Lifetime Modelling of High Temperature Corrosion Processes

Proceedings of an EFC Workshop 2001

European Federation of Corrosion
Publications
NUMBER 34

Lifetime Modelling of High Temperature Corrosion Processes

Proceedings of an EFC Workshop 2001

Edited by
M. Schütze, W. J. Quadakkers and J. R. Nicholls

Published for the European Federation of Corrosion
by Maney Publishing on behalf of The Institute of Materials

MANEY
publishing

Book Number B772
Published in 2001 by Maney Publishing
on behalf of The Institute of Materials
1 Carlton House Terrace, London SW1Y 5DB

Maney Publishing is the trading name of
W. S. Maney & Son Ltd

ISBN 1-902653-47-5

Typesetting by
spiresdesign

Made and printed in Great Britain

Contents

Part 2 – General Lifetime Prediction Approaches 337

European Federation of Corrosion Publications
Series Introduction

The EFC, incorporated in Belgium, was founded in 1955 with the purpose of promoting European co-operation in the fields of research into corrosion and corrosion prevention.

Membership is based upon participation by corrosion societies and committees in technical Working Parties. Member societies appoint delegates to Working Parties, whose membership is expanded by personal corresponding membership.

The activities of the Working Parties cover corrosion topics associated with inhibition, education, reinforcement in concrete, microbial effects, hot gases and combustion products, environment sensitive fracture, marine environments, surface science, physico–chemical methods of measurement, the nuclear industry, computer based information systems, the oil and gas industry, the petrochemical industry, coatings, automotive engineering and cathodic protection. Working Parties on other topics are established as required.

The Working Parties function in various ways, e.g. by preparing reports, organising symposia, conducting intensive courses and producing instructional material, including films. The activities of the Working Parties are co-ordinated, through a Science and Technology Advisory Committee, by the Scientific Secretary.

The administration of the EFC is handled by three Secretariats: DECHEMA e. V. in Germany, the Société de Chimie Industrielle in France, and The Institute of Materials in the United Kingdom. These three Secretariats meet at the Board of Administrators of the EFC. There is an annual General Assembly at which delegates from all member societies meet to determine and approve EFC policy. News of EFC activities, forthcoming conferences, courses etc. is published in a range of accredited corrosion and certain other journals throughout Europe. More detailed descriptions of activities are given in a Newsletter prepared by the Scientific Secretary.

The output of the EFC takes various forms. Papers on particular topics, for example, reviews or results of experimental work, may be published in scientific and technical journals in one or more countries in Europe. Conference proceedings are often published by the organisation responsible for the conference.

In 1987 the, then, Institute of Metals was appointed as the official EFC publisher. Although the arrangement is non-exclusive and other routes for publication are still available, it is expected that the Working Parties of the EFC will use The Institute of Materials for publication of reports, proceedings etc. wherever possible.

The name of The Institute of Metals was changed to The Institute of Materials with effect from 1 January 1992.

The EFC Series is now published by Maney Publishing on behalf of The Institute of Materials.

A. D. Mercer
EFC Series Editor,
The Institute of Materials, London, UK

EFC Secretariats are located at:

Dr B A Rickinson
European Federation of Corrosion, The Institute of Materials, 1 Carlton House Terrace, London, SW1Y 5DB, UK

Mr P Berge
Fédération Européene de la Corrosion, Société de Chimie Industrielle, 28 rue Saint-Dominique, F-75007 Paris, FRANCE

Professor Dr G Kreysa
Europäische Föderation Korrosion, DECHEMA e. V., Theodor-Heuss-Allee 25, D-60486, Frankfurt, GERMANY

Preface

Resistance to environmental corrosion is a major property required for components which have to operate at high temperatures, e.g. in steam power plants, chemical and petrochemical plants, waste incineration plants, engines and gas turbines. The resistance of the materials to high temperature corrosion relies on the formation of protective surface scales which prevent direct access of the corrosive environment to the free surface of the material. Extensive research efforts in recent years have led to the development of a large number of materials which possess suitable resistance to corrosion in specific applications.

It is important to note that in nearly all applications resistance to environmental corrosion at high temperatures does not mean that the material is 'immune' against the aggressive environment. High Temperature Corrosion resistance is a non-specific definition which in fact means that the rates of corrosion are reduced to such an extent that the lifetime of the component is sufficient to allow a safe and economically feasible plant design and operation. This philosophy implies that the development of models which allow long term extrapolation of materials degradation by high temperature corrosion and prediction of component life are of crucial importance for reliable plant operation especially in cases where materials are subjected to severe service conditions in respect to temperature, stress and/or corrosivity of the environment.

A two-day workshop held at DECHEMA in Frankfurt/Main on 22 and 23 February 2001 addressed these important aspects of high temperature materials application. It was organised by the working party "Corrosion by Hot Gases and Combustion Products" of the European Federation of Corrosion and followed the tradition of EFC workshops in various areas of high temperature corrosion which have been held since the beginning of the 1980s. The aim of the workshop was to present the state of the art in the development of lifetime modelling of high temperature oxidation and corrosion processes and the incorporation of these models in component design.

The present volume contains the written contributions to this workshop. In their papers the authors report results from recent research projects and thus provide a comprehensive survey of the present knowledge and understanding of this technologically important topic. The first eight papers came out of a joint European project with the acronym LEAFA which was especially devoted to the lifetime extension of alumina-forming alloys. These eight papers at the same time represent the final report of this project. The other papers come from various projects in Europe and the USA. The contribution of each author to this monograph is gratefully acknowledged and recognised in the authorship of individual papers: thanks are also due to all colleagues who helped by carefully refereeing the papers.

M. Schütze   *Karl-Winnacker-Institut der DECHEMA e.V., Frankfurt, Germany*
Chairman of EFC Working Party "Corrosion by Hot Gases and Combustion Products"

W. J. Quadakkers   *Forschungszentrum Jülich, Institute for Materials and Processes in Energy Systems, Jülich, Germany*

J. R. Nicholls   *Cranfield University, United Kingdom*

Previous EFC publications from the WP on 'Corrosion by Hot Gases and Combustion Products':

EFC No. 14   Guidelines for Methods of Testing and Research in High Temperature Corrosion

EFC No. 27   Cyclic Oxidation of High Temperature Materials

Part 1

Investigation and
Modelling of Specific
Degradation Processes

1.1 Alumina formers

1

Life Extension of Alumina-Forming Alloys — Background, Objectives and Achievements of the BRITE/EURAM Programme LEAFA

J. R. NICHOLLS and M. J. BENNETT

SIMS, Cranfield University, Cranfield, Bedford MK43 0AL, UK

ABSTRACT

During the last decade or so increased technological interest in the FeCrAlRE (where RE is a reactive element) class of alloys has necessitated a corresponding upsurge in research activity to provide engineering data backed by improved scientific understanding. In Europe most work has been undertaken by a succession of three major multi-partnered collaborative programmes within Community framework actions. The third, and recently completed, three year programme has been LEAFA (Life Extension of Alumina Forming Alloys), undertaken under the BRITE/EURAM action, by ten partners from Industry, Research Institutes and Universities from six Member Countries.

The LEAFA project focussed on the life-limiting corrosion process, that results in the chemical failure of commercial FeCrAlRE alloys. Also a series of high purity model FeCrAlY alloys has been studied to establish the role of important minor constituent elements, Zr, P, Ti, V and Ca. Full lifetime testing was carried out in oxidising environments, to develop a life prediction model, while an initial appraisal was undertaken also on the potential role of more aggressive gas contaminants (S, N, Cl and H_2O bearing species) likely to be present in many industrial applications.

The background to the LEAFA project, its objectives, work programme and achievements together with the implications of the project on a European dimension will be outlined in this paper. This will provide an introduction to the seven companion papers detailing the main scientific results derived by the project as presented at the EFC Workshop 'Life Modelling of High Temperature Corrosion Processes' (EFC Event No. 248) at DECHEMA e.V. Frankfurt, Germany on 22 and 23 February 2001 (EFC publication No. 34), published by The Institute of Materials, London, 2001.

1. Introduction and Background

Over the last decade or so, numerous innovative industrial developments have emerged necessitating improved metallic component performance at ultra-high temperatures (≥1000°C) to meet the challenges of enhanced plant reliability and lower costs, together with compliance with ever more stringent regulatory, safety and environment requirements. The technologies involved are wide ranging and include the automotive, power generation and aerospace industries, as well as the manufacture of domestic appliances and industrial plant. Figure 1 illustrates a number of component parts of a combustion chamber, manufactured from PM2000 (courtesy of U. Miller, Plansee) for use at ultra-high temperatures.

The main class of alloys having potential for such applications are the FeCrAlRE alloys (where RE is a reactive element, such as yttrium, zirconium, hafnium, etc.) as their corrosion resistance derives from the formation of a highly protective alumina scale. Substantial research work was undertaken on these alloys through the 1960s to the 1980s but was aimed primarily at the development of commercial alloy compositions, with emphasis on the role of the reactive element. However, most of the supporting corrosion studies were of short term duration (typically <200 h) and were focussed on the development/growth of the protective alumina scale. Little knowledge was generated regarding the lifetime governing processes, especially the breakdown/failure of the protective scale. Such modes of failure have proved to be catastrophic, limiting the use of chromia protected alloys (e.g. [1–3]), widely employed for industrial applications at lower temperatures (≥900°C). Therefore, the recent increasing technological interest in the FeCrAlRE class of alloys, for use at ultra high temperatures, has necessitated a corresponding upsurge in worldwide research activity into the lifetime of these alloys, with the main emphasis being directed to the definition of life-limiting processes.

Since European companies (such as VDM-Krupp, Kanthal-Sandvik, Ugine-Savoie, Special Metals and Plansee) are world leaders in the production of these alloys, as would be expected, a considerable proportion of this research work has been and is being carried out in Europe. The main thrust has stemmed from a succession of three major multi-partnered collaborative programmes within the European Community Framework Actions. Each programme has focussed on separate and distinct, but at the same time inter-related, objectives and has been undertaken by

Fig. 1 *Parts of a combustion chamber manufactured from PM2000 (courtesy of U. Miller, Plansee).*

grouping partners best equipped to achieve them. The first study, within the COST 501/2, W.P.4 (Heat-Exchanger) Action, examined the long term (>20 kh) oxidation behaviour of one alloy group; the commercial high strength ODS FeCrAlRE alloys. This programme established that there were two modes of failure causing the initial deterioration and the ultimate final loss of scale protectiveness (respectively mechanical and chemical failure modes). It led also to the development of preliminary alloy life-prediction models [4,5]. Mechanical failure of scales formed on commercial FeCrAlRE alloys fabricated by all the main manufacturing routes (conventional melting, powder metallurgy, as well as mechanical alloying) was investigated in the second programme 'IMPROVE' under the BRITE/EURAM Action (Project BE 7972 (1994)). This enhanced the scientific understanding of, and modelled the basic processes underlying, scale cracking/spallation and also of the subsequent rehealing of scales on further environmental exposures [6–8]. A major technological conclusion to emerge was that it was essential to consider failure in the context of the whole system, alloy substrate as well as surface rather than purely as a surface phenomenon.

The third in this related series of programmes is the current 'LEAFA' (Life Extension of Alumina Forming Alloys) Project (No. BE-97-4491), which has concentrated on the ultimate, and life-governing, scale failure process: chemical failure, which leads to breakaway and non-protective oxidation. Figure 2 illustrates the breakaway oxidation of a PM2000 foil (120 µm thick) following cyclic oxidation at 1300°C. This SEM image shows iron oxide formation in the breakaway region, with partial rehealing of the scale. This focussed, concerted, fundamental scientific three year programme, undertaken also under the BRITE/EURAM Action, commenced on 1st January 1998. Ten partners, from six member countries of the European Community, have been involved from Universities, Research Institutions and Industry, namely Cranfield University (Project Technical Co-ordinator), T. U. Clausthal, University of Liverpool, FZJ (Julich), Armines, CEC-JRC-IAM,

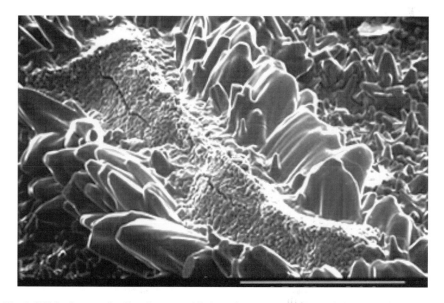

Fig. 2 *SEM micrograph of breakaway oxidation of a 125 µm PM2000 foil at 1300°C.*

Krupp-VDM, Kanthal AB, Plansee and Diffusion Alloys Ltd. The Partners were selected on the basis of proven expertise and the availability of appropriate facilities to undertake this complex project. In addition many had worked harmoniously and successfully together on a range of previous CEC funded projects.

The LEAFA project objectives, the scientific and technical approach, and achievements together with the current status of the dissemination of scientific information arising and of the technological implications for European Industry are detailed. The overall objective, therefore for this paper is to provide an introduction to the seven companion papers [9–15] detailing the main scientific results derived by the Project, as presented at the EFC Workshop 'Lifetime Modelling of High Temperature Corrosion Processes' (EFC Event No. 248) at DECHEMA e.V. Frankfurt, Germany on 22nd and 23rd February 2001 (this volume).

2. Objectives and Commitments of the LEAFA Projects

At the project outset this scientific programme had four main technical objectives, as follows:

1. The development of a life prediction model for the chemical failure, in oxidising environments, of a range of commercial FeCrAlRE alloys, fabricated by all the major metallurgical routes, by European manufacturers. This model would cover all critical variables, including temperature, thermal cycling conditions, and alloy component geometry. Such modelling would be supported by lifetime test data and detailed mechanistic understanding.

2. Preliminary extension of this model to indicate the probable behaviour in industrial environments containing other corrodents, such as Cl, S, H, N and C bearing gases.

3. Improved understanding of the role of potentially critical minor alloy constituents/ contaminants (such as Zr, P, V, Ca and Ti), which could affect oxidation life-times of commercial FeCrAlRE alloys, through studies of the behaviour of high purity model alloys,

4. Preliminary development of potential lifetime improvement procedures, with particular emphasis on enhancement of the alloy Al content by diffusion coatings.

Additionally the project had four further major commitments, to:

(i)   Disseminate rapidly and effectively the unique information generated by the Project to the Scientific Community within Europe, but also worldwide, for peer comment.

(ii) Facilitate the comprehensive and immediate technological exploitation of the Project results by European Industry to maximise the economic, social and environmental benefits thereby arising within the European Community.

(iii) Ensure maximum training opportunities for young research graduates, and

(iv) Enhance collaboration between relevant European Industrial Research Institutes and University Laboratories involved in this important area of scientific research and development.

3. Scientific and Technical Approach

The LEAFA project was undertaken essentially in two phases, the first examined chemical failure under oxidising conditions and the second evaluated the role of gas contaminants on chemical failure (Fig. 3). The work programme was split into nine tasks, the first five (Critical analysis of existing information; Materials selection, procurement, characterisation; Definition of test methodologies and test environments; Data generation (involving lifetime testing and mechanistic understanding); Development of a life prediction model) comprised Phase 1, scheduled to be undertaken during Project Years 1 and 2. The remaining four tasks (Data generation in bioxidant environments, and mechanistic understanding; Extension of the life prediction model; Life improvement procedures; Model validation and practical implications) formed the basis of Phase 2 undertaken mainly during the Project third year.

The project was co-ordinated by meetings of Partners at four to six month intervals and by the generation of comprehensive progress reports as required by the CEC.

4. Achievements

For clarity of presentation, and to facilitate correlation of achievement against objective, the main technical results will be summarised in the context of each task.

4.1. Task 1

A critical review of existing data at the Project inception not only confirmed the main project objectives, the alloys to be tested, the data required and the approaches to be pursued but also highlighted the necessity to increase work on several topics already identified particularly the role of trace additions, introduced into the alloy as part of the manufacturing process, on oxidation behaviour.

4.2. Task 2

Materials procurements/fabrication and characterisation. The four commercial FeCrAlRE alloys (Aluchrom YHf, Kanthal AF, Kanthal APM and PM2000), which formed the core of the test programme, were manufactured in the form either of 0.5–2.0 mm thick sheet or 30–200 μm thick foil (Table 1) and were prime products of the Project Industrial Partners. Fabrication of these alloys covered all the major metallurgical routes employed, namely, conventional melting (M), powder metallurgy (P.M.) and mechanical alloying (M.A.).

Phase 1: Chemical Failure Under Oxidising Conditions

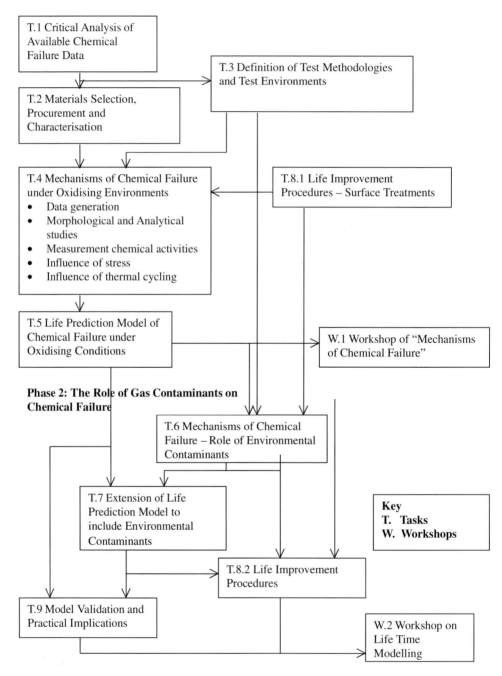

Fig. 3 *The LEAFA Technical Programme flow diagram.*

The compositions of these commercial alloys were determined and are detailed in ref. [9].

For studies on the role of P, V, Ca, Ti and Zr, seven high purity alloys, were fabricated having the same basic composition (Fe + 20%Cr + 5% Al + 500 ppm Y, referred to as MRef.) but with defined separate additions of these trace elements, see reference [12]. The analysed compositions of the model alloys indicated that the Cr, Al and Y contents were close to the planned values. The S, O, N levels were acceptable but below detection limits, while the C levels were comparable (90–98 ppm) in six alloys but slightly lower (74 ppm) in M–P due to a fabrication error. The introduction of P did not prove easy experimentally, which caused the variation in levels in MRef., M+V, M+Ca, M+Ti, M+Zr and a lower value, 236 ppm, in alloy M+P than planned (400 ppm). The additions of V, Ti and Zr were close to the concentrations aimed for, while Ca proved difficult to add to M+Ca and the concentration (<5 ppm) analysed was again below the detection level.

4.3. Task 3

The test methodologies, environments and test matrices to generate the data required under the Project Phases 1 and 2 fully exploited the facilities available to laboratories of the Partners. The recommendations of TESTCORR (the European guidelines for high temperature corrosion testing) [16,17] were followed as closely as was compatible with LEAFA requirements. All the Phase I tests were undertaken in air, on the same material batches and used the same methodology (specimen production and cleaning, exposure

Table 1. Materials produced for the LEAFA Project

Alloy	Sheet thickness, mm	Foil thickness, μm	Manufacturer route	Fabrication*
Aluchrom YHf	1.0	30, 50, 80, 200	Krupp VDM	M
Kanthal AF	0.5, 1.0, 2.0	50, 70	Kanthal AB	M
Kanthal APM	0.5, 1.0, 2.0	–	Kanthal AB	P.M.
PM 2000	0.5, 1.0, 2.0	125	Plansee	M.A.
Model Alloys	1.0, 2.0, (6mm Ø rods)	70, 90, 100	Alloys made by Armines. Foils produced from rod by Kanthal AB. Foils rolled from sheet by Krupp VDM.	M.

*See page 5.

procedures, gravimetric measurements). Tests were carried out at 1100, 1200 and 1300°C (and in some instances also at 1350°C). Depending on the anticipated exposure duration the tests were either isothermal/continuous or discontinuous, with thermal cycles of 20, 100 or 300 h duration. In all tests the cooling and heating rates were those of the furnace. Description of the oxidation behaviour of these alloys necessitated measurement of the formation and development of the protective oxide scale and of its subsequent failure mechanically (in some instances) and chemically, in all cases [11].

In many of these test programme packages the times to breakaway proved to be longer than predicted at the project onset, but were carried on to failure, nevertheless, such that representative alloy lifetimes could be measured and documented. This then fully justified why it was imperative to carry out full life testing to underpin life-time prediction modelling. The overall test programme status at the project completion was that although a few specimens still have not failed, even after extended exposures (< 15 000 h), such data also proved valuable for the modelling work.

Seven gas compositions (Table 2) were used in the corresponding Phase 2 test programme undertaken [15] using facilities readily available in the laboratories of seven partners. Also indicated in Table 2 are the major technologies applications having these representative environments. In most laboratories the Phase 2 tests could only be commenced on completion of the Phase 1 experiments, with the consequence that the Phase 2 test programme had to be limited, in the main, to1000 h exposures. All alloys

Table 2. Gas composition for the test environments of the LEAFA Phase 2 test programme

Environment	Contaminant levels	Application
Air + H_2O	2.5 vol.% H_2O	General effect of humidity.
N_2 + H_2 + H_2O	95 vol.% N_2 + 5 vol.% H_2 + ~630 vppm H_2O (dew point > –25°C)	Shield gas for furnace heaters.
N_2 + NO	5000 ppm NO	Major contaminant in automotive exhaust gas.
Air + SO_2/SO_3	3000 ppm SO_2	Major contaminant in coal fired power station environments.
Air + HCl	50 ppm HCl	Major contaminant in waste incinerator and biogas combustion exhaust gases.
Exhaust gas	N_2 + 9.2 vol.% H_2O + 10.2 vol.% CO_2 + 4.6 vol.% CO + 2300 ppm C_3H_8	Automotive catalytic converters.
Combustion gas	N_2 + 15 vol.% O_2 + 4 vol.% H_2O + 3.4 vol.% CO_2	Gas turbines in power plants.

were tested in all environments at 1200°C and also at either 1100 or 1300°C. The test methodologies were basically the same as employed in the Project Phase 1 but with the additional necessity to maintain and monitor the more complex gaseous test environments, again in accordance with TESTCORR guidelines [16] where feasible. At the project completion all the tests had been carried out as planned.

4.4. Task 4

This was the principal Project Phase 1 task concerned with the mechanisms of chemical failure in oxidising environments. Most of the results, concerning the behaviour of commercial FeCrAlRE alloys, the mechanisms of chemical failure, the mechanism and modelling of oxide spallation, the critical role of minor elemental constituents in the model alloys and the influence of surface geometry on commercial alloy lifetimes are detailed respectively in ref. [9–13].

4.5. Task 5

The development of a life prediction model of chemical failure of FeCrAlRE alloys in oxidising environments is described elsewhere in this volume [14].

4.6. Task 6

Data obtained concerning the behaviour of FeCrAlRE alloys in bioxidant environments, i.e. during the LEAFA project Phase 2, are presented and discussed elsewhere in this volume [15].

4.7. Task 7

The life prediction model has been extended, to include the impact of environmental contaminants.

4.8. Task 8

Diffusion coating procedures have been developed as a route to enrich the aluminium surface levels of FeCrAlRE alloy components.

As a direct consequence of the wide ranging technologies encompassed by the current research, the potential benefits stemming from the life extension of commercial alumina-forming FeCrAlRE alloys in high temperature corrosion environments, cannot be over emphasised. It is, therefore, crucial that in-service operation of alloys has to be optimised to prolong component life and enhance plant reliability, thereby increasing efficiency. The substantial body of data produced by the LEAFA project have enabled the formulation of detailed recommendations for 'Best Practice' and component design criteria.

4.9. Task 9

The life prediction model has been validated by comparison with technological experience.

With the completion of these nine tasks all the main objectives of the LEAFA project have been achieved and can be summarised as follows:

(i)   Establishment of a life-time database on the oxidation behaviour of commercial FeCrAlRE alloys at temperatures in the range 1100–1400°C [9].

(ii)   Modelling of oxide spallation from FeCrAlRE alloys [11].

(iii)   Development of a mechanistic understanding of the chemical failure of these FeCrAlRE based alloys [10].

(iv)   Investigation of the effect of component geometry on the chemical failure of commercial FeCrAlRE alloys [13].

(v)   Investigation of the role of critical trace elements on the high temperature oxidation behaviour, using model FeCrAlY alloys [10,12].

(vi)   A preliminary study of the role of bioxidants on the lifetime of the commercial FeCrAlRE family of alloys [15] and finally

(viii) Development of a life-prediction model to predict the chemical failure of these high temperature FeCrAlRE based alloys [14].

5. Benefits at a European Dimension

The first major commitment was the rapid and effective dissemination of the unique scientific information arising from the LEAFA Project initially among partners, which was accomplished through the regular Project meetings and subsequently to the Scientific Community, primarily with Europe but also worldwide to advance the frontiers of scientific understanding and for peer review/comment. Twenty-one publications will have arisen from the project at its completion. Many of the papers have been presented at one of four major International Workshops/Conferences. Two of these provided ideal forums for the two Workshops (W1, and W2) required of the Project (see Fig. 3). The first Workshop formed part of the International Conference, *Microscopy of Oxidation—4*, held in Cambridge, UK on 20–22 September, 1999, while the second Workshop (W2) formed part of the EFC Workshop on 'Lifetime Modelling of High Temperature Processes' held at DECHEMA e.V., Frankfurt am Main, Germany on 22 and 23 February 2001. W1 papers were published in the Journal *Materials at High Temperatures* 2000, **17**, (1), while the W2 papers [9–15] appear also in the present volume.

The LEAFA project had immediate and direct relevance to several CEC priority themes, in particular Safety and Improved Economics of Industrial Components and Emission Reduction Technology (and thereby, by implication, also Citizen Safety and Environmental Protection). The LEAFA project Industrial Partners already have appraised the technological implications of this work to current and future activities (and thus the potential value for money benefits regarding these CEC priority themes). At the completion of the research work several major factors are clearly apparent regarding:

(i) improved alloy composition and quality control, which will result in a reduction in batch to batch variable behaviour, and therefore lower manufacturing costs.

(ii) improved FeCrAlRE component engineering design criteria, with assured lifetime performance.

(iii) improved alloy marketability through enhanced customer relations/ confidence and thereby, competitiveness.

(iv) potential for recycling high cost components, and

(v) development of new alloys.

Overall economically, the conservative estimates of the LEAFA Industrial Partners would indicate that the combined benefits to the European Community annually during the next five years are likely to be at least an order of magnitude greater than the original investment made in the LEAFA research project.

The remaining commitments of the LEAFA project also have been fulfilled in that training opportunities were provided for five research graduates; three presented doctoral theses based on project studies. Several researchers have widened their experience based as a consequence of working for extended periods in the laboratories of other Partners. Finally the project has enhanced collaboration between relevant European Industrial, Research Institutes and University Groups involved in this important area of scientific research and development, which in turn has led also to the instigation of several other novel research initiatives.

6. Conclusions

The LEAFA project was a major research undertaking and its success in meeting all the central objectives and commitments stemmed from multi-partner collaboration exploiting fully the facilities and dedicated expertise existing within these European laboratories.

Finally, as expressed by Kanthal AB, Industry must never rest on its laurels to ensure it can meet all future technological challenges for improved safety, reliability and economics for the benefits of all communities worldwide. There has been a substantial leap forward resulting from the LEAFA project which must be built on by future research programmes leading to the development of the next generation of superior high temperature alloys and component design.

7. Acknowledgements

We are immensely grateful for the substantial contributions made by all our Partners, which ensured the success of the LEAFA Project. We wish also to thank our CEC Scientific Officer, Dr. Lothar Schmidt for his guidance and support throughout. Finally, this Project could not have been undertaken without the financial commitment of the CEC BRITE/EURAM Action, Project BE97-4491.

References

1. M. J. Bennett and D. P. Moon, in *The Role of Active Elements in the Oxidation Behaviour of High Temperature Metals and Alloys* (E. Lang, ed.). Elsevier Applied Science, 1989, 111–129.

2. M. J. Bennett and R. C. Lobb, *Oxid. Metals*, 1991, **35**, 35–52.

3. H. E. Evans, A. T. Donaldson and T. C. Gilmour, *Oxid. Metals*, 1999, **52**, 379.

4. M. J. Bennett, R. Perkins, J. B. Price and F. Starr, in *Materials for Advanced Power Engineering* (D. Coutsouradis *et al.*, eds). Kluwer Academic Publishers, 1994, **2**, 1553–1562.

5. W. J. Quadakkers and M. J. Bennett, *Mater. Sci. Technol.*, 1994, **10**, 126–131.

6. BRITE/EURAM IMPROVE Programme, Project No. BE7972 (1994)) Final Report (1997), CEC Brussels.

7. J. P. Wilber, J. R. Nicholls and M. J. Bennett, in *Microscopy of Oxidation—3* (S. B. Newcomb and J. A. Little, eds). Published by The Institute of Materials, London 1997, UK, pp.207–220.

8. J. P. Wilber, M. J. Bennett and J. R. Nicholls, in *Cyclic Oxidation of High Temperature Materials* (M. Schütze and W. J. Quadakkers, eds). Publication No. 27 in European Federation of Corrosion Series. Published by The Institute of Materials, London, 1999, pp.137–147.

9. R. Newton *et al.*, *Lifetime Modelling of High Temperature Corrosion Processes* (M. Schütze, W. J. Quadakkers and J. R. Nicholls, eds). Publication No. 34 in European Federation of Corrosion Series. This volume, pp.15–36.

10. H. Al-Badairy, *et al.*, *Lifetime Modelling of High Temperature Corrosion Processes* (M. Schütze, W. J. Quadakkers and J. R. Nicholls, eds). Publication No. 34 in European Federation of Corrosion Series. This volume, pp.50–65.

11. H. E. Evans and J. R. Nicholls, *Lifetime Modelling of High Temperature Corrosion Processes* (M. Schütze, W. J. Quadakkers and J. R. Nicholls, eds). Publication No. 34 in European Federation of Corrosion Series. This volume, pp.37–49.

12. D. Naumenko, *et al.*, *Lifetime Modelling of High Temperature Corrosion Processes* (M. Schütze, W. J. Quadakkers and J. R. Nicholls, eds). Publication No. 34 in European Federation of Corrosion Series. This volume, pp.66–82.

13. G. Strehl, *et al.*, 'Life Modelling of High Temperature Corrosion Processes', Eds. M. Schütze, W. J. Quadakkers and J. R. Nicholls. This Volume pp.107–122.

14. J. R. Nicholls, *et al.*, *Lifetime Modelling of High Temperature Corrosion Processes* (M. Schütze, W. J. Quadakkers and J. R. Nicholls, eds). Publication No. 34 in European Federation of Corrosion Series. This volume, pp.83–106.

15. A. Kolb-Telieps, *et al.*, *Lifetime Modelling of High Temperature Corrosion Processes* (M. Schütze, W. J. Quadakkers and J. R. Nicholls, eds). Publication No. 34 in European Federation of Corrosion Series. This volume, pp.123–134.

16. TESTCORR – Final Code of Practice for Discontinuous Corrosion Testing in High Temperature Gaseous Atmospheres, Contract SMT4-CT95-2001, European Commission, 2001.

17. J. R. Nicholls, Discontinuous Measurements of High Temperature Corrosion, in *Guidelines for Methods of Testing and Research in High Temperature Corrosion* (H. J. Grabke and D. B. Meadowcroft, eds). Publication No. 14 in European Federation of Corrosion Series. Published by The Institute of Materials, London, 1996, pp.11–36.

2

The Oxidation Lifetime of Commercial FeCrAl(RE) Alloys

R. NEWTON, M. J. BENNETT, J. P. WILBER, J. R. NICHOLLS, D. NAUMENKO*,
W. J. QUADAKKERS*, H. AL-BADAIRY†, G. TATLOCK†, G. STREHL§,
G. BORCHARDT§, A. KOLB-TELIEPS¶, B. JÖNSSON**, A. WESTERLUND††,
V. GUTTMANN, M. MAIER and P. BEAVEN††

SIMS, Cranfield University, Cranfield, Bedfordshire, MK43 0AL, UK
*Forschungszentrum Jülich, IWV-2, 52425, Jülich, Germany
†University of Liverpool, Liverpool, L69 3GH, UK
§Technische Universität Clausthal, D-38678 Clausthal-Zellerfeld, Germany
¶Krupp-VDM, D-58778, Werdohl, Germany
**Kanthal AB S-73427, Hallstahammar, Sweden
††CEC-Joint Research Centre, 1755 ZG Petten, The Netherlands

ABSTRACT

Reliable, lifetime data are a prerequisite for the development of any model aiming at predicting the onset of the ultimate high temperature corrosion degradation mode of FeCrAlRE alloys, namely chemical failure. Thus, the main data generation task of the LEAFA project has been to obtain lifetime data on the oxidation behaviour, in air, of four commercial alloys, Aluchrom YHf, Kanthal AF, Kanthal APM and PM2000. These alloys contained different reactive elements (RE), yttrium, as metal and in the ODS strengthened PM2000 as yttria, zirconium and hafnium either separately or in combination but their aluminium contents were ~5 mass%. The alloys were manufactured by the main fabrication routes, conventional melting, powder metallurgy and mechanical alloying, and as a consequence had a range of mechanical properties, with the ODS alloy being the most creep resistant.

The alloys, in the form either of foils (thickness 30–125 µm) or sheets sections (0.2–2.0 mm thick), were oxidised either isothermally or discontinuously, with thermal cycles of 20, 100 or 300 h duration, at 1100, 1200 and 1300°C. Detailed measurements were made of the scale growth rates, and also of the mechanical failure of scales, through determination of the critical thicknesses for the onset of spallation and of the continuing spall rates.

Chemical failure occurred when the aluminium content of the alloy fell below a value $[Al_{crit}]$, such that the protective alumina scale could no longer be formed/sustained. Breakaway oxidation then ensued leading to the formation of non-protective iron–chromium oxides. Values of both the time to breakaway ($t_{B/O}$) and of $[Al_{crit}]$ were determined. Critical parameters affecting the respective oxidation behaviours of the alloys included the substrate alloy mechanical properties and thickness.

1. Introduction

The last decade or so has witnessed increasing use of the FeCrAlRE (where RE is a reactive element) family of alloys for industrial components for service at ultra-high temperatures (≥1000°C). A major attraction of these alloys for such applications centres on their outstanding corrosion resistance, emanating from the protection afforded by the alumina scales formed on them. However, 'nothing in life lasts forever' and scale protectiveness can eventually break down, by several mechanisms, the ultimate being by chemical failure leading to catastrophic non-protective oxidation. As this would have major economic and safety implications technologically, it is imperative to be able to predict oxidation induced component life times. The development of a life prediction model [1] for industrial applications has been the principal objective of the European collaborative focused fundamental project, LEAFA; (Life Extension of Alumina Forming Alloys) [2] carried out as part of the BRITE/EURAM Action. Reliable life time oxidation data are an essential prerequisite for the model development and this was the main data generation task of the LEAFA project. Some of the basic oxidation information, (i.e. scale growth and spallation kinetics) obtained at 1300°C were presented at the first Workshop disseminating information derived from the project [3]. For completeness these results have been updated and are reproduced again here. This paper, therefore, will detail all the data generated by the project and will discuss the critical parameters and understanding that have emerged from the LEAFA project relevant to the development of a life-prediction model [1].

2. Experimental

2.1. Materials

The four commercial FeCrAlRE alloys (Aluchrom YHf, Kanthal AF, Kanthal APM and PM2000), tested in the form either of 30–125 µm thick foils or of 0.2–2.0 mm thick sheets, were prime products of the Project Industrial Partners. Fabrication of these alloys covered all the major metallurgical routes employed, namely, conventional melting (M), powder metallurgy (P.M.) and mechanical alloying (M.A.) (Table 1).

The compositions of these commercial alloys were determined by chemical analysis (mainly using AAS, ICPMS, GDMS, colourimetry and oxidation (C/S) or melting (O/N)), (Tables 2 and 3). A problem was encountered with the aluminium analysis

Table 1. Materials produced for the LEAFA Project

Alloy	Foil (nominal thickness, µm)	Sheet (nominal thickness, mm)	Manufacturer	Fabrication route
Aluchrom YHf	30, 50, 80	0.2, 1.0	Krupp VDM	M
Kanthal AF	50, 70	0.5, 1.0, 2.0	Kanthal AB	M
Kanthal APM	–	0.5, 1.0, 2.0	Kanthal AB	P.M.
PM 2000	125	0.5, 1.0, 2.0	Plansee	M.A.

of PM2000 but was resolved by repeat determinations at Plansee, Armines (MSE), Kanthal and FZJ from which convergent values were derived (Table 2). Accurate values of the initial aluminium contents of the alloys were essential as these were used in the calculations of the critical parameter $[Al_{crit}]$ (the aluminium content remaining in the alloy at the onset of chemical failure), as will be described in Section 3.

2.2. Test Metholodgy

Oxidation test matrices to generate the data required and appropriate cross-checks were set up to exploit fully the facilities available in the laboratories of seven Project partners. These matrices comprised 15 programme packages involving over 300 specimens. The recommendations of TESTCORR (the European guidelines for high temperature corrosion testing) [4] were followed as closely as was compatible with LEAFA requirements. All tests were undertaken using 2 cm × 1 cm (nominal) coupons in either laboratory or synthetic air on the same material batches and used the same methodology (specimen production, surface grinding on 1200 grit paper and cleaning, measurement of specimen dimensions, exposure procedures, gravimetric measurements). Experiments were carried out at 1100, 1200 and 1300°C. Depending on the anticipated exposure duration the tests were either isothermal/continuous or discontinuous, with thermal cycles of 20, 100 or 300 h duration. In all experiments the heating and cooling rates (0.10°C s^{-1}) were those of the furnace. Description of the oxidation behaviour of these alloys necessitated gravimetric measurements of the formation and development of the protective oxide scale and of its subsequent failure mechanically (in some instances) and chemically, in all cases.

Table 2. Aluminium and chromium contents of the commercial FeCrAl(RE) alloys

Alloy/thickness	Percentage of aluminium and chromium in the commercial alloys, mass%					
	Armines (AAS)*		Manufacturer (AAS)		FZJ (ICPMS)**	
	Al	Cr	Al	Cr	Al	Cr
Aluchrom YHf 1 mm	5.5	19.7	5.6	20.25		
Kanthal AF 1 mm	5.2	20.8	5.15	21.06		
Kanthal APM 1 mm	5.8	21.0	5.86	21.14		
PM2000 0.5 mm	4.0	18.9	3.8	18.95	4.00	20.00
PM2000 1mm	5.3	18.1	5.35	17.85	5.40	19.60
PM2000 2mm	5.4	19.5	5.40	19.45	5.40	20.90

*Atomic Absorption Spectroscopy. **Inductively coupled plasma mass spectroscopy.

Additionally, in some discontinuous experiments at 1300°C, with thicker Kanthal AF, Kanthal APM and PM2000 alloy sections, the specimen dimensions were measured after at least the first five cycles (each of 100 h duration) but thereafter, as most specimens had become distorted, they were only photographed.

Table 3. Analyses of minor constituents of commercial FeCrAlRE alloys

Alloy	Aluchrom YHf		Kanthal AF	Kanthal APM	PM2000		
Thickness, mm	1		1	1	0.5	1	2
Source	MSE	VDM	MSE	MSE	MSE	MSE	MSE
Y, ppm	460	500	340	<0.1	3700	3700	3700
C, ppm	2100	2300	280	290	260	300–600	100
S, ppm	1.3	20	1.5	<1.0	20	30	28
O, ppm	<10	–	<10	<1.0	3000	3100	2800
N, ppm	40	40	150	160	150	300	60
P, ppm	130	130	140	170	–	–	24
Zr, ppm	540	100	580	1100	<10	<10	14
V, ppm	860	300	200	360	–	–	77
Ti, ppm	98	100	940	260	4500	1800	4500
Cu, ppm	110	200	330	240	–	–	70
Ca, ppm	12	10	0.95	1.1	16	15	31
Hf, ppm	310	500	3.1	1.1	10	<5	0.05
Mn, ppm	1800	2500	610	800	380	430	310
Si, ppm	2900	–	1900	4000	–	–	240
Nb, ppm	<50	<30	<50	<50	–	–	<50
Mg, ppm	78	90	17	22	–	–	30
Mo, ppm	100	100	58	41	–	–	30

N.B. Value not supplied: MSE = Armines, St. Etienne, France.

In many of these test programme packages the time to breakaway proved to be longer than predicted at the project onset. However, tests were carried out to failure, such that representative alloy lifetimes could be measured and documented. This then fully justified why it was imperative to carry out full life testing to underpin lifetime prediction modelling. The overall test programme status at the project completion was that over 98% of all the specimens had been oxidised either to chemical failure or had been stopped deliberately just before failure for examination. Although the remaining specimens still had not failed, even after extended exposures (> 16 000 h), such data also proved valuable for the modelling work.

Following oxidation surface morphologies and the nature of the oxidation were examined using a range of conventional surface analytical procedures, including optical and scanning microscopy, EDX, EPMA and XRD.

3. Results

For clarity of presentation the respective basic oxidation data for commercial FeCrAlRE alloy foils (30–125 μm thick) and thicker sections (0.2–2.0 mm)* will be described separately. Although this division was somewhat arbitary, it does enable the data to be shown with the maximum possible detail bearing in mind, for example, that the exposure durations for foils were often ≥200 h and for sections up to 16 000 h. However, as will be discussed later all the data taken together provide a logical and complete understanding of the critical parameters controlling the oxidation behaviours of these alloys.

The increase in the gross mass gain (a measure of the oxygen uptake) with exposure time for Aluchrom YHf, Kanthal AF and PM2000 foils during isothermal/continuous and discontinuous (20 h cycles) oxidation, in air, at 1200°C, are shown in Figs 1, 2 and 3 respectively. These indicate that the kinetics of oxidation (i.e. of alumina scale formation and growth) were similar whether the exposure was continuous or discontinuous and were such that the rate of oxidation was fastest initially, and subsequently decreased increasingly with time. There did not appear to be any significant differences between the overall oxidation resistances of these alloys.

The corresponding gravimetric data (gross mass gain vs time) for all thicker sections (nominal thicknesses in range 0.2–2.0 mm), together with that of some foils, are shown in Fig. 4(a) for Aluchrom YHf at 1100, 1200 and 1300°C and in Figs 5 and 6 for Kanthal AF, Kanthal APM and PM2000 at 1200 and 1300°C respectively. In these discontinuous exposures the thermal cycle was either 20, 100 or 300 h depending on the test duration and the ability to define the time to breakaway ($t_{B/O}$) with most reasonable accuracy.

The main conclusions to emerge from these results, as with the foils described above, were that at least up to the onset of scale mechanical failure (see below) the oxidation resistances of the alloys were comparable and that neither thermal cycle frequency nor alloy thickness exerted any significant influence on scale growth

---

*NOTE: The thicknesses quoted in the text are the nominal values for the alloy foil/sheet produced. The experimentally measured thicknesses of the individual specimens are used however, in the production of the Figs.

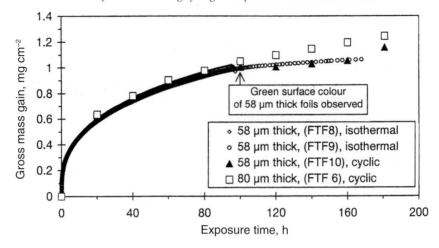

Fig. 1 *Gross mass gain data for 58 μm and 80 μm thick Aluchrom YHf foils during isothermal and cyclic (20 h heating/2 h cooling) oxidation in air at 1200°C.*

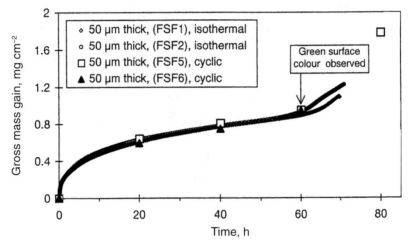

Fig. 2 *Gross mass gain data for 50 μm thick Kanthal AF foils during isothermal and cyclic (20 h heating/2 h cooling) oxidation in air at 1200°C.*

kinetics. As with the foils, the rate of oxidation of the thicker sections was fastest initially, decreasing with time according to a sub-parabolic, then essentially paralinear relationship until the mass gain increased rapidly due to breakaway oxidation (designated B/O in Figs 4(a), 5 and 6), also discussed below. Up to breakaway the gravimetric data for all alloys may be fitted best to a power law relationship, $[\Delta m]^n = Kt$. A plot of n against volume/surface area (mm) (i.e. essentially 0.5 thickness) for all relevant information from the present and previous programmes [5–7] (Fig. 7) revealed that with increasing thickness n gradually decreased from an average value of ~3.5 for most foils (<125 μm thick) to values between 2.0–2.5 for the thicker alloy sections (0.2–2.0 mm thick). The explanation could derive from the concurrent deformation of substrates by creep relaxation resulting from oxidation induced stresses.

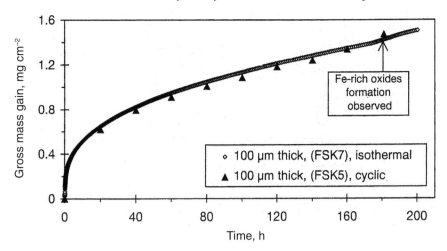

Fig. 3 *Gross mass gain data for 100 μm thick PM2000 foils during isothermal and cyclic (20 h heating/2 h cooling) oxidation in air at 1200°C.*

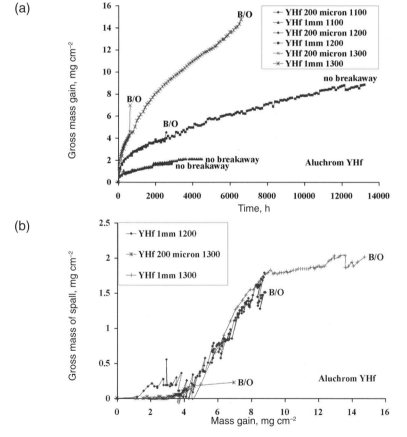

Fig. 4 *Gross mass gain (a) and spallation (b) data for the discontinuous oxidation of Aluchrom YHf in air at 1100, 1200 and 1300°C.*

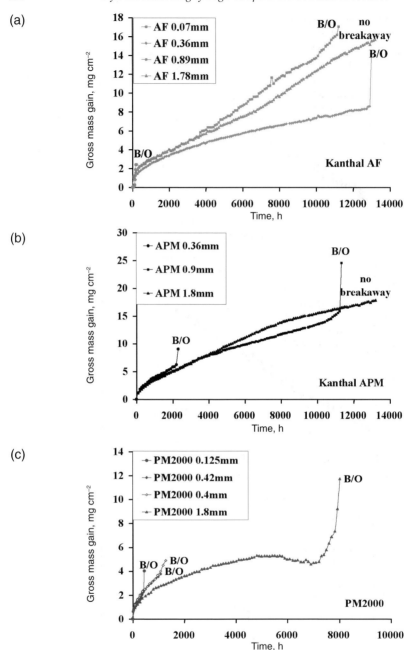

Fig. 5 *Gross mass gain data for the discontinuous oxidation of (a) Kanthal AF, (b) Kanthal APM and (c) PM2000 in air at 1200°C.*

The foils deformed extensively [3] but could not be measured easily. However, it was possible to assess this phenomenon on Kanthal AF, Kanthal APM and PM2000 thicker sections by measurement of their extension after each of the initial five 100 h oxidation cycles at 1300°C [3], assuming, of course, that any substrate damage arising

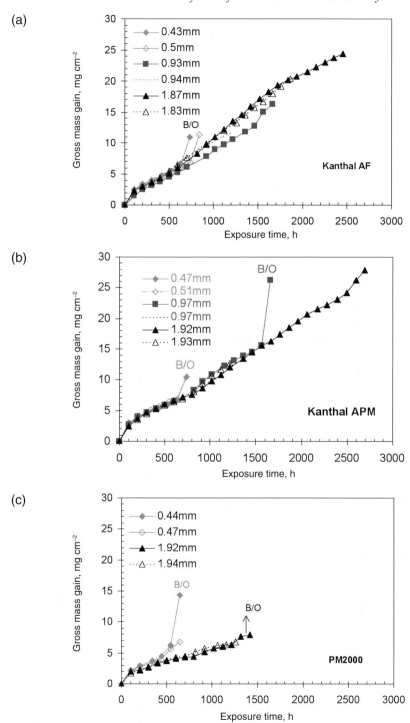

Fig. 6 *Mass gain data for the discontinuous oxidation of (a) Kanthal AF, (b) Kanthal APM and (c) PM2000 in air at 1300°C.*

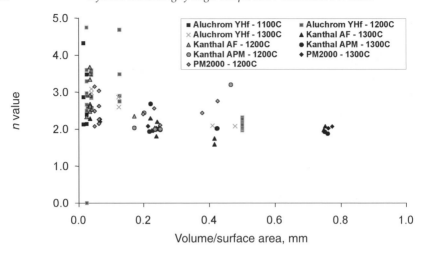

Fig. 7 *Variation with the volume/ surface area ratio (i.e. 0.5 thickness) of the parameter n in a power law relationship describing the mass gain data during the oxidation of a range of commercial FeCrAl(RE) alloys in air at 1100–1300°C.*

through oxidation did not affect alloy creep properties significantly. Additionally, it should be recognised, that since specimens often bent, the length values given in Fig. 8, at best, were semi-quantitative, especially for specimens of the weakest alloy, Kanthal AF. Both Kanthal alloys elongated linearly with oxidation time by extents, which were lowest on 2 mm thick samples and highest on 0.5 mm samples. By contrast PM2000 was more resistant to oxidation-induced deformation, which could only be measured on the 0.5 mm thick alloy, for 2 mm thick material it was within measurement error.

Turning next to mechanical failure, when the stresses generated by scale growth and particularly also on cooling, could no longer be relieved by substrate creep relaxation, interaction with defects developed within the scale during its growth caused mechanical failure (cracking and spallation) [8]. The extents of spallation as a function of the gross mass gain for Aluchrom YHf oxidised in air at 1200 and 1300°C are shown in Fig. 4(b). The corresponding data for Kanthal AF, Kanthal APM and PM2000 at 1200°C and 1300°C are detailed in Figs 9 and 10 respectively. For all four alloys spallation was initiated at a critical mass gain (i.e. corresponding to a critical oxide thickness). At 1300°C these values increased as the thickness of the specimen decreased from 2.0 to 0.2mm and were alloy dependent, being smallest on PM2000 and highest on Kanthal AF and Aluchrom YHf (i.e. consistent with the elongation measurement at the same temperature). As examples, the critical values for 2 mm thick PM2000 were calculated to be 10 μm, while the corresponding values for 2 mm thick Kanthal AF and 1 mm thick Aluchrom YHf were 20 μm. These calculations were made assuming a mass gain of 1 mgcm^{-2} results in a scale thickness of 5 μm [8]. In contrast such variations with alloy and thickness were less obvious for the corresponding results at 1200°C, as all the critical scale thicknesses appeared to fall within the range 7–10 μm.

On re-exposure to oxidant at both oxidation temperatures, 1200°C and 1300°C, alumina reformed on those areas from which scale had spalled, whilst existing scale

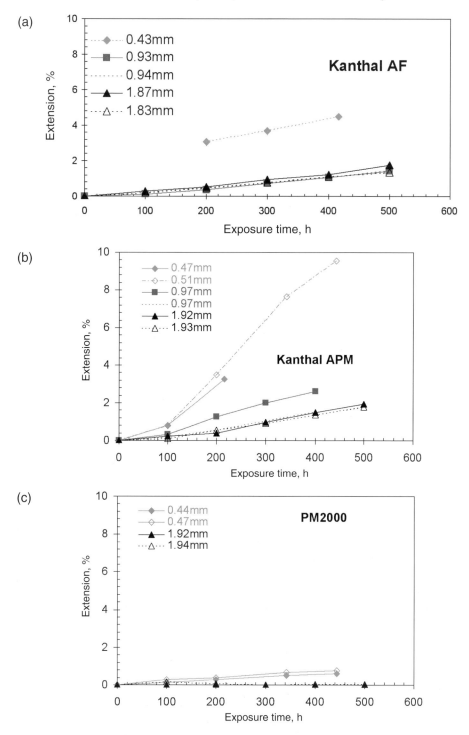

Fig. 8 *Extension of (a) Kanthal AF, (b) Kanthal APM and (c) PM2000 as a result of cyclic oxidation in air at 1300°C.*

(a)

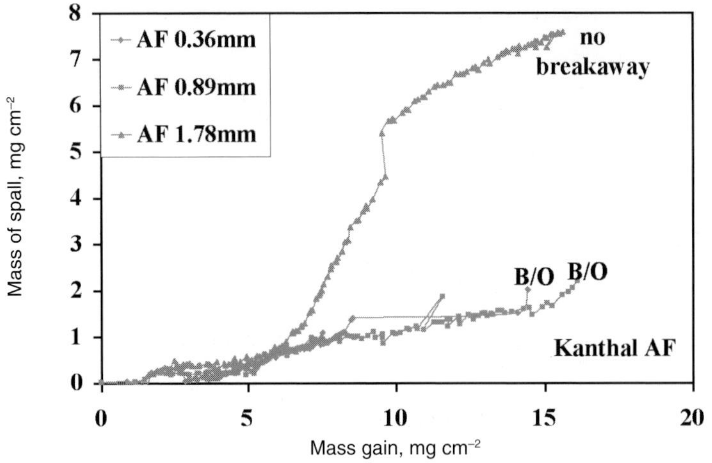

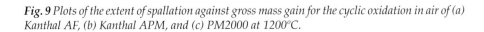

(b)

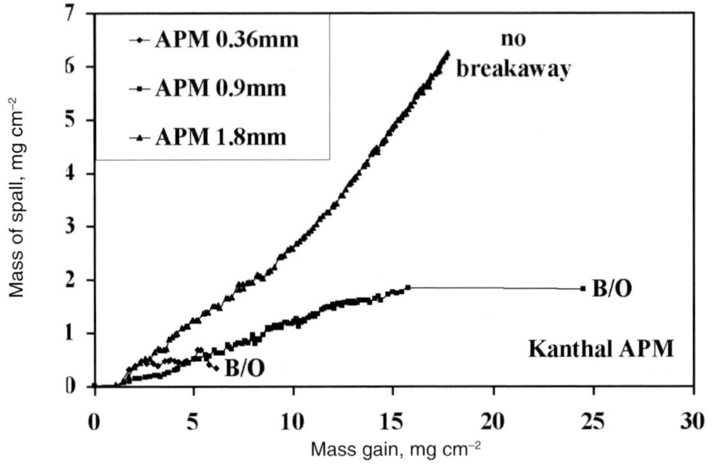

(c)

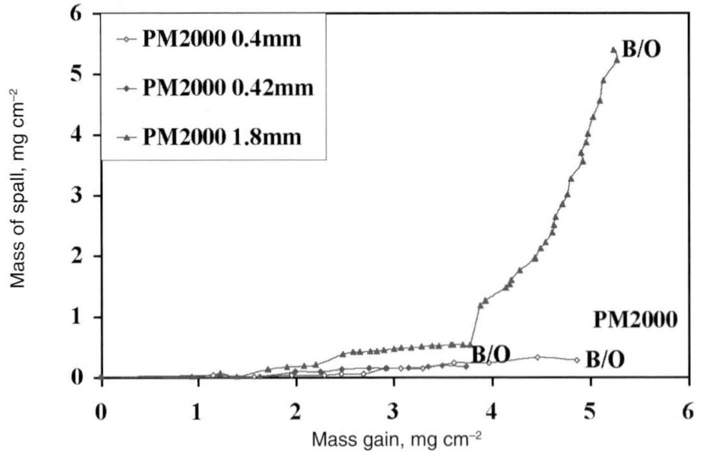

Fig. 9 *Plots of the extent of spallation against gross mass gain for the cyclic oxidation in air of (a) Kanthal AF, (b) Kanthal APM, and (c) PM2000 at 1200°C.*

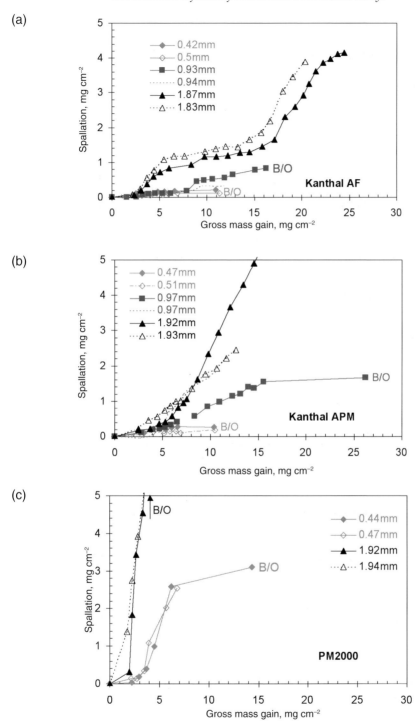

Fig. 10 Plots of the extent of spallation against gross main gain for the cyclic oxidation of (a) Kanthal AF in air of Kanthal AF, (b) Kanthal APM and (c) PM2000.

continued to grow. During each successive cooling period of the discontinuous exposure cycle further spallation occurred such that the rate of spallation accumulation was higher the stronger (highest PM2000, lowest Aluchrom YHf) and the thicker the alloy. It should be noted that no oxide spallation occurred from 0.2mm and 1 mm thick Aluchrom YHf at 1100°C, from 0.2 mm thick Aluchrom YHf at 1100°C and 1200°C or from 30–125(m thick Aluchrom YHf Kanthal AF, Kanthal APM and PM2000 foils at 1100, 1200 and 1300°C because the scale thicknesses generated were below the critical values for spallation.

Chemical failure of these FeCrAlRE alloys resulted in the inability of the protective alumina scale to be sustained (i.e. to reform/self-heal) once the substrate aluminium level fell below a critical value, defined for parallel sided coupon specimens as $[Al_{crit}]$. This led to non-protective/breakaway oxidation eventually, with total oxidation of the alloy to (FeCr) oxides, as described previously e.g. [3,5,8–10], and shown in a further example, Fig. 11.

For further discussion of this life-limiting process it is again easier to describe separately the behaviour of foils and of thicker sections. There is also an underlying physical difference, as described below, in that $[Al_{crit}]$ is thickness dependent.

With Kanthal AF and Aluchrom YHf foils (≤ 80 µm thick), when all the Al was essentially consumed (i.e. by definition chemical failure) chromia formed as an underlying layer, as was witnessed by a surface colour change from grey to green [10] (Figs 1, 2 and 12). This afforded pseudo-protection for a period until non-protective Fe/Cr oxide formation ensued as indicated by a rapid mass increase (Figs 2, 3, 4(a) and 5(a). The 100–125 µm PM2000 foil showed a slight variant in behaviour, in that during oxidation the surface colouration changed from grey to red [10] (Fig. 3), due to iron oxide formation within the scale and this was taken as an

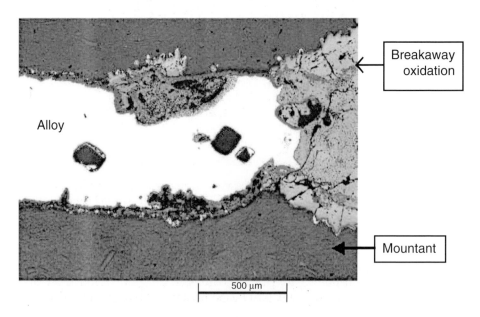

Fig. 11 *Scanning electron micrograph of a transverse cross section of 1 mm Kanthal APM following 11300 h oxidation in air at 1200°C.*

Table 4. Pseudo-protection periods (h) for Aluchrom YHf, Kanthal AF and PM2000 foils oxidised in air at 1100, 1200 and 1300°C

Alloy	Thickness, μm	Psuedo protection period, h		
		1100°C	1200°C	1300°C
Aluchrom YHf	30	288, 480, 576	N.T.	N.T.
	50	1056, 1152, 1248	70,100	N.T.
	80	$Al_{crit} \geq$ zero at B/O	140, 200	30
Kanthal AF	50	N.D.	20, 40	N.M.
	70	N.D.	60	2600
PM2000	125	200	100	N.M.

N.T. = Not tested.
N.D. = Not determined.
N.M. = Not measurable.

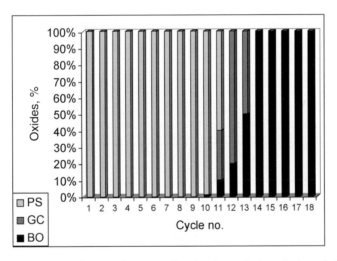

Fig. 12 *Schematic diagram showing the extent of α-alumina scale degradation, of alloy Kanthal AF at 1200°C in laboratory air (PS = Protective Scale, GC = Green Colour and BO = Breakaway Oxidation. Each cycle was of 20 h duration, so time = 20× cycle no).*

indication of the onset of chemical failure. This again preceeded complete scale breakdown and non-protective oxidation Fig. 5(c).

The times to chemical failure/breakaway oxidation ($t_{B/O}$) for Kanthal AF and Aluchrom YHf foils ($\leq 80\,\mu$m) were taken at the point of total Al consumption, shown, as described, by the surface colouration change (grey–green), by a change in the instantaneous oxidation rate constant (Fig. 13) or by calculation from the mass gain data assuming the oxygen uptake was associated entirely with alumina formation.

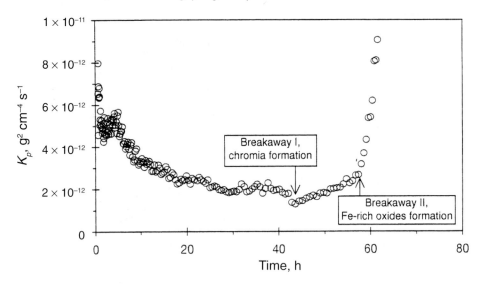

Fig. 13 *Parabolic growth rate constant vs time for 50 μm thick Kanthal AF, foil during isothermal oxidation, at 1200°C, in synthetic air.*

The corresponding $t_{B/O}$ for PM2000 foils also has been described above. The onset of chemical failure for these foil samples, will be compared with data on thicker sections later in this paper. However, before doing this it is interesting to assess the periods of psuedo-protection for foils at 1100°C, as well as 1200 and 1300°C, which are summarised in Table 4. It is recognised that the definition of the period (i.e. between $t_{B/O}$ as defined and the onset of rapid mass gain associated with non-protective iron oxide formation) is somewhat arbitary and therefore the values are at best qualitative. Nevertheless, the data indicated that the pseudo–protection period decreased with increasing oxidation temperature, increased with substrate thickness but then diminished for thickness ≥ 125 μm probably associated with a change in failure mechanism and finally was alloy dependent (being longest on Aluchrom YHf, and shortest on PM2000).

Turning now to all the thicker sections (0.2–2 mm) of these four FeCrAlRE alloys, since the exposures were discontinuous there was a time interval between successive gravimetric measurements/visual observation of the specimens. As a consequence values for $t_{B/O}$ were taken as the mean of the last time in the protective oxidation regime and the first time in the regime of accelerating oxidation (Figs 4, 5 and 6). In these tests no periods of pseudo-protection, as observed in the behaviour of thinner foils of some of these alloys, could be discerned. Also, despite the extended exposure periods the following seven specimens still had not gone into breakaway oxidation, 0.24, 0.51 and 1.02 mm thick Aluchrom YHf after 9312 h at 1100°C, 0.90mm thick Aluchrom YHf after 16 000 h at 1200°C, 1.78 mm thick Kanthal AF after 16 000 h at 1200°C, 0.97 mm thick Kanthal APM after 13 000 h at 1200°C and 1.90mm thick Kanthal APM after 16 000 h at 1200°C.

Values of $t_{B/O}$ for the oxidation in air of foils only at 1100°C and of both foils and thicker sections at 1200°C are plotted against thickness in Fig. 14, and the best estimate lines shown were for the Aluchrom YHf, 1000°C, the Aluchrom YHf, 1200°C and the

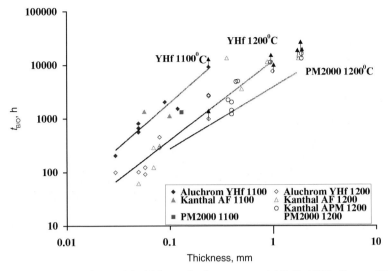

Fig. 14 *Dependence of* $t_{B/O}$ *with thickness for four commercial FeCrAlRE alloys oxidised in air at 1100°C and 1200°C (↑ denotes time of last measurement as sample has not yet failed).*

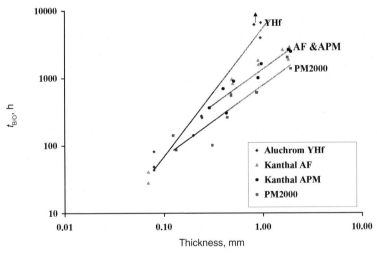

Fig. 15 *Dependance of* $t_{B/O}$ *with thickness for four commercial FeCrAlRE alloys oxidised in air at 1300°C. (↑ denotes time of last measurement as sample has not yet failed).*

PM2000, 1200°C data. The corresponding values for foils and thicker sections at 1300°C are plotted in a similar manner in Fig. 15. The main general trends to emerge, as expected, were that for all alloys, at each temperature, $t_{B/O}$ increased with thickness, a consequence of the aluminium available and for each thickness decreased with increasing temperature, arising from the increased rate of aluminium consumption by oxidation.

For foils the $t_{B/O}$ vs thickness relationship, at all three temperatures, was essentially alloy independent. In contrast, the corresponding relationship for thicker sections, determined at both 1200°C and 1300°C, was alloy dependent. All the $t_{B/O}$ values for Aluchrom YHf thicker sections (i.e. up to ~1 mm thick) followed the same relationship as for the foils. However, the corresponding $t_{B/O}$ values for the thicker sections of the two Kanthal alloys and particularly for PM2000 were lower, with the differences in behaviour between the alloys widening with increasing thickness. The onset with thickness of the alloy dependence essentially coincided with that of scale mechanical failure. The severity of this synergistic interaction between mechanical failure and chemical failure was consistent with the observations described above regarding mechanical failure — i.e. the propensity to scale mechanical failure increased with increasing substrate alloy creep resistance (PM2000 being the highest, Aluchrom YHf the lowest) and with alloy thickness.

Two procedures were used to derive experimentally values of [Al$_{crit}$]. In the first, values of [Al$_{crit}$] (calculated) were computed from the gravimetric mass gain data at 1100–1300°C using the last gross mass gain measurement in the protective oxidation regime (i.e. prior to the onset of breakaway) presupposing that the mass gain was associated entirely with oxidation of aluminium. These data are plotted against thickness in Fig. 16. The second series of values, [Al$_{crit}$] (analysed), were measurements by EDX/EPMA of the aluminium contents of samples immediately following the onset of breakaway, which usually occurred at corners/edges. Analyses were made at regions of specimens well away from the breakaway area and where the alloy was still protected by an alumina scale. Because only a limited number of [Al$_{crit}$] (analysed) values for parallel sided coupon specimens were available additional data, derived from the previous COST 501/2 project, for PM2000 oxidised at 1100°C and for another ODS alloy, ODM751 (but with a similar composition to PM2000) at 1200(C, were

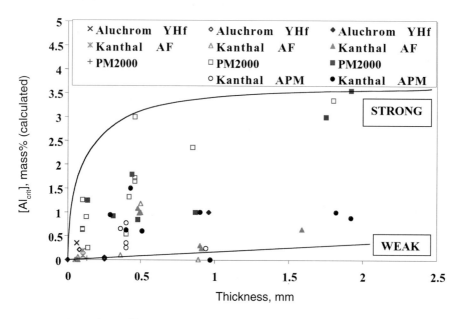

Fig. 16 *Dependence of [Al$_{crit}$] (calculated) with thickness for four commercial FeCrAlRE alloys oxidised in air at 1100–1300°C. (The upper and lower boundary lines are drawn solely to delineate the data set.)*

included in the plot of these values again as a function of thickness in Fig. 17. Inevitably there were noticeable scatters in both data sets but the general trends were that $[Al_{crit}]$ (calculated) values were higher than those of $[Al_{crit}]$ (analysed), that there was no significant temperature effect but that both series of values for the more creep resistant (i.e. stronger) alloys (PM2000 and ODS 751) tended to be higher than those for the weaker alloys (Aluchrom YHf, Kanthal AF and Kanthal APM). This alloy dependence could be affected also by thickness, particularly as shown by the PM2000 $[Al_{crit}]$ (calculated), data set, which would suggest $[Al_{crit}]$ (calculated) increased with thickness. It could be argued that, particularly for the two Kanthal alloys and Aluchrom YHF, another trend could be that $[Al_{crit}]$ (calculated) increased with thickness from 0.05 to 0.5 mm then for thicker sections (up to 2 mm) decreased to a much lower value.

4. Discussion

A considerable body of lifetime oxidation data has been generated for four commercial FeCrAlRE alloys over a temperature range (1100–1300°C) of technological significance. These data have enabled the development of a life-prediction model [1] backed now by a general overall understanding of the oxidation behaviour of the FeCrAlRE class of alloys to emerge from this study that is substantiately in accord with, but which also augments, previous observations [1–3,8,9].

For each alloy the kinetics of the initial protective alumina scale formation and growth were independent both of alloy thickness and whether the exposures were continuous or discontinuous. Any modest differences between the oxidation rates

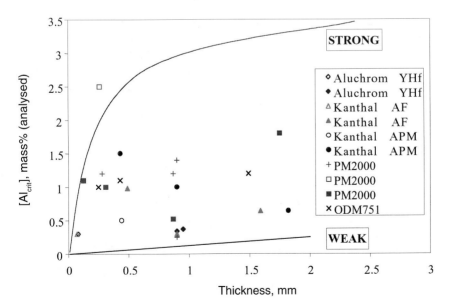

Fig. 17 *Dependence of $[Al_{crit}]$ (analysed) with thickness for five commercial FeCrAlRE alloys oxidised in air at 1100–1300°C. (The upper and lower lines embracing the data are the same as shown in Fig. 16.)*

of the individual alloys probably derived from effects (e.g. on transient oxidation, diffusion processes) originating from the various reactive elements (RE) and levels of other minor alloy constituents that differed between these alloys.

During growth defects developed within the alumina scale and probably also at the scale–substrate interface. Once a critical defect concentrations was reached, i.e. at a critical scale thickness, interaction between this porosity and the applied stress resulting from scale growth and also on cooling due to the CTE (coefficient of thermal expansion) mismatch between the scale and substrate caused mechanical failure (cracking and spallation) of the scale [8]. The actual residual stress acting on the scale was reduced by substrate creep deformation by an extent depending on the creep properties of the alloy. In accordance with the elongation measurements/ substrate deformation observations the critical scale thickness for spall initiation decreased with increasing alloy thickness and higher alloy creep resistance.

On subsequent exposure, the spall regions were re-oxidised, while intact protective scale continued to thicken so that at successive cooling stages at the end of each exposure cycle, the scale again failed mechanically. As substrate stress relief continued to be important the rates of continuing spallation followed similar patterns to those of spall initiation, being higher the thicker the alloy section and the more creep-resistant the alloy.

The ultimate and limiting criterion for the FeCrAlRE alloy oxidation lifetimes was chemical failure of the protective alumina scale. Non-protective, breakaway oxidation then ensued leading to through section oxidation of all the metallic constituents. Chemical failure was initiated when the aluminium concentration in the alloy fell below a critical value needed to sustain/reheal the protective alumina scale. Aluminium diffusion in these alloys was rapid and thereby, not controlling [5]. In confirmation of previous observations [3, 5,10] the onset of breakaway ($t_{B/O}$) was shown experimentally to be governed by three critical factors. The first was the initial quantity of aluminium available in the alloy (i.e. the aluminium reservoir), which was a function of the aluminium content (typically about 5 mass% for these alloys) and thickness, as confirmed, as $t_{B/O}$ increased with thickness (Figs 14 and 15). The second factor was the rate of aluminium consumption through oxidation, determined by the scale growth kinetics, which was temperature dependent. This was also alloy dependent as aluminium consumption was exacerbated further by scale mechanical failure resulting in spallation. The final factor, was the critical Al value [Al_{crit}], for the initiation of breakaway, which depended on the alloy and thickness. Unfortunately it was not possible to determine this parameter precisely experimentally. As a consequence two procedures were employed to evaluate [Al_{crit}] (calculated) and [Al_{crit}] (analysed) (see Figs 16 and 17 respectively) both of which were subject to uncertainty. The former values based on the last oxygen uptake measurement prior to breakaway onset did not take account of the oxidation of any alloy constituent (such as Fe, Cr, Ti, RE) other than aluminium. These values probably overestimated [Al_{crit}] and as such were conservative. The other series of [Al_{crit}] (analysed) values were determined following the onset of breakaway. These also could be subject to error, arising from that of measurement and that the available free aluminium in the alloy could be less than measured if internal oxidation and/or nitridation had occurred to react with aluminium. As a consequence these values were low and underestimated [Al_{crit}].

In accordance with recent related work concerned with the chemical failure of protective chromia scales on alloys at lower temperatures [11] two failure modes would appear to be involved. The first, Intrinsic Chemical Failure (INCF), was governed entirely by aluminium consumption, such that [Alcrit] was either zero or a low value (< 0.3mass%). It controlled the lifetimes of thin (≤ 125 μm) foils of all alloys and also of the thicker Aluchrom YHf sections, as witnessed by the continuity of the $t_{B/O}$ versus thickness relationship at 1200°C and 1300°C over the entire thickness range. A second failure mode, Mechanically Induced Chemical Failure (MICF), occurred under those circumstances where mechanical processes caused scale failure on cooling. It resulted in lower times to breakaway both indirectly, through scale spallation enhancing the rate of aluminium consumption and directly, by initiating chemical failure at cracks through the scale, which were unable to reheal on further exposure. This failure mode controlled the lifetimes of the thicker Kanthal AF, Kanthal APM and PM2000 sections and occurred at a somewhat higher [Al$_{crit}$] value with possibly a step function increase to ≈ 1.5mass%. This synergistic interaction between mechanical failure and chemical failure processes was influenced by the alloy creep properties and thickness — and as a consequence was alloy dependent, as observed experimentally with PM2000 being the more severely affected FeCrAlRE alloy.

5. Conclusions

1. Lifetime data have been obtained for the oxidation, in air at 1100–1300°C, of four commercial FeCrAlRE alloys (Aluchrom YHf, Kanthal AF, Kanthal APM and PM2000), with a range of thicknesses (30–125 μm thick foils and 0.2–2.0 mm thicker sections).

2. A detailed scientific understanding has been developed of the oxidation behaviour of these alloys, involving the formation and development of an alumina protective scale and its subsequent failure by mechanical and chemical processes.

3. Chemical failure, the ultimate life-limiting process, leading to catastrophic non-protective oxidation, resulted from the inability of the protective alumina scale to be sustained/rehealed once the substrate aluminium level fell below a critical value [Al$_{crit}$].

4. Two chemical failure modes were involved. Intrinsic Chemical Failure was governed entirely by aluminium consumption, such that [Al$_{crit}$] was either zero or a low value (≤ 0.3mass%). It controlled the lifetimes of thin (≤ 125 μm) foils of all alloys and essentially also of the thicker Aluchrom YHf sections. Mechanically Induced Chemical Failure occurred under those circumstances where mechanical processes caused scale failure on cooling (i.e. by cracking and spallation) and resulted in lower times to breakaway. This failure mode controlled the lifetimes of the thicker Kanthal AF, Kanthal APM and PM2000 sections and occurred at a higher [Al$_{crit}$] value, typically ≈ 1.5 mass%. Its impact

was influenced by the alloy creep properties and thickness and thereby, was alloy dependent.

6. Acknowledgements

This study formed part of the BRITE/EURAM LEAFA Project funded by the European Community (Project No. BE-97-4491). The authors are grateful to their Partners for the supply of the alloys tested, for the chemical analyses of the alloys and for their scientific input in discussing these results.

References

1. J. R. Nicholls, *et al.*, EFC Publication No. 34. This volume, pp.83–106.
2. J. R. Nicholls and M. J. Bennett, EFC Publication 34. This volume, pp.3–14.
3. J. P. Wilber, M. J. Bennett and J. R. Nicholls, *Mater. High Temp. 2000*, **17**, 125–132.
4. First Draft Code of Practice for Discontinuous Corrosion Testing in High Temperature Gaseous Atmospheres, European Commission Project SMT4–CT95–2001, TESTCORR, ERA Technology (1996).
5. M. J. Bennett, R. Perkins, J. B. Price and F. Starr, in *Materials for Advanced Power Engineering* (D. Coutsouradis *et al.*, Eds). Kluwer Academic Publishers **2**, 1994, 1553–1562.
6. J. P. Wilber, M. J. Bennett and J. R. Nicholls, in *Materials for Advanced Power Engineering* (J. Lecomte-Beckers *et al.*, eds). Forschungszentrum Jülich GmbH, Germany, **2**, 1998, 835–846.
7. Final Report European BRITE/EURAM Project "How to Improve the Failure Resistance of Alumina Scales on High Temperature Materials" (Project Number BE7972-1994) CEC Brussels, April 1997.
8. J. P. Wilber, J. R. Nicholls and M. J. Bennett in *Microscopy of Oxidation—3* (S. B. Newcomb and J. A. Little, eds). The Institute of Materials, London, 1999, pp.207–220.
9. H. Al-Badairy, G. J. Tatlock and M. J. Bennett, *Mater. High Temp.*, 2000, **17** 101–107.
10. H. Al-Badairy, *et al.*, EFC Publication No. 34. This volume, pp.50–65.
11. H. E. Evans, A. T. Donaldson and T. C. Gilmour, *Oxid. Met.*, 1999, **52**, 379–402.

3

Prediction of Oxide Spallation from an Alumina-Forming Ferritic Steel

H. E. EVANS and J. R. NICHOLLS*

School of Metallurgy and Materials, The University of Birmingham, Birmingham, B15 2TT, UK
*School of Industrial and Manufacturing Science, Cranfield University, Cranfield, Bedford, MK43 0AL, UK

ABSTRACT

Finite element methods have been used to model alumina spallation through the growth of a wedge crack along the oxide/metal interface of a 0.38 mm-thick sample of Kanthal APM (Fe–23Cr–6Al mass%). The test conditions examined involved a period of isothermal oxidation at temperatures of 1000, 1100 or 1200°C followed by cooling at a constant rate of 10^4 °C/h. Separate runs were performed to evaluate the influence of cooling rate from 1100°C. A consistent observation has been that creep relaxation within the alloy reduces the rate of crack propagation and increases the temperature drop to initiate spallation. The effects are most marked when cooling from a high temperature or at low rates and effective interface fracture energies for such conditions are high. This complex behaviour is described in terms of a 3-dimensional spallation map. It is shown how this map can also be used to define domains in which chemical failure can also arise.

1. Introduction

Aluminium-containing ferritic steels are widely used at high temperatures in applications such as automobile catalyst supports and furnace windings [1] and increasingly as heat exchangers and turbine components [2]. Their success depends not only on the formation of a protective alumina layer but also on its resistance to spallation during temperature changes. This is particularly important since this class of alloy is often used in thin sections, e.g. as little as 50 μm thickness [1], and repeated spallation and re-formation of the alumina layer can accelerate the rate of reduction of the alloy's residual aluminium content [2]. With continued aluminium depletion, there will come a time when re-formation of the spalled alumina layer will not be possible and non-protective iron-rich oxides will form. This type of breakaway oxidation has been termed generally [3] Mechanically Induced Chemical Failure (MICF). Of course, continued growth of the alumina layer will also result in aluminium depletion even in the absence of spallation. In this case, however, chemical failure will arise, at very low levels of aluminium, by reduction of the alumina layer by chromium from the alloy. This process has been termed [3] Intrinsic Chemical Failure (InCF).

The purpose of this paper is to provide the results of numerical calculations of the critical temperature drop, ΔT_c, required to initiate oxide spallation from, specifically, Kanthal APM (Fe–23Cr–6Al mass%) and to relate these predictions to chemical failure.

As has been apparent in previous studies [4–6], the use of finite element methods is necessary since (non-linear) alloy creep has a significant influence on the kinetics of wedge crack growth along the oxide/metal interface. As a consequence, ΔT_c is likely to vary in a complex manner with cooling rate and oxidation temperature. These aspects will be considered in this present paper.

2. The Finite Element Model

The finite element (FE) model is essentially the same as that used previously [4–6] and envisages oxide spallation to occur by the growth of an interfacial wedge crack nucleated during cooling by the formation of a shear crack within the adherent surface oxide layer, as shown schematically in Fig. 1. The oxide layer is taken to be initially stress-free but experiences compressive stressing during cooling as a result of thermal mismatch strains between the oxide and metal. In Fig. 1, the wedge is driven along the interface to the left-hand side by sliding along the surface of the shear crack which is, here, inclined at 45°. The associated out-of-plane displacement generates tensile stresses normal to the oxide/metal interface and these favour the propagation of the wedge crack along the interface to the left-hand side of the shear crack.

The detailed FE mesh over this spallation zone is shown in Fig. 2. The elements used are axisymmetric and the spallation zone is consequently circular and, in this case, of 8 μm radius. This is centred within a spall cell of 80 μm radius, so ensuring that edge effects are negligible. The half-thickness of the underlying alloy is 190 μm and the results presented in this paper refer to a surface oxide thickness layer of 5 μm. The left-hand edge of the mesh shown in Fig. 2 is the centre line of the (axisymmetric) model and all lateral displacements for these edge nodes are constrained to be zero for reasons of symmetry. A feature of the model is the use of interfacial elements [7] of high aspect ratio representing the oxide/metal interfacial zone and the shear crack. Both sets of these elements are 0.1 μm thick. In general, they can have mechanical and physical properties different from the adjacent phases and so can represent discrete phases or cracked zones. In earlier studies [4] on an austenitic steel, the interfacial zone was amorphous silica and was, in reality, of similar thickness to the interfacial elements. In this case, the interfacial zone was treated as a distinct deformable phase within the model. For the present alumina-forming steel there is a true oxide/metal interface and the 0.1 μm thick elements used in the model

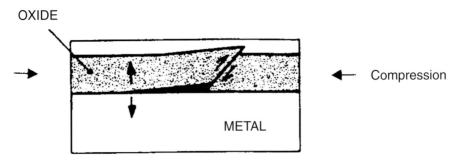

OXIDE

Compression

METAL

Fig. 1 A schematic diagram of the wedging process during cooling.

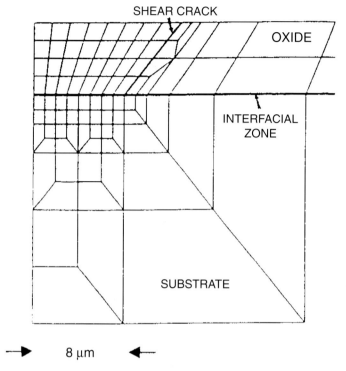

Fig. 2 *Details of the finite-element mesh in the vicinity of the wedge crack.*

will be a less-good approximation to this. Unfortunately, thinning these elements can lead to instabilities and difficulties with convergence and, so, the compromise approach has been to limit within-element deformation and fracture to those elements within the spall zone.

The presence of the interfacial elements permits the growth kinetics of the wedge crack to be modelled. This occurs incrementally by a length of 1 μm (the length of an interfacial element to the left of the shear crack in Fig. 2) whenever the maximum tensile stress within an interfacial element exceeds a pre-set fracture value, σ_f, and there is a reduction in total strain energy when this element becomes non load-bearing. For the present alloy, it is assumed that the interfacial zone has the same elastic properties as alumina and that σ_f has a value of 1700 MPa. This is the same approach as used previously [5,6] in other alumina-forming alloys but, clearly, there must be uncertainty over the actual value of the fracture stress. The following argument can be used, however, to show that the value chosen is not entirely arbitrary. It is recognised that, since the interface ahead of the wedge crack is free of defects, the fracture stress must correspond to a cleavage value for the wedge crack. Within the FE model, the so-called crack-tip stress is actually monitored at a distance, r, 0.33 μm ahead of the crack and, so, has a much smaller value than that existing near the actual crack tip. For a similar crack in bulk alumina, the appropriate stress field would vary as $r^{-1/2}$ and the critical stress of 1700 MPa at 0.33 μm would correspond to an approximate cleavage stress of $E_{ox}/10$ at approximately a molecular diameter from the actual crack tip, where E_{ox} is the Young's modulus.

For the computations undertaken here, the shear crack is assumed to pre-exist, i.e. it penetrates to the top of the oxide/metal interfacial zone (but not to the underlying alloy) even at the oxidation temperature. This assumption is intended to produce conservative estimates of ΔT_c but these will not be unduly pessimistic in view of the large, microns-long, lenticular voids known to form in some alumina layers [8,9]. Shear displacements on this crack surface are taken to be frictionless. The alumina layer and interface are taken to be creep-rigid elastic solids but alloy creep is incorporated into the model. For all phases, the elastic properties are assumed to be temperature invariant but this simplification has negligible effect on the predictions. The material parameters for the oxide phases and alloy substrate used in the computations are given in Table 1. The creep rate of the Kanthal APM alloy is described, from the manufacturer's data sheets, as:

$$\dot{\varepsilon} = 1.3 \times 10^{-35} \sigma^{5.5} \exp\left(\frac{-28484}{T}\right), s^{-1} \tag{1}$$

where stress, σ, is in Pa and temperature, T, in K.

3. Results and Discussion

3.1. Crack Kinetics

This influence of alloy creep can be appreciated from Fig. 3 which shows the kinetics of growth of the interfacial wedge crack during cooling from temperatures of 1200, 1100 and 1000°C at a constant rate of 10^4 °C/h. The striking feature is that crack growth is inhibited during the early stages of cooling because alloy creep permits stress relaxation at the crack tip with associated blunting. At intermediate temperatures, when such relaxation rates described through eqn (1) are much reduced, the rate of stress build-up at the crack tip increases and crack propagation continues. These trends are shown in Fig. 4 where, as described earlier, the out-of-plane interface stress is monitored at a position 0.33 µm ahead of a 1-µm long wedge crack. It can be seen that the rate of stress relaxation during the early stages of cooling from the

Table 1. Elastic properties of alumina and Kanthal APM

Material	Young's Modulus, GPa	Poisson's Ratio	Thermal expansion coefficient, $\times 10^6$ K^{-1}	Fracture stress, MPa
Kanthal APM	150	0.30	16.0	Not applicable
Alumina layer	387	0.27	7.9	Not applicable
Oxide/metal interface	387	0.27	7.9	1700

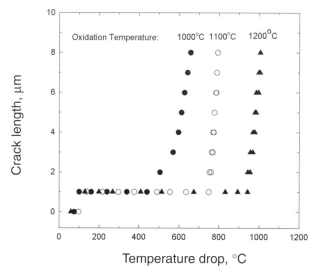

Fig. 3 *Kinetics of wedge crack growth under a 5-mm thick alumina layer on Kanthal APM during cooling at a rate of 10^4 °C/h from various temperatures.*

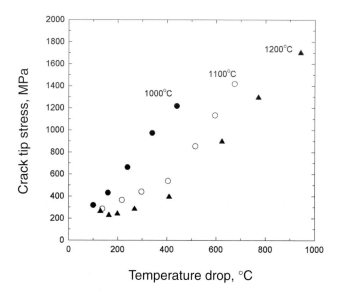

Fig. 4 *The development of out-of-plane interfacial stress ahead of a 1-μm wedge crack in APM during cooling from various oxidation temperatures.*

highest temperature, 1200°C, is sufficiently rapid that the interface stress actually reduces in value. The effect of the strong temperature dependence of creep rate (eqn 1) on stress relaxation ensures that cooling from high temperatures, at a given oxide thickness, is beneficial in that a larger critical temperature drop, ΔT_c, arises. Actual values, for the case considered of a 5 μm alumina layer, are given in Table 2.

Table 2. *Computed values of* ΔT_c *and* γ_F *for spallation of a 5 μm alumina layer from Kanthal APM*

Oxidation temp., °C	Cooling rate, °C h⁻¹	ΔT_c, °C	γ_F, J.m⁻²
1200	10^4	1005	66.0
1100	10^6	613	34.8
1100	10^4	793	58.3
1100	10^2	1130	118.3
1000	10^4	657	40.0

When the wedge crack is small (1 μm), the tensile and shear stresses along the interface ahead of it are of similar value and, in that sense, the crack is of mixed mode. However, as the sliding displacement along the oxide shear crack increases, i.e. as temperature continues to fall, even small wedge cracks acquire dominantly tensile characteristics. This is demonstrated in Fig. 5 for a 3 μm crack during cooling from 1200°C where it can be seen that the tensile stress close to the crack tip is an order of magnitude larger than the shear stress. Nevertheless, whereas this shear stress decays gradually with distance, the tensile stress rapidly falls and, indeed, becomes compressive a few microns ahead of the crack.

The influence of cooling rate on crack kinetics from an oxidation temperature of 1100°C follows similar trends to that found for the effect of oxidation temperature,

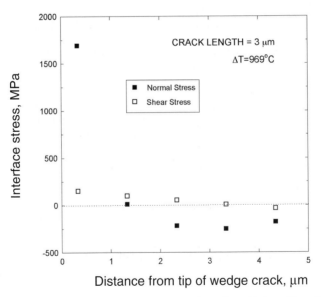

Fig. 5 *The distribution of tensile and shear stress along the oxide/metal interface ahead of a 3-μm wedge crack during cooling from 1200 °C at* 10^4 *°C/h.*

as shown in Fig. 6. Again, the striking feature is that crack growth is inhibited during the early stages of cooling. The extent of this period of zero crack growth increases as the cooling rate reduces and relaxation of crack-tip stresses by substrate creep is favoured. Again, at intermediate temperatures, when relaxation rates become insignificant, crack propagation continues. The consequence is that slow cooling rates increase the spallation resistance, at least for the simple cooling transient considered here. The computed values of ΔT_c, the critical temperature drop for oxide spallation, are given in Table 2. From these it can be appreciated that fast cooling at 10^6 °C/h from 1100°C will initiate spallation during the transient at a temperature drop of 613°C whereas no spalling will occur during slow cooling.

3.2 The Spallation Map

The process of oxide spallation by wedge cracking can be predicted using the concept of a critical strain energy [10,11]. This envisages that spallation occurs when the strain energy within the oxide layer exceeds that required to produce fracture at the oxide/metal interface. The concept is, of course, widely used in fracture mechanics but first seems to have been considered in the context of oxide spallation by U. R. Evans [12]. Assuming negligible stress in the oxide layer at the oxidation temperature and biaxial stressing during cooling, it can be shown that the critical temperature drop to initiate spallation is given as [10]:

$$\Delta T_c = \left[\frac{\gamma_F}{\xi E_{ox}(\Delta\alpha)^2(1-\nu_{ox})} \right]^{\frac{1}{2}} \quad (2)$$

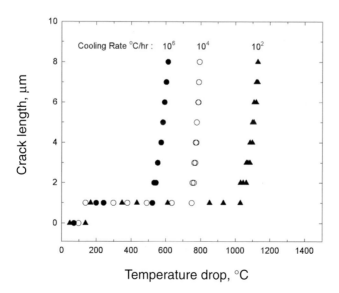

Fig. 6 Kinetics of wedge crack growth under a 5 μm-thick alumina layer on Kanthal APM during cooling from 1100 °C at various rates.

Here, ξ is the oxide thickness, E_{ox} its Young's modulus, ν_{ox} its Poisson's ratio and $\Delta\alpha$ $(= \alpha_{metal} - \alpha_{ox})$ is the difference in thermal expansion coefficients. An important parameter is γ_F, the effective fracture energy, which in the presence of substrate creep can take values well in excess of those expected for elastic behaviour. A merit of the present FE modelling approach is that γ_F can be evaluated from eqn (2) since values of ΔT_c are predicted over a range of experimental conditions. These deduced values of γ_F are included in Table 2 and demonstrate the variation that occurs with cooling rate and oxidation temperature. They serve to emphasise that no single unique value of the effective fracture energy can be ascribed to the spallation process.

Nevertheless, for a given oxidation temperature and cooling conditions this concept of a critical strain energy for spallation is known to be successful in explaining the spallation behaviour of chromia from an austenitic steel [10]. Specifically, the critical temperature drop to initiate spallation was found, experimentally, to be proportional to the parameter $\xi^{-1/2}$, demonstrating that a constant value of effective fracture energy (6 Jm^{-2} in this case) existed over a range of oxide thicknesses. The same result was found [4] using the same FE code as currently. However, these results are intuitively reasonable only if the stored strain energy within the oxide layer can be released instantaneously at the critical temperature, ΔT_c, and that negligible release of energy to the cracking process occurs prior to this stage. It has been argued [13] that creep relaxation and the insertion of a period of zero crack growth during cooling provides a mechanism whereby this condition can be approximated. As can be appreciated from Figs 3 and 6, the presence of this quiescent period of crack growth ensures that most of the crack propagation stage, and release of energy, occurs in most cases over a temperature range comprising, typically, <10% of the total temperature drop. Exceptions arise at the lowest oxidation temperature (1000°C) and at the fastest cooling rate (10^6 °C/h) at 1100°C where, in each case, creep processes are more limited.

This mechanistic basis for the critical strain energy criterion provides confidence in its use to create a spallation map for the present system. The procedure is simply to use the deduced values of γ_F from Table 2 to evaluate values of ΔT_c for a wide range of oxide thicknesses for each of the oxidation temperatures used in the computations. The overall effect of these parameters for a cooling rate of 10^4 °C/h. is represented in 3-dimensional space in Fig 7. Beneath this surface, oxide spallation will not occur and the map can be used to define permissible temperature drops to achieve this condition. As example, for an oxidation temperature of 1200°C, it is predicted that spallation will not occur during cooling to room temperature for oxide thicknesses <4.6 μm but that, for a 50 μm thick layer, the temperature drop must not exceed 356°C.

This predicted behaviour agrees well with measured data [14]. As part of the LEAFA studies, Kanthal APM samples of thicknesses 0.36, 0.9 and 1.8 mm were oxidised at 1200°C by the Jülich team (these data are reported in detail elsewhere in this volume [14]). The samples were furnace cooled every 100 h and weighed. Net and gross mass gains were recorded allowing the mass of spall to be calculated. The furnace cooling rate was close to 10^2 °C/h. The critical mass gains at onset of spallation were measured as 1.45 mg cm^{-2} at 0.36 mm thickness, 1.50 mg cm^{-2} at 0.9 mm thickness and 1.47 mg cm^{-2} at 1.8 mm thickness. This shows that the critical thickness for the onset of spallation, when cooling from 1200°C to room temperature, is 7.5–7.8 μm (corresponding to a mass gain of 1.45–1.50 mg cm^{-2}) and that this critical oxide thickness is independent of substrate thickness for samples thicker than 0.36 mm.

In this modelling work, a sample thickness of 0.38 mm has been used (similar to the thinner Kanthal APM samples evaluated above). For a 5 μm thick alumina scale the critical temperature drop is 1005°C at a cooling rate of 10^4 °C/h from 1200°C (see Table 2). Cooling at a slower rate would increase this critical temperature drop, while increasing the oxide thickness would act to counteract this, reducing the necessary temperature drop. Thus, it can be appreciated that a measured critical thickness to spall of 7.5–7.8 μm, when furnace cooled, is consistent with the current predictions.

3.3. Chemical Failure

For operating conditions above the surface shown in Fig. 7, oxide spallation will occur during cooling and these specimens or components will be susceptible to Mechanically Induced Chemical Failure (MICF) as outlined in the Introduction. This type of chemical failure will arise when the residual aluminium content of the alloy has been reduced below a critical value, C_h, so that re-formation of the alumina layer will not occur on reheating to the oxidation temperature. The magnitude of C_h is not well established but experimental work [15] on the oxidation of ternary FeCrAl alloys at the lower temperature of 800°C indicates that a minimum concentration of aluminium of 2.4 mass% was needed to form alumina. This value tends to an upper bound to the residual concentrations of aluminium at chemical failure recently found [14,16] after oxidation at 1300°C but the spread of measured values of C_h was large, however, with a minimum of 0.53 mass%. Critical residual values of aluminium in the range 1.2 to 2.0 mass% have also recently been reported [17] after exposure at 1350°C. For present purposes, a value for C_h of 2.0 mass% will be adopted but it is recognised that this is likely to be a conservative estimate and lead to underestimates of the critical oxide thickness, ξ_{MICF}, above which MICF could occur.

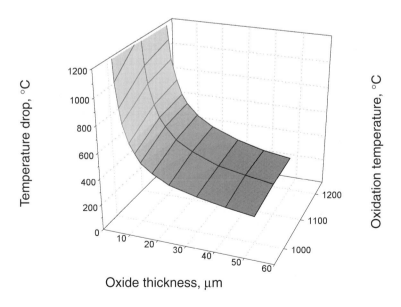

Fig. 7 *A 3-dimensional spallation map for Kanthal APM cooled at 10^4 °C/h.*

At the elevated oxidation temperatures considered here, aluminium diffusion rates within the alloy are high [18] and the depletion profile will be flat [19], i.e. aluminium may be considered to be depleted uniformly across the substrate cross section. Since the aluminium removed will be incorporated into the oxide layer, the residual aluminium content, C (mass fraction), within the alloy can be related to the thickness, ξ, of this alumina layer according to [17]:

$$C = \frac{(C_o \rho_m x_m - 2C_{ox}\rho_{ox}\xi)}{(\rho_m x_m - 2C_{ox}\rho_{ox}\xi)} \tag{3}$$

where C_o is the initial aluminium concentration in the alloy (mass fraction), C_{ox} is the mass fraction of aluminium in the oxide, ρ_m and ρ_{ox} are the corresponding densities, x is the metal thickness and ξ is the thickness of the alumina layer. Equating C to the critical value C_h and rearranging gives the critical oxide thickness above which the sample is susceptible to MICF as:

$$\xi_{MICF} = \frac{\rho_m x_m (C_o - C_h)}{2C_{ox}\rho_{ox}(1 - C_h)} \tag{4}$$

Using $C_o = 0.06$, $C_h = 0.02$, $C_{ox} = 0.54$ (for alumina), $x_m = 0.38 \times 10^{-3}$ m, $\rho_m = 7.2 \times 10^6$ gm^{-3}, $\rho_{ox} = 4.1 \times 10^6$ gm^{-3} gives this critical oxide thickness as 25.2 µm. This value is insensitive to oxidation temperature although, of course, the time taken to reach this oxide thickness will increase as temperature decreases.

Measurements of the critical mass gain at onset of breakaway oxidation have been made [14]. At 1200°C, this critical value is 6.2 mgcm^{-2} for Kanthal APM, at a sample thickness of 0.36 mm, i.e. at approximately the same thickness as used in the present modelling study. A mass gain of 6.2 mgcm^{-2} is equivalent to an oxide thickness of 32 µm. The present estimated critical oxide thickness for MICF of 252 µm is somewhat less than the value found [14] but is consistent with it in that it provides a conservative estimate, as intended.

MICF will arise only if oxide spallation occurs and this critical condition can be incorporated into the spallation map of Fig. 7. For simplicity however, this will be demonstrated for the single oxidation temperature of 1200°C as shown in Fig. 8. Here the solid line represents the 1200°C section taken from Fig. 7 and shows how the critical temperature drop, ΔT_c, for spallation decreases with increasing oxide thickness. Above this line, spallation will occur during cooling but this broad region is subdivided by the vertical line A–A which represents the boundary of the MICF domain located at an oxide thickness of 25.2 µm. To the left of A–A, i.e. at thinner oxide thicknesses, areas of the alloy surface exposed by spallation will re-form an alumina layer even though they may also be associated with outer layers of iron-rich oxides, i.e. a form of oxidation consisting of healed iron-rich pits. To the right of A–A however, such healing will not occur and spallation will result in MICF on re-exposure to temperature.

A similar construction can be undertaken to define the domain of Intrinsic Chemical Failure (InCF). This process will arise in specimens which have not experienced spallation but for which the aluminium concentration has been reduced to such an extent that reduction of the alumina layer by residual chromium can occur. This

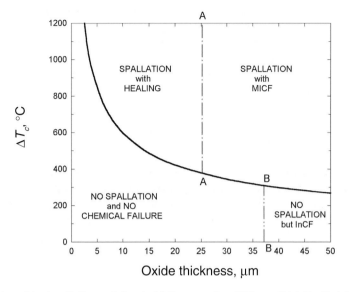

Fig. 8 *A combined spallation and chemical failure map for a 0.38-mm thick Kanthal APM specimen oxidised at 1200°C.*

requires the residual aluminium concentration to be essentially zero [20,21]. The critical oxide thickness, ξ_{InCF}, at which this will occur can then be obtained from eqn (4) by equating C_h to zero. For the sample of 190 μm half thickness considered here, ξ_{InCF} = 37.1 μm. As for the case of MICF above, this critical value will form the boundary between a safe region and one in which chemical failure will occur. This boundary is shown in Fig. 8 as the vertical line (B–B) beneath the spallation line since InCF will occur without spallation. To the left of this line, neither oxide spallation nor chemical failure will occur but InCF will arise to the right of this line but, again, oxide spallation will not occur.

4. Conclusions

Finite element methods have been used to model alumina spallation through the growth of a wedge crack along the oxide/metal interface of a 0.38 mm–thick sample of Kanthal APM (Fe–23Cr–6Al mass%). The test conditions examined involved a period of isothermal oxidation at temperatures of 1000, 1100 or 1200°C followed by cooling at a constant rate of 10^4 °C/h. Separate runs were performed to evaluate the influence of cooling rate from 1100°C.

A consistent observation was that the propagation kinetics of the wedge crack were strongly influenced by creep relaxation processes within the alloy. These were manifested by periods of zero crack growth during the early stages of cooling as the rate of increase of stress in the vicinity of the crack tip was reduced by stress relaxation. This effect was most marked when creep was favoured, i.e. during cooling from high temperatures or during cooling at low rates. The overall effect was that the critical temperature drop to initiate spallation was significantly larger in such cases.

An additional effect of the dissipation of strain energy into substrate creep was that fracture energies for spallation were appreciably higher than might be expected from elastic deformation.

The dependence on oxide thickness and oxidation temperature of the critical temperature drop to initiate spallation has been described by a 3-dimensional map which defines a surface below which oxide spallation will not occur. This map is appropriate only to the (essentially) isothermal exposure conditions considered in this work. It can also incorporate domains within which chemical failure associated with aluminium depletion can occur. Of note is that the domain for Mechanically Induced Chemical Failure (MICF) exists only above the spallation surface and for alumina thicknesses greater than approximately 25 µm. This is considered to be a conservative lower estimate of the critical oxide thickness for MICF. The domain for Intrinsic Chemical Failure (InCF) exists only below the spallation surface and for oxide thicknesses greater than approximately 37 µm.

5. Acknowledgements

This work was undertaken as part of the Life Extension of Alumina Forming Alloys (LEAFA) project funded by the EC under contract number BRPR-CT97-0562.

References

1. J. Klöwer, *Mater. Corros.*, 1998, **49**, 758–763.
2. W. J. Quadakkers and M. J. Bennett, *Mater. Sci. Technol.*, 1994, **10**, 126–131.
3. H. E. Evans, A.T. Donaldson and T. C. Gilmour, *Oxid. Met.*, 1999, **52**, 379–402.
4. H. E. Evans, G. P. Mitchell, R. C. Lobb and D. R. J. Owen, *Proc. Roy. Soc. (Lond.) A*, 1993, **440**, 1–22.
5. H. E. Evans, A. Strawbridge, R. A. Carolan and C. B. Ponton, *Mater. Sci. Eng. A.*, 1997, **225**, 1–8.
6. H. E. Evans and M. P. Taylor, *Surf. Coat. Technol.*, 1997, **94–95**, 27–33.
7. J. M. A. Cesar de Sa and D. R. J. Owen, "Computer Aided Modelling of Grain Boundary Failure Mechanisms", Report No. C/R 449/83, Department of Civil Engineering, University College Swansea, UK, 1983.
8. J. P. Banks, D.D. Gohil, H. E. Evans, D. J. Hall and S. R. J. Saunders, in *Materials for Advanced Power Engineering—II* (D Coutsouradis *et al.*, eds). Kluwer Academic Publishers, Dordrecht, The Netherlands, 1994, pp.1543–1552.
9. V. Guttman, F. Hukelmann, P. A. Beaven and G. Borchardt in *Cyclic Oxidation of High Temperature Materials* (M. Schütze and W. J. Quadakkers, Eds). Publication No. 27 in European Federation of Corrosion Series. Published by The Institute of Materials, London, 1999, 17–32.
10. H. E. Evans and R.C. Lobb, *Corros. Sci.*, 1984, **24**, 209–222.
11. J. Robertson and M. I. Manning, *Mater. Sci. Technol.*, 1990, **6**, 81–91.
12. U. R. Evans, *An Introduction to Metallic Corrosion*. Edward Arnold, London, 1948.
13. H. E. Evans and A. Strawbridge in *Fundamental Aspects of High Temperature Corrosion* (D. A. Shores *et al.*, eds). The Electrochemical Society, Pennington, NJ, 1997, pp.1–15.
14. R. Newton *et al.* This volume, pp.15–36
15. P. Tomaszewicz and G. R. Wallwork, *Oxid. Met.*, 1983, **20**, 75–109.
16. J. P. Wilber, M. J. Bennett and J. R. Nicholls, *Mater. High Temp.*, 2000, **17**, 125–132.
17. H. Al-Badairy and G. J. Tatlock, *Oxid. Met.*, 2000, **53**, 157–170.

18. A. Heesemann, E. Schmidke, F. Faupel, J. Klöwer and A. Kolb-Telieps, *Scr. Mater.*, 1999, **40**, 517–522.
19. I. Gurrapa, S. Weinbruch, D. Naumenko and W. J. Quadakkers, *Mater. Corros.*, 2000, **51**, 224–235.
20. P. Niranatlumpong, C. B. Ponton and H. E. Evans, *Oxid. Met.*, 2000, **53**, 241–258.
21. G. Strehl, *et al.*, *Mater. High Temp.*, 2000, **17**, 87–92.

4

Mechanistic Understanding of the Chemical Failure of FeCrAl-RE Alloys in Oxidising Environments

H. AL-BADAIRY, G. J. TATLOCK, H. E. EVANS*, G. STREHL[†],
G. BORCHARDT[†], R. NEWTON[§], M. J. BENNETT[§], J. R. NICHOLLS[§],
D. NAUMENKO[¶] and W. J. QUADAKKERS[¶]

Materials Science and Engineering, Department of Engineering, The University of Liverpool,
Liverpool L69 3GH, UK
*School of Metallurgy and Materials, The University of Birmingham, Birmingham B15 2TT, UK
[†]Institut für Metallurgie, Technische Universität Clausthal, D-38678 Clausthal-Zellerfeld, Germany
[§]School of Industrial and Manufacturing Science, Cranfield University, Bedford MK43 0AL, UK
[¶]Forschungszentrum Jülich, 1WV-2 Postfach 1913, 52425 Julich, Germany

ABSTRACT

Chemical failure in high temperature alloys, such as Fe–20Cr–5Al with reactive element (RE) additions, occurs when the protective alumina scale is unable to be maintained, leading to catastrophic breakaway oxidation. Voluminous oxides of iron and chromium are then produced and any remaining metal is rapidly consumed by oxidation. The onset of chemical failure is controlled by a large number of factors including alloy composition, microstructure, specimen thickness, oxide scale growth and cracking, thermal history and stress levels within the scale. In this study the oxidation of commercial and specially prepared model alloys has been investigated at temperatures up to 1300°C in laboratory air. Two distinct types of failure have been identified. Firstly, in thin samples (~200μm), Intrinsic Chemical Failure (InCF) can occur, where the growth of the protective scale depletes the Al level in the alloy to such an extent that chromia formed by alumina reduction is thermodynamically favourable. The samples adopt a greenish tinge as chromia is formed either below or within the existing alumina scale and this is followed by classical breakaway oxidation.

The second type of failure is mechanically induced by cracking and / or spalling of the oxide scale. If this occurs when the aluminium concentration in the alloy is below a critical level C_{NOSH} (NOSH = NO-Self-Healing), then the protective alumina scale cannot reform and the alloy goes into breakaway oxidation via a Mechanically Induced Chemical Failure (MICF) mechanism. Repeated spallation and regrowth of the oxide scale during the early stages of exposure lead to more rapid degradation, since the aluminium level is quickly reduced below the C_{NOSH} value.

This paper illustrates that the colour changes of the oxide, as oxidation progresses, form a useful guide to the degradation mechanism for a given alloy. This colour change is correlated with the instantaneous parabolic rate constant from thermogravimetric studies and demonstrates that the instantaneous rate constant changes from a parameter that decreases with time to one that increases with time at this transition, precisely pinpointing the onset of breakaway. Results from mechanically weak model alloys have been contrasted with much stronger oxide dispersion strengthened commercial alloys such as PM2000.

1. Introduction

The oxide scales that form on metal alloys used in high temperature industrial applications are very important in determining the suitability of these alloys for such applications. In particular, chromia- and alumina-forming alloys are the most common commercial alloys used in high temperature applications due to the slow growth, protective properties and thermodynamic stability of the chromia and alumina scales. For example, alumina-forming ferritic stainless steels, of approximate composition Fe–20Cr–5Al, are used extensively in high temperature furnace components and catalyst supports [1–3]. This is due to the formation of a protective α-alumina scale when oxidised above 1000°C. In reality, FeCrAl alloys are used in complex gaseous environments as well as in laboratory air. For example, oxidation of such alloys in synthetic exhaust-gas atmosphere has been conducted by Sigler [4,5]. He concluded that the spallation behaviour of alumina-forming Fe–Cr–Al alloys in air and in a synthetic exhaust gas (N_2 + CO_2 +H_2O) is influenced mainly by the presence of alloying elements and the control of sulfur. However, highly volatile elements such as Ca and Mg, appear to lose their sulfur-controlling ability by diffusing out of the metal matrix into the scale. In his investigation it was found that Zr provided additional improvements to scale adhesion by growing extensive oxide pegs beneath the alumina scale. He showed that in synthetic exhaust gas (N_2 + CO_2 + H_2O) spallation in localised areas was caused by H_2O in the atmosphere.

The protection that the α-alumina scale provides during use at high temperatures can be maintained, providing that the aluminium level in the alloy substrate does not drop below a critical value. In thin foils, degradation often occurs with the formation of a chromia-rich layer within or at the base of the protective alumina scale [6] before the alloy goes into the more classical pattern of breakaway corrosion with the formation of voluminous iron rich oxide.

Extensive investigations on the prediction of breakaway oxidation for alumina-forming alloys have been carried out, [7–12]. These have demonstrated that the oxidation-induced lifetime of components at high temperatures is dependent upon the initial aluminium reservoir (a function of alloy content and material thickness), the rate of aluminium consumption by oxide scale formation and the critical aluminium content for the onset of breakaway. For thicker samples (greater than ~0.2 mm) the scale thickens above the critical value for spallation, which in turn enhances the rate of aluminium consumption by oxidation and shortens the time to the onset of breakaway.

Advances in current mechanistic understanding, resulting from the vast amount of data generated during the LEAFA (Life Extension of Alumina Forming Alloys in High Temperature Environments) project, are presented. These recent developments concern primarily the more precise definition of the relationship between the substrate alloy and the alumina scale designed to protect the alloy from rapid reaction with its environment. More details of the role of bioxidant atmospheres and the influence of minor elements on the oxidation behaviour of FeCrAl alloys are described elsewhere [13,14].

2. Experimental Procedure

2.1. Raw Materials and Sample Preparation

Samples from the commercial alloys Kanthal AF, Kanthal APM , Aluchrom YHf and PM2000 were supplied by our industrial partners, Kanthal AB, Sweden, Krupp VDM, Germany and Plansee, Germany respectively. The thickness of the samples studied varied from 50 μm to 2 mm and all the samples were chemically cleaned prior to oxidation. This involved ultrasonically cleaning for 20 min in methanol, followed by iso-propyl alcohol and analar iso-propyl alcohol. The final cleaning step involved the use of a reflux condenser with analar iso-propyl alcohol. At lease two reflux cycles were used. The chemical composition of the selected alloys is shown in Table 1. The oxidation temperature during this study ranged between 1200 and 1300°C and the oxidising environments were laboratory air and moist air (air + 3.2vol.% water vapour). The oxidation type was cyclic oxidation in three different laboratories, University of Liverpool, UK, Research Centre Jülich, and Technical University Clausthal, Germany, respectively.

2.2. Sample Examination

Initial observations of some of the samples after each oxidation cycle were made to record the occurrence and extent of breakaway oxidation. Also gross mass gain was recorded for each oxidised sample. Topographical investigation was conducted using Scanning Electron Microscopy (SEM). Electron Microprobe Analysis (EPMA) was also used during this study.

3. Results

Our preliminary observations showed that samples which were prepared from thin foils of the mechanically weak alloys Kanthal AF and Aluchrom YHf deformed badly during oxidation at high temperature, but there was little or no deformation in the

Table 1. Alloy composition

Alloy	Fabrication route	Chemical composition, mass%						
		Cr	Al	Y	Ti	Zr	Hf	Y$_2$O$_3$
PM2000	M.A.	20.0	5.5	–	0.5	–	–	0.5
Aluchrom YHf	Wr	20.0	5.3	0.06	<0.05	<0.1	<0.1	–
Kanthal AF	Wr	20.0	5.0	0.08	0.08	0.06	–	–
Kanthal APM	P.M.	21.0	5.9	–	0.02	0.1	–	–

M.A. = Mechanically alloyed; Wr = Wrought; P.M. = Powder metallurgy.

case of samples that were prepared from the stronger PM2000 ODS alloy. For example, Fig. 1 shows a sample of PM2000 after 60 h oxidation at 1300°C. The coupon exhibited little distortion, but the start of breakaway oxidation may be observed at the corners of the sample. In contrast, a badly deformed sample of Aluchrom YHf oxidised at 1300°C for 40 h is shown in Fig. 2.

Fig. 1 *Optical micrograph showing the appearance of non-protective oxide at the corner of the sample, while the rest of the coupon remains undistorted. Alloy PM2000 oxidised at 1300°C for 60 h in laboratory air (20 h cycle).*

Fig. 2 *Optical micrograph showing the badly deformed AluchromYHf alloy after oxidation at 1300°C for 40 h in laboratory air (20 h cycle).*

The effect of the scale growth rate on alloy lifetime is illustrated in Fig. 3. For the 0.5 mm thick specimen of the commercial wrought alloy Kanthal AF, breakaway occurred after 13000 h cyclic oxidation (100 h cycles) at 1200°C. In contrast, the specimen of another alloy Kanthal APM with a comparable Al-reservoir (due to similar initial Al-content and thickness) failed under the same experimental conditions after only 3200 h. The mass change data up to 500 h exposure for the two alloys show almost identical scale growth rates during these short term exposures. However, it can be seen that the difference in lifetime between the two materials can be explained by increased scale growth on Kanthal APM after 500 h exposure, represented by an increase in both gross mass change (where the mass of any spalled oxide is included) and net mass change (where only the mass of the sample is measured). Hence, not only alloy composition but also processing route and creep strength may be important in the prediction of oxidation lifetime, and these points are returned to in the discussion. It has also been reported [15] that large variations in growth kinetics may be encountered, not only between samples of alloys of different origin, but even between batches of samples from the same commercial alloy.

Metallographic investigations showed that scale spallation occurred on most samples once the oxide thickness had reached a critical value, but thinner foils never reached this condition. Hence for example, a 2mm thick sample of PM2000 oxidised at 1300°C in air showed extensive spallation after 1517 h, as illustrated in Fig. 4. In thin (<200 µm) foils, however, metallographic cross-sections of weak alloys showed extensive convolutions of the oxide scale (see Fig. 5). This occurred typically after creep deformation of the substrate on repeated heating and cooling. For thin foils of the strong alloys such as PM2000 a network of fine cracks was generated rather than convoluted scales.

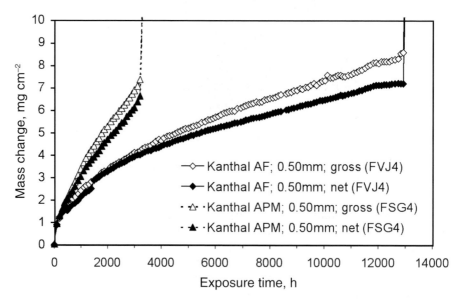

Fig. 3 *Mass change data during 100h cycles oxidation of 0.5 mm thick coupons of commercial FeCrAl alloys Kanthal APM and Kanthal AF at 1200°C in air. Rapid mass gains indicate breakaway.*

Fig. 4 Optical image of alloy PM2000 (2mm thick) showing scale spallation, after oxidation at 1300°C for 1517h in laboratory air (isothermal oxidation).

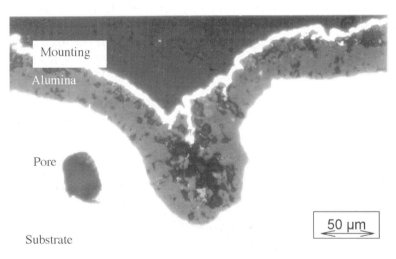

Fig. 5 Metallographic cross-section of model ODS alloy DAJ with 0.2 mass% Y_2O_3 showing buckling of alumina scale.

These differences in failure mechanism could also be linked to the specimen surface colouration. From observations and analysis of PM2000, AF and YHf, a grey surface colour was associated with α-alumina, a green colour showed evidence of chromia followed by black iron oxide. In the case of AF and YHf alloys, chromia formed before total failure of the protective scale as shown in Fig. 6. In the case of PM2000, a much stronger alloy at high temperatures, the change of surface colouration was very different. It went from grey to red and black, suggesting a totally different failure mechanism sequence, regardless of the oxidising environment. Figures 7 and 8 show the extent of each type of oxide as the tests continued and provided a unique insight

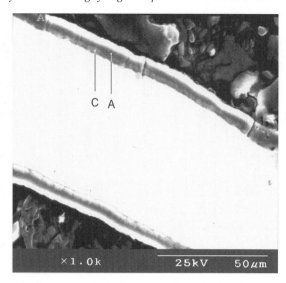

Fig. 6 *Aluchrom YHf 50 mm foil oxidised at 1200°C for 100 h in air. Underneath the alumina scale (A), a chromia layer (C) is visible. The outer layer appears bright due to charging of the sample.*

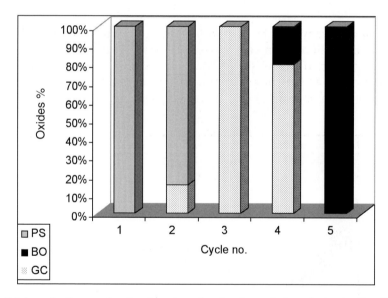

Fig. 7 *Schematic diagram showing the extent of α-alumina scale degradation in alloy Aluchrom YHf at 1300°C in laboratory air (20 h cycle); PS = Protective Scale, GC = Green Colour, BO = Breakaway Oxidation.*

into the progress of the oxide from protective to non-protective oxidation. All the data are presented for the laboratory air tests, but testing in moist air gave similar results, except that the initial scale growth was slightly faster in air + 3.2vol.% water vapour.

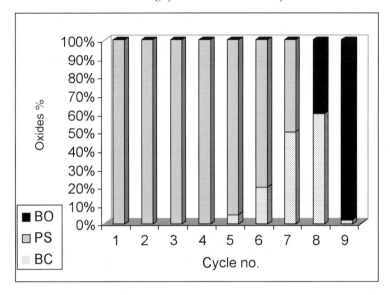

Fig. 8 Schematic diagram showing the extent of α-alumina scale degradation. Alloy PM2000 at 1300°C in laboratory air (20h cycle), PS = Protective Scale, BC = Brown Colour, BO = Breakaway Oxidation.

Colour changes may also be linked back to the thermogravimetric data. For example, the onset of a green colouration can be correlated with changes in the instantaneous oxidation rate constant. Figure 9 shows that the rate constant suddenly increases with time as the chromia formation starts and this can also be used to pinpoint the onset of breakaway oxidation, as first suggested in [15].

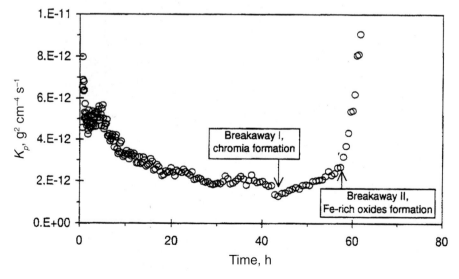

Fig. 9 Parabolic Growth Rate Constant for 50 mm thick Kanthal AF foil during isothermal oxidation at 1200°C in air.

The aluminium content of the alloy at any time is clearly important in the rehealing of any region under spalled oxide or cracks in the scale. Figure 10, for example, shows a situation where iron oxides have started to form in cracks within the aluminium oxide on PM2000 but complete breakaway has been prevented by the further growth of an aluminium and titanium rich oxide. EPMA was used to measure the residual aluminium concentration in the alloy at breakaway and it was found that in thin foils of weak alloys where chromia growth is prevalent, the aluminium concentration is usually very low (<0.1 mass%.). In the stronger alloys, however, the alumina concentration is usually much higher at breakaway (1–1.5 mass%). Figure 11 shows a series of Al measurements across two ODS alloys which contain different amounts of the strengthening yttria dispersion, 0.5% in one case and 0.2% in the other. The EPMA data show that the mean values of aluminium content vary from 0.5 to 1.8 between the two alloys, with the 0.2% yttria alloy behaving more like a weak alloy. However, these values must be treated with some caution since they do not correspond to the residual levels of aluminium available. Internal precipitates of aluminum nitrides and oxides are often present in the matrix close to the breakaway region (Gurappa *et al.* [16]), and a contribution to the aluminium signal from these would explain the scatter in the EPMA results. Finally, the XRD investigation during this study showed that the bulk oxide when the whole sample had gone into breakaway oxidation (no metal left) was an oxide solid solution of the type $(Fe,Cr)_2O_3$ regardless of the oxidising environment.

4. Discussion

Given that FeCrAlRE alloys rely on the protection afforded by an α-alumina scale against the rapid corrosion rates that lead to the ultimate failure of the component,

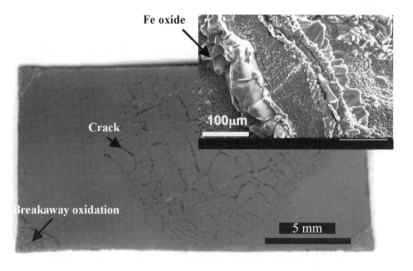

Fig. 10 *Optical and SEM images showing the formation of cracks on PM2000 oxidised at 1300°C for 120 h in dry cylinder air (20 h cycle).*

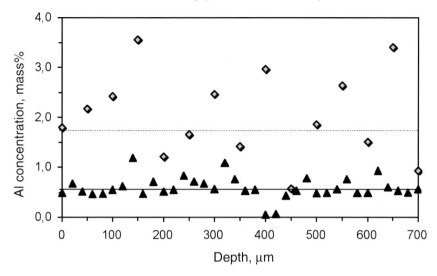

Fig. 11 *EPMA measurements of residual Al content in two FeCrAl ODS alloys after cyclic oxidation until breakaway at 1200°C in air. Alloy DAH contain 0.5%wt Y_2O_3-dispersoid; alloy DAJ contains 0.2%wt Y_2O_3-dispersoid. Dashed and solid lines represent mean values of the measured data for DAH and DAJ respectively.* ▲ *0.2%yttria; DAJ* ◆ *0.5%yttria;DAH*

the establishment of an adequate scale is essential. This scale forms by reaction of the alloy with its environment. The two requirements for the formation of an alumina scale are sufficient aluminium and oxygen. The partial pressure of oxygen must be sufficiently high to form an α-alumina scale. The kinetics and thus the time to breakaway depend on the partial pressure of oxygen. For example, greater lifetimes were achieved when the partial pressure of oxygen was reduced, but still sufficient to form a protective α-alumina scale [13]. The mechanism proceeds in the same manner as for a purely oxidising environment, but with decreased rate. Thus, the partial pressure of oxygen was observed to have no marked effect on the mechanism.

The availability of aluminium is dependent on the substrate. Commercial alloys such as PM2000, Aluchrom YHf, Kanthal APM or Kanthal AF possess sufficient aluminium to form a continuous oxide layer of α-alumina. The intrinsic structure, properties and growth of this scale influence its ability to protect the substrate. The more protective the scale, the greater the lifetime of the component. Any deterioration in the protection afforded by the oxide layer will involve further reaction of the substrate beneath. The mechanism by which protection is gradually lost, and after which more rapid oxidation ensues, is determined by both the chemical and mechanical properties of this layer. They relate in turn to the consumption of aluminium in the alloy for the formation of the scale. For continued protection sufficient aluminium must remain available to preserve this layer of aluminium oxide.

This investigation revealed that thin samples (<200 µm) had gone into breakaway oxidation with no occurrence of scale spallation and that the breakaway always occurs at the corners of the sample where the volume to surface ratio is minimised (see Fig. 1). However, in thicker samples spallation occurred before breakaway oxidation as shown in Fig. 4 (PM2000 alloy, 2 mm thick).

In general spallation occurs when a specific scale thickness is reached. Values for the critical scale thickness to spallation vary between 12 and 20 µm depending on the alloy. In [17] the material thickness needed to provide enough aluminium to form a scale of a certain thickness $d_{alumina}$ has been calculated, assuming that all the aluminium from the alloy is transformed to alumina.

$$d_{alumina} = c_0 \frac{\rho_{alloy}}{\rho_{alumina}} \left(1 + \frac{3}{2} \frac{M_o}{M_{Al}}\right) \frac{V}{A} \tag{1}$$

The density of the ρ_{alloy} is = 7.15 gcm^{-3} and that of alumina $\rho_{alumina}$ 3.98 gcm^{-3}. The molar masses of the elements involved are M_o = 16 and M_{Al} = 27 g mol^{-1}, respectively. The aluminium mass fraction in the alloy can be given by C_0 in first approximation. The volume to surface ratio V/A can be related to the alloy thickness using eqn (2), if the thickness of the specimen is much smaller than the width and length, which is in general true for sheet material.

$$V = d_{alloy} \frac{A}{2} \tag{2}$$

Combining eqn (1) and (2) with the numbers in the text gives the desired relationship.

$$d_{alumina} = 0.05 \frac{7.15 \text{ g cm}^{-3}}{3.98 \text{ g cm}^{-3}} \left(1 + \frac{3}{2} \frac{16 \text{ g mol}^{-1}}{27 \text{ g mol}^{-1}}\right) \frac{d_{alloy}}{2} = 0.0848 \cdot d_{alloy} \tag{3}$$

If an average critical thickness to spallation of 17 µm is assumed, using eqn (3), a material thickness (d_{alloy}) of 200 µm can be evaluated. Materials with a thickness of 200 µm and less are referred to as foils, because they are so thin that the oxide does not spall before the complete consumption of the scale forming element, aluminium. As the amount of aluminium in the alloy decreases, a point is reached where the activity of aluminium at the metal oxide interface is so low that the partial pressure of oxygen rises and the oxidation of chromium starts. This effect has been reported in several publications [6,18,19], while predictions of scale spallation are described elsewhere [20].

This study also showed that the onset of chemical failure (breakaway oxidation) is controlled by a large number of factors including alloy composition, microstructure, specimen thickness, oxide scale growth and cracking, thermal history and stress levels within the scale. Hence any mechanistic model of chemical failure must in itself be quite complex. At one extreme, for example, in thin samples that may deform to accommodate any extra stresses generated, Intrinsic Chemical Failure (InCF) [21] is observed. Here the growth of the protective scale depletes the Al level in the alloy to such an extent that chromia formation is thermodynamically favourable and the samples adopt a greenish tinge as chromia is formed either below or within the existing alumina scale. If local equilibrium is assumed, then thermodynamic calculations [22] would suggest that the aluminium level needs to drop below 0.2 ppm before chromia is formed. In practice, other kinetic factors must also be taken into account. Hence, experimentally, chromia is formed when the aluminium content drops below ~0.1 mass%. Clearly the composition of the alloy will affect the growth of both scales, and the role of minor elements in this process is discussed in detail in [14]. However, once the chromia has formed, as illustrated in Fig. 6 and 7 where a chromia layer is visible underneath the alumina scale, this is one indication of chemical failure. The alumina scale cannot reform if damaged either by the growth of the chromia layer or by other external factors, and breakaway oxidation then commences. Also the thickness of the alumina layer will be reduced by its continued reduction by chromium and, again, non-protective conditions will develop.

For thicker samples, stresses may lead to cracks in the alumina scale that must be healed by the growth of new Al-rich oxide. Once the Al level drops below a critical value, this cannot occur and the alloy goes into breakaway degradation, as shown in Fig. 10 where cracks have formed during oxidation of PM2000 at 1300°C in laboratory air and less protective oxides, mainly Fe oxide, were present. This indicates that the sample has gone into failure with the formation of a brownish coloured oxide which comprises Al and Fe oxides as revealed by EDX, instead of a Cr-rich oxide (greenish colour) which was the case during the failure of Kanthal AF and Aluchrom YHf alloys. The critical Al level which is required to sustain a protective alumina scale in PM2000 seems to have an upper limit of 2 mass%.

The critical Al concentration or C_{NOSH} (NOSH — NO Self Healing) is a fundamental property of the specific alloy and does not depend on sample geometry, stress state, etc. It depends only on the alloy composition. A scale may remain protective, even when the Al concentration drops below this critical value but the sample will fail if anything disrupts the protective alumina. While the amount of aluminium in the commercial alloys prior to oxidation (C_0) is normally 5–6 mass%, different values of the aluminium level at breakaway (C_B) have been reported in the literature for alloys with very similar chemistry. For ODS alloy MA956 at 1000 to 1200°C, a C_B of 1.2 to 1.5 mass% was measured by EPMA [23], however, it was reported to be <2 mass% during oxidation at 1350°C [24]. The recent results with thin wrought FeCrAl foils at 1200°C [15] have shown that the Al may be depleted essentially to a near zero value before formation of breakaway oxides starts. It has been proposed that the value of C_B depends on the protective properties (gas tightness and adherence) of the alumina scale [25]. The scale adherence has been recently related to the alloy creep, which can facilitate the relaxation of the oxidation-induced stresses [21]. This means that C_B can depend on the alloy creep strength. The experimental evidence for this suggestion is represented by EPMA measurements of the residual Al-contents in Fig. 10. The model FeCrAl ODS alloy DAH (0.5% yttria dispersion) has a higher C_B of ~1.75% than another model alloy DAJ (low yttria content of 0.2% and thus smaller creep resistance), which has C_B of ~0.5 mass%. Some variation in Al-content across the specimen thickness apparently occurred due to the presence of internal precipitates of aluminium nitrides and/or oxides which are commonly observed in the breakaway affected alloy matrix [16]. The metallographic cross section of the weak alloy DAJ (Fig. 6) shows a convoluted scale, which typically occurred due to the creep deformation of the substrate on repeated heating and cooling.

5. Mechanistic Models

All these different mechanisms can be represented on a series of diagrams in which the Al concentration is plotted with time. Figure 12(a) represents intrinsic chemical failure (InCF). Breakaway occurs after the Al concentration has dropped well below C_{NOSH} and $C_{Cr_2O_3}$ — where Cr_2O_3 has started to form below or within the Al_2O_3 scale. Figure 12(b) corresponds to the case where the Al level is reduced in the alloy as the alumina grows (without spalling) until the level is below C_{NOSH}. Any disruption of the scale by internal or external factors then causes MICF and breakaway. If the sample geometry is such that breakaway occurs in one part of the sample while the rest is still protected by a layer of alumina, then it is possible to get a situation where breakaway occurs in some regions just as Cr_2O_3 starts to form and this special case is represented in Fig. 12(c). Finally, although spallation of the protective alumina scales formed during the early stages of oxidation does not necessarily lead immediately to breakaway oxidation, spalling can give rise to mechanically assisted depletion of the Al concentration until it drops below C_{NOSH} which then gives a situation where any internally or externally generated disruption will lead to MICF and breakaway; this is shown in Fig. 12(d).

Both C_{NOSH} and $C_{Cr_2O_3}$ are independent of sample geometry, oxide thickness, thermal history and stresses which may have developed in the alloy or oxide. C_{NOSH} is the minimum Al concentration for re-formation of an alumina layer. It can be thought of as a kinetic boundary and is likely to depend on alloy composition, microstructure, diffusion rates and possibly temperature. On the other hand, $C_{Cr_2O_3}$ is a thermodynamic boundary and represents the conditions that Cr in the alloy can reduce alumina. The critical value will depend on alloy composition and temperature. Measured residual values of Al concentration, C_B, in the alloy can vary between $C_{Cr_2O_3}$ and C_{NOSH} but higher values of C_B are favoured in creep-resistant alloys and thick samples.

This framework now allows us to explore all the cases of breakaway and to look at exceptional situations, for example, where geometry or thermal treatment can lead to selective oxidation at certain positions on a sample, which can in turn lead to changes in alloy composition (depletion of Cr for example) and hence unusual breakaway behaviour. The influence of sample geometry on the lifetime of FeCrAl alloys is discussed by Strehl *et al.* [17].

6. Conclusions

- The study showed that breakaway oxidation was always initiated at the corners and edges of the samples.

- No scale spallation was observed during the oxidation of thin foils, whereas the scale spalled prior to total breakaway oxidation in the case of thicker samples.

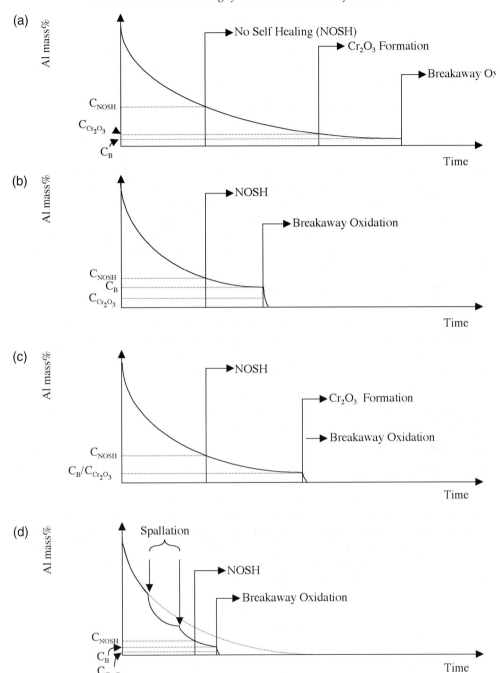

Fig. 12 *Schematic diagram illustrating the different mechanisms of failure of alumina scale which form on FeCrAl alloys. (a) Intrinsic chemical failure, (b) mechanically induced chemical failure without spallation, (c) failure due to simultaneous chromia formation and breakaway oxidation and (d) mechanically induced chemical failure after spallation.*

- The surface colouration evaluation showed that progress of the oxidation from protective to non-protective is different in PM2000 from that in AF and YHf alloys. In the case of PM2000, a brownish colour was formed instead of a greenish colour prior to the formation of the voluminous iron oxide, which was the case during the oxidation of AF and YHf.

- The critical aluminium level is a quantity that may be related to the mechanism of failure and in turn the time to breakaway oxidation. Intrinsic chemical failure (InCF) is accompanied by a depletion of the aluminium level to near zero. In the critical residual Al concentration ($C_{Cr_2O_3}$) represents a thermodynamic boundary at which Cr can reduce aluminia.

- Residual Al values recorded when failure occurs by mechanically induced chemical failure (MICF) are much higher than $C_{Cr_2O_3}$. This critical Al concentration, C_{NOSH} (NOSH — NO Self Healing), is a kinetic boundary and represents the minimum Al concentration for re-formation of an alumina layer. It will depend on alloy composition, microstructure, diffusion rates and possible temperature. A typical value is 2 mass%. A scale may remain protective, even when the Al concentration drops below this critical value; but the sample will fail if anything disrupts the protective alumina.

7. Acknowledgements

The authors would like to thank the European commission for financial support of this work, through contract no. BRPR-CT-97-0562. We are also grateful to our industrial partners for the supply of the alloys tested.

References

1. D. R. Sigler, Oxidation resistance of alumina-coated Fe–20Cr alloys containing rare earth or yttrium. *Oxid. Met.*, 1993, **40**, 295–312.
2. Itoh I., *et al.*, Development of ferritic stainless steel foil as metal support for automotive catalytic converter. Nippon Steel Technical Report, 1995, **64**, 69–74.
3. T. Ohashi and T. Harada, High-temperature oxidation of Fe–Cr–Al–Si alloys extruded into honeycomb structure. *Oxid. Met.*, 1996, **6**, 235–255.
4. D.R. Sigler, Adherence behavior of oxide grown in air and synthetic exhaust gas on Fe–Cr–Al alloys containing strong sulfide-forming elements, Ca, Mg, Y, Ce, La, Ti and Zr, *Oxid. Met.*, 1993, **40**, 555–583.
5. D. R. Sigler, The oxidation behavior of Fe–20Cr alloy foils in a synthetic exhaust-gas atmosphere, *Oxid. Met.*, 1996, **46**, 335–363.
6. G. Strehl, *et al.*, Cyclic oxidation of thin FeCrAl(RE) foils in air: SEM and EPMA investigations, in *Cyclic Oxidation of High Temperature Materials* (M. Schütze and W. J. Quadakkers, eds). EFC Publication No. 27, published by The Instsitute of Materials, London, 1999, 82–94.

7. W. J. Quadakkers and M. J. Bennett, Oxidation induced lifetime limits of thin-walled, iron based alumina forming oxide dispersion strengthened alloy components, *Mater. Sci. Tech.*, 1994, **10**, 126–131.

8. W. J. Quadakkers and K. Bongartz, The prediction of breakaway oxidation for alumina forming ODS alloys using oxidation diagrams. *Werkst. Korros.*, 1994, **45**, 232–241.

9. M. J. Bennett, Recent studies and current problems concerning the high temperature oxidation behaviour of alumina forming iron base alloys. *Solid State Phenomena*, 1995, **41**, 235.

10. M. J. Bennett, A. H. Harker, R. Perkins, J. B. Price and J. A. Desport, The oxidation behaviour of alumina forming ferritic oxidation dispersion strenghthened alloys. *Proc. 1st Mexican Workshop and 2nd Int. Symp. on Metallic Corrosion* (L. Maldonado and M. Pech, EDS). Mexico, DF UNAM Facultad du Quimica Press, pp.100–118, 1995.

11. W. J. Quadakkers, T. Malkow and H. Nickel, The Effect of major and minor alloying elements on the oxidation limited life of FeCrAl-Base Alloys, in *Proc. 2nd Int. Conf. on Heat-Resistant Materials*, Gatlinburg, TN, 1995, pp.19–96.

12. W. J. Quadakkers, K. Schmidt, H. Grubmeier and E. Wallura, Composition, structure and protective properties of alumina scales on iron-base oxide dispersion strengthened alloys, *Mater. High Temp.*, 1992, **10** (1), 23–32.

13. A. Kolb-Telieps *et al.*, The role of bioxidant corrodents on the life time behaviour of FeCrAlRE alloys. This volume, pp.123–134.

14. D. Naumenko, *et al.*, critical role of minor elemental constituents on the life time oxidation behaviour of FeCrAlRE alloys. This volume, pp.66–82.

15. D. Naumenko, L. Singheiser and W. J. Quadakkers, Oxidation limited life of FeCrAl based alloys during thermal cycling, in *Cyclic Oxidation of High Temperature Materials*, (M. Schütze and W.J. Quadakkers, eds). EFC Publication No. 27, published by The Institute of Materials, 1999, pp.287–305.

16. I. Gurappa, S. Weinbruch, D. Naumenko and W. J. Quadakkers, *Mater. Corros.*, 2000, **51**, 224.

17. G. Strehl, *et al.*, The influence of sample geometry on the life time of FeCrAl alloys. This volume, pp.107–122.

18. G. Strehl *et al.*, The effect of aluminium depletion on the oxidation behaviour of FeCrAl foils, *Mater. High Temp.*, 2000, **17**, 87–92.

19. F.H. Stott and N. Hiramatsu, Breakdown of protective scales during the oxidation of thin foils of Fe–20Cr–5Al alloys at high temperatures, *Mater. High Temp.*, 2000, **17**, 93–100.

20. H. E. Evans and J. R. Nicholls, Prediction of oxide spallation from an alumina-forming ferritic steel. This Volume, pp.37–49

21. H. E. Evans, A.T. Donaldson and T. C. Gilmour, *Oxid. Met.*, 1999, **52**, 379–402.

22. G. Strehl, Private communication.

23. W. J. Quadakkers and K. Bongartz, *Werkst. Korros.*, 1994, **24**, 232.

24. H. Al-Badairy and G. J. Tatlock, The application of a wedge-shaped sample technique for the study of breakaway oxidation in Fe–20Cr–5Al based alloys, *Oxid. Met.*, 2000, **53**, 157–170.

25. H. Al-Badairy and G. J. Tatlock, The influence of the moisture content of the atmosphere on alumina scale formation and growth during high temperature oxidation of PM2000, *Mater. High Temp.*, 2000, **17**.

5

Critical Role of Minor Element Constituents on the Lifetime Oxidation Behaviour of FeCrAl(RE) Alloys

D. NAUMENKO, W. J. QUADAKKERS, V. GUTTMANN*,
P. BEAVEN*, H. AL-BADAIRY†, G. J. TATLOCK†, R. NEWTON§,
J. R. NICHOLLS§, G. STREHL¶, G. BORCHARDT¶, J. LE COZE**,
B. JÖNSSON†† and A. WESTERLUND††

Forschungszentrum Jülich, IWV-2 Postfach 1913, 52425 Jülich, Germany
*Joint Research Centre, PO Box 2, 1755 ZG Petten,The Netherlands
†Materials Science & Engineering, Department of Engineering, University of Liverpool,
Liverpool L69 3GH, UK
§School of Industrial and Manufacturing Science, Cranfield University,
Bedford MK43 0AL, UK
¶Institut für Metallurgie, Technische Universität Clausthal,
D-38678 Clausthal-Zellerfeld, Germany
**Ecole des Mines de Saint-Etienne, S M S - M H P, 158 Cours Fauriel, Cedex 2,
42023 Saint-Etienne, France
††Kanthal AB, Box 502, S-734 27 Hallstahammar, Sweden

ABSTRACT

In the present study the effect of a number of common alloying additions and impurities, i.e. Zr, Ti, V, Ca, P on the oxidation limited lifetime of FeCrAlY alloys in the temperature range of 1100-1300°C has been studied. The elements in concentrations ranging from 5 to 300 ppm were added to an alloy with a base composition of (mass%) Fe–20Cr–5Al–0.05Y with a sulfur content below 10 ppm. The long term tests revealed that P had a detrimental effect on the oxidation resistance, especially at 1300°C. It has also been found that minor additions of carbide forming elements such as Ti and Zr to Y-containing FeCrAl alloys are of vital importance for improved long term cyclic oxidation performance. Comparison of the obtained results with data from commercial materials of similar composition revealed that the fundamental findings derived for high purity alloys can be substantially changed by interaction of the mentioned minor alloying additions with common alloy impurities such as C and N.

1. Introduction

During high-temperature oxidation of FeCrAl-components, the alloy Al content is continuously depleted due to formation of the protective alumina scale. This depletion becomes significantly enhanced, if under thermal cycling conditions spallation of the alumina scale occurs, thus consuming the Al for scale re-healing. After a critical Al-depletion has occurred, a rapid breakaway oxidation of Fe and Cr commences, which can be considered as the life limit of the FeCrAl components [1]. Hence, the oxidation limited lifetime of FeCrAl-alloy based components is determined primarily by the rate of consumption of the alloy Al-reservoir [1]. The latter factor in turn crucially depends on the minor alloying constituents including intentionally added reactive elements (RE), e.g. Y, La, Zr, etc. [2], as well as common alloy impurities, such as sulfur [3]. The commonly observed effects of the reactive elements on the oxidation behaviour include: significant improvement of the oxide scale adherence, prevention of the deleterious effect of sulfur and change in the scale growth mechanism [4–6]. However, most of the scientific understanding of the above phenomena is based on relatively short term oxidation data and, therefore, is not always relevant to the lifetime operation of FeCrAl components. Also, commercial FeCrAl-based materials frequently contain a combination of reactive elements and a number of unavoidable metallic and non-metallic contaminants. Thus, a lot of important effects, related to interaction of the reactive elements with each other as well as with common alloy impurities have hardly been considered.

The current study deals with the effects of a number of common alloying additions/impurities, i.e. Zr, Ti, V, Ca and P on the long term cyclic oxidation of FeCrAl alloys at temperatures up to 1300°C and exposure times exceeding 15 000 h. Some additional short time oxidation tests followed by extensive characterisation of the formed oxide scales using a variety of surface analysis techniques were performed to elucidate the effects of these minor elements on the oxidation mechanisms.

The obtained experimental results are discussed mainly in terms of alloy chemical composition and microstructure. First, in Sections 3.1 and 3.2 the mass change data obtained during the cyclic oxidation tests is correlated with the observed morphological features of the oxide scales and alloy microstructure. Further analytical studies carried out to elucidate the mechanistic role of the minor elements, which appeared to have the most significant influence on the oxidation of the model FeCrAlY alloys, i.e. P, Ti and Zr will be discussed in detail in Sections 3.3 to 3.5.

2. Experimental

The model alloys with a basic composition Fe+20Cr+5Al+0.05Y (mass%), doped with P, Ca, V, Ti and Zr in concentrations ranging from 5 to 300 ppm were produced by induction melting in cold crucible under high purity argon. The levels of known detrimental impurities such as sulfur were kept as low as technologically possible. The chemical composition of the studied alloys, determined by ICP-MS and GDMS, are given in Table 1.

The model alloy ingots were rolled down to obtain: (a) sheets of 1 mm thickness and (b) thin foil strips with a thickness of 70 and 90 μm. The foil specimens were oxidised in the as-rolled condition, whereas from the 1 mm thick sheets 20 × 10 mm coupons were machined and ground to 1200 grit surface finish. All specimens were ultrasonically cleaned in an ethanol bath prior to the oxidation exposures.

The life time cyclic oxidation tests were performed in resistance heated furnaces at temperatures varying from 1100 to 1300°C in laboratory air. The cyclic conditions

Table 1. *Chemical composition of the studied model FeCrAl alloys*

	M Ref	M + P	M + V	M + Ca	M + Ti	M + Zr	M - P
Cr (mass%)	19.9	19.9	19.7	19.7	19.7	19.7	19.8
Al (mass%)	5.0	4.96	4.88	4.80	4.91	4.86	4.90
Y (ppm)	510	500	500	520	480	490	500
C (ppm)	92	96	98	97	90	90	74
S (ppm)	<10	<10	<10	<10	<10	<10	<10
O (ppm)	<10	<10	<10	<10	<10	<10	<10
N (ppm)	<10	<10	<10	<10	<10	<10	<10
P (ppm)	86	236	83	64	78	78	2.4
Zr (ppm)	0.64	-*	-*	-*	-*	290	0.64
V (ppm)	4.1	-*	83	-*	-*	3.3	3.9
Ti (ppm)	15	-*	-*	-*	220	12	15
Cu (ppm)	4.7	-*	-*	-*	-*	5.0	4.5
Ca (ppm)	<0.05	-*	-*	<5	-*	<0.05	<0.05
Hf (ppm)	<0.005	-*	-*	-*	-*	<0.005	<0.005
Mn (ppm)	1.3	-*	-*	-*	-*	1.5	1.4
Si (ppm)	20	-*	-*	-*	-*	18	17
Nb (ppm)	<50	-*	-*	-*	-*	<50	<50
Mg (ppm)	<0.005	-*	-*	-*	-*	<0.005	<0.005
Mo (ppm)	0.89	-*	-*	-*	-*	0.85	0.81

(-*) not analysed, but should be similar to the values of alloys M Ref, and M-P.

were either 20 h cycles (for the thin foils) or 100 h cycles (for the 1 mm thick sheets), using air cooling.

The oxidised specimens were characterised using optical metallography, X-ray diffraction (XRD), scanning electron and transmission electron microscopy (SEM and TEM) coupled with energy dispersive X-ray analysis (EDX), Auger electron spectrometry (AES), secondary ion mass spectrometry (SIMS) and Rutherford backscattered electron spectrometry (RBS). The specific aspects of application of the above techniques for studying the alumina scales and the underlying metal substrates were described elsewhere [7,8].

(a)

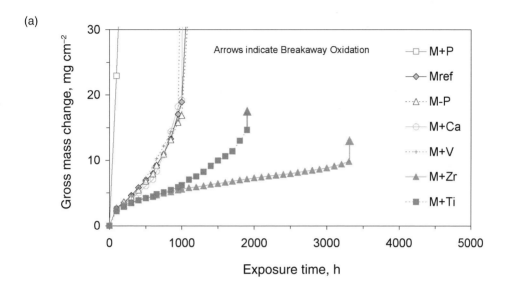

(b)

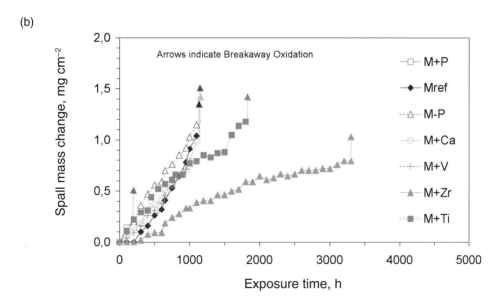

Fig. 1 *Mass change data of studied model alloys during lifetime cyclic oxidation (100 h cycles) at 1300°C in air: (a) total oxygen uptake related to adherent scale plus spalled oxide; (b) mass change of spalled oxide.*

3. Results and Discussion

3.1 Lifetime Oxidation Behaviour

During oxidation at 1300°C the shortest time to breakaway for the 1 mm thick specimens was observed for alloy M+P, which already failed after 2 cycles, i.e. 200 h.

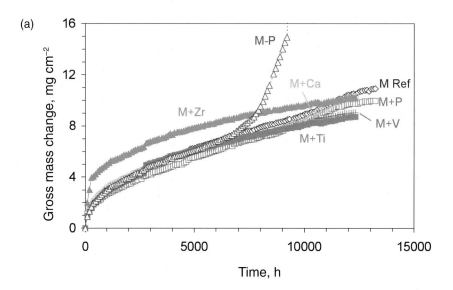

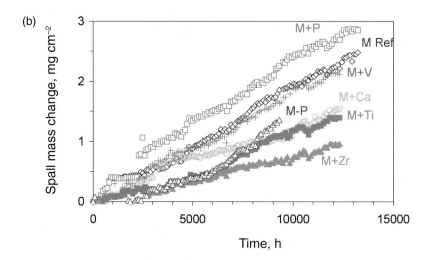

Fig. 2 *Mass change data of studied model alloys during lifetime cyclic oxidation (100 h cycles) at 1200°C in air: (a) total oxygen uptake related to adherent scale plus spalled oxide; (b) mass change of spalled oxide.*

The longest lifetimes at 1300°C were found for alloy M+Zr (3800 h) and M+Ti (1800 h) (Fig. 1). All the other alloys failed after very similar times, i.e. around 1000 h.

At 1200°C the P addition, which was very detrimental at 1300°C, showed only a minor effect. The oxidation rate of alloy M+P was similar to those of most of the other materials (Fig. 2). Until the maximum exposure time was reached (15 000 h) only one of the materials showed breakaway oxidation, i.e. alloy M-P after around 9000 h. Although the other alloys did not yet reach breakaway, it can be concluded that also at this temperature Ti and especially Zr had a positive effect on oxidation resistance. The scales on the Ti- and Zr-containing alloys appeared to be less susceptible to spalling than those formed on the other alloys (Fig. 2b). At 1200°C alloy M+Zr exhibited a much higher initial oxidation rate than all other alloys. However, after around 500 h of exposure the oxidation rate of M+Zr decreased and became similar to those of the other materials (Fig. 2a). It is noteworthy that, in spite of a large scale thickness of around 50 µm, after 15 000 h cyclic oxidation, alloy M+Zr maintained a very adherent, non-cracked scale with only minor oxide spallation at edges and corners (Fig. 3).

An interesting observation during the cyclic oxidation exposures was formation of broccoli-like nodules protruding through the alumina scale on the surfaces of some of the tested pieces. The formation of broccolies occurred not in the beginning of the cyclic tests, but after extended exposure time, which was shorter at 1300°C and longer (several thousands of hours) at 1200°C. Alloys M-P, M+V and M+P

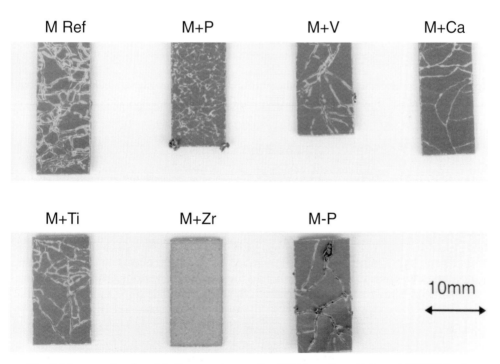

Fig. 3 *Macrographs of model alloys specimens after 8000 h cyclic oxidation at 1200°C in air (compare with mass change data in Fig. 2)*

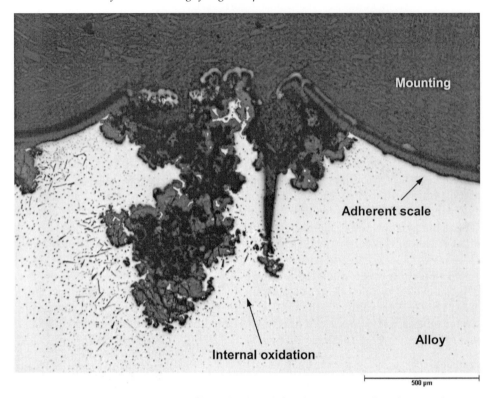

Fig. 4 *Metallographic cross-section of broccoli-like nodule. Alloy M-P, 9300 h cyclic air oxidation at 1200 °C.*

appeared to be the most susceptible to this effect, in contrast to M+Ti and M+Zr, which showed no protrusions at all (Fig. 3).

3.2 The Broccoli effect

Extensive metallographic and SEM studies on a number of samples revealed that the effect, which is macroscopically apparent as a broccoli-like nodule, is related to a type of local scale disruption resulting from severe, local internal alumina formation and the associated volume increase (Fig. 4). As shown in a larger magnification in Fig. 5, the craters typically developing in the substrate were partially filled with alumina remnants (note: sometimes also chromia has been observed). These craters were surrounded by an irregularly shaped alumina layer of non-uniform thickness.

This effect, which occurred on alloys M Ref, M+P, M-P, M+Ca and M+V after relatively long exposure times at 1200°C, could not be observed during the SEM studies of these alloys after short term oxidation. However, in the latter case Y- and Cr-rich phases were found to be precipitated at the alloy grain boundaries (Fig. 6). The EDX analyses of the Y-rich pegs at the scale/metal interface and the internal Y-rich precipitates in Fig. 6(a) revealed that these were Y/Al-mixed oxides, presumably of the type $Y_3Al_5O_{12}$. The EDX line scan across the Cr-rich phase on the alloy grain

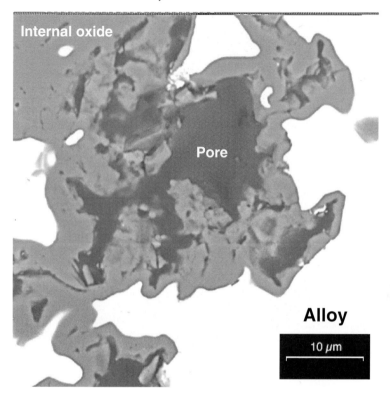

Fig. 5 *Advanced stage of internal oxidation associated with the broccoli effect (SEM image). Cracking of the oxide at various locations leads to the penetration of the oxidation front.*

boundary (Fig. 6b) indicated the presence of chromium carbide precipitates. After medium exposure times such as 2000 h at 1200°C, frequently crack initiation in the vicinity of the Y-rich pegs was observed (Fig. 7).

Taking into account all the above observations, the broccoli effect is likely to be triggered by an internal oxidation of the alloy grain boundaries. Scale cracking in the vicinity of the Y/Al-oxide pegs, which apparently serve as stress concentrators during temperature cycling, allows oxygen ingress to the alloy grain boundary decorated with chromium carbides. Rapid oxidation of the carbides at high temperature results in release of gaseous carbon oxides. The pressure of these gaseous products appears to be high enough to deform the mechanically weak alloy substrate at 1200°C, leading to the typical sponge-like internal alumina morphology after longer exposure times (Figs 4 and 5). The effective area available for oxidation becomes significantly increased resulting in a higher oxidation rate (see e.g. alloy M-P in Fig. 2a) with the consequence of a faster depletion of the alloy Al-reservoir and shorter times to breakaway for the alloys susceptible to the broccoli effect (Fig. 1 and 2).

An essential point with respect to the broccoli effect concerns the advance of the internal oxidation, which appeared to be mechanically induced probably according to the following mechanism. The internally growing oxides at the beginning of their generation were found to contain growth defects, such as pores and microcracks. Upon cooling from the oxidation temperature, compressive stresses act on these

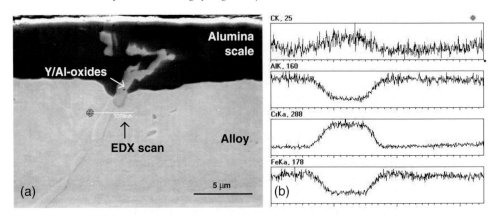

Fig. 6 *SEM studies of oxide scale on model alloy M Ref after 100 h isothermal oxidation at 1200 °C in air: (a) backscattered electron image of scale/alloy cross-section; (b) EDX line scan data across alloy grain boundary for C, Al, Cr and Fe at scan position marked in (a).*

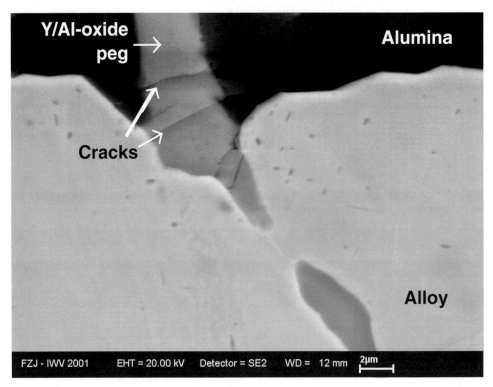

Fig. 7 *SEM backscattered electron image of cross-sectioned scale on M Ref after 2000 h cyclic oxidation at 1200 °C in air.*

internal oxides, leading to local through scale cracking. During the next oxidation cycle the oxide growth stresses will cause enlargement of the initial defects. In addition the forming new oxides will penetrate the substrate in the places, where the through cracking took place. The latter will prevent crack closure during subsequent sample cooling, while new through cracks are expected to form and part of the oxides breaks off. Repeating of the above processes acting during each oxidation cycle enables the internal oxidation to advance. It is important to stress, that the occurrence of the broccolis is not identical with the chemical breakaway failure of the FeCrAl materials. It only causes an enhanced Al-consumption, thus shortening the alloy life time.

3.3. The Role of Phosphorus

With regard to the influence of phosphorus, several samples have been investigated after oxidation at different temperatures (1100, 1200 and 1300°C) and times (5–24 h), as shown in Table 2. All the samples were bent within the analysis chamber of an Auger electron microscope during this investigation and the observations made are summarised also in Table 2. TEM cross-section samples were prepared from some of the oxidised samples.

The investigations showed that the scale mainly spalled and fractured at the emergent grain boundaries of the metal. Phosphorus was found to be present at the metal/oxide interface, as indicated in Fig. 8 and Table 2 but not in all cases. The difficulty in locating the phosphorus at the metal/scale interface may be due to several reasons. For example, in several cases the oxide may have fractured in a cohesive manner. Thus, some phosphorus, which might have segregated to the metal/scale interface could not be detected. Alternatively, phosphorus might have been incorporated into the scale as P_2O_5 (which is the most stable form of phosphorus oxide) or, in the case of element segregation, might have been lost to the atmosphere, since the boiling point of phosphorus is low (280°C). Transmission Electron Microscopy indicated that all the alumina scales had a columnar grain structure at the metal/scale interface (Fig. 9b), whereas at the oxide/gas interface the grain structure was equiaxed (Fig. 9a). This type

Table 2. *AES-investigation of the role of phosphorus*

Alloy	Thickness, μm	Temp., °C	Time, h	Summary of observations
M2	90	1100	5	The scale fractured in thin lines. No analysis was carried out.
M2	90	1100	15	The scale fractured, but no P was detected.
M2	90	1100	25	The scale fractured, P was detected.
M2	90	1300	24	The sample fractured during bending.
M2	300	1300	24	The scale fractured but no P was detected; the scale probably fractured in a cohesive manner or at the Al_2O_3/Cr_2O_3 interface.
M2	240	1200	24	The scale fractured, P was detected.

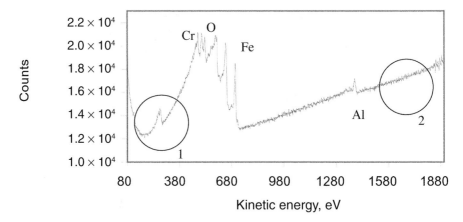

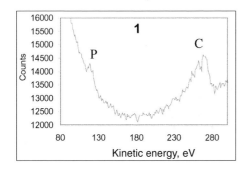

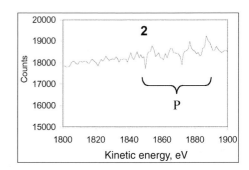

Fig. 8 *Auger electron spectra taken from scale/metal interface (region of spalled oxide) of model alloy M+P. Oxide scale was formed during 15 h oxidation at 1100°C in air.*

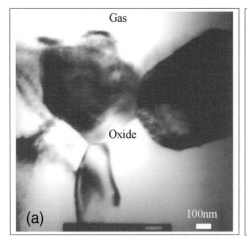

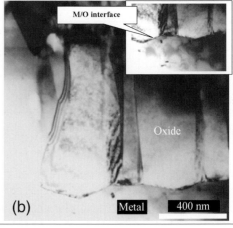

Fig. 9 *TEM cross sectional image of the model alloy M+P oxidised in air for 15 h at 1100°C: (a) near the scale/gas interface; (b) near the scale/metal interface.*

of scale structure has been typically observed on RE-containing FeCrAl alloys [2]. No P-rich phases were found on the scale grain boundaries in the present TEM study.

The effect of phosphorus on the oxidation of FeCrAlY alloys appeared to be very temperature sensitive. At 1300°C P had a clear detrimental effect on the growth rate of the alumina scale, resulting in a fast Al-depletion and short times to failure of the material. At lower temperatures, i.e. 1100 and 1200°C this effect is not pronounced, however, the scale adherence seems to be lower on the P-doped alloy than on the other model alloys probably due to the P-segregation to the scale/metal interface. As far as it is known to the authors, the P-effect has not been explicitly described in the literature [9]. More detailed oxidation tests and analytical studies are necessary in order to clarify the mechanisms responsible for the deleterious role which P can play in the high temperature oxidation of FeCrAl alloys.

3.4. The Role of Titanium

The most important effect of Ti additions in the high purity model alloys at 1200 as well as at 1300°C was the suppression of the broccoli type local oxides. As discussed in Section 3.2 the reason for the broccoli-oxides is probably related to locally enhanced oxidation promoted by Cr-carbide formation on alloy grain boundaries at oxidation temperature and/or during specimen cooling. Titanium additions apparently getter the carbon impurity by formation of thermodynamically stable Ti-carbides, thus preventing Cr carbide formation on alloy grain boundaries. In order to get a better mechanistic understanding of the effect of Ti on the scale growth mechanisms, the model alloys M Ref and M+Ti were exposed isothermally for short times at 1200°C with subsequent analyses of the surface scale by RBS. Up to 20 h oxidation a slightly higher concentration of Y was detected on the surface of the alumina scale on alloy M+Ti, than on MRef. However, after longer times, i.e. after 50 h and 100 h exposure, this effect appeared to vanish and even the opposite effect occurred (Fig. 10).

The latter observations are different from those, made recently with two model FeCrAl ODS alloys (a Ti-free and a 0.4%Ti-doped), in which yttrium was present as an oxide dispersion [10]. The results in ref. [10] strongly indicated a positive effect of Ti in the ODS alloys with a steady time dependence, by promoting incorporation of yttrium into the alumina scale and enhancing its mobility at the scale grain boundaries. A possible reason why Ti is not as beneficial in the wrought alloys as in the ODS alloys, might be that in the former case Ti appeared to promote internal oxidation due to formation of mixed Y/Ti/Al-oxides at the alloy grain boundaries (Fig. 11). Therefore, in the wrought alloys Ti may decrease the mobility of Y, which is necessary for the latter element to become incorporated into the scale.

3.5. The Role of Zr

As in the case of the model alloy M+Ti, the Zr addition to the FeCrAlY base composition completely prevented formation of the broccoli-type oxides. The mechanism of this effect can be explained as being similar to that of Ti-doping, in terms of the formation of thermodynamically stable and homogeneously distributed carbo nitrides within the alloy matrix. To study the effect of zirconium on the scale growth mechanism, a foil of model alloy M+Zr has been oxidised at 1300°C for 60 h,

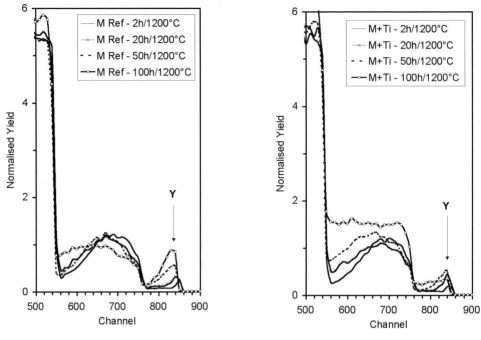

Fig. 10 *RBS spectra near Y high energy edge taken from oxide surface scales formed on the model alloys M Ref and M+Ti after various oxidation times at 1200°C.*

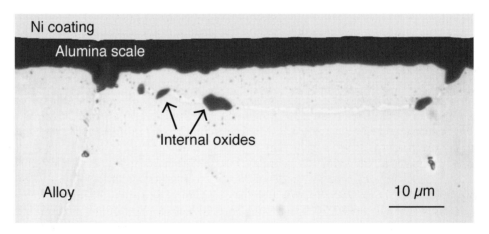

Fig. 11 *Metallographic cross-section of scale formed on alloy M+Ti after 100 h isothermal oxidation at 1200°C in air, showing internal Y/Ti/Al-rich oxide precipitates in the alloy matrix.*

i.e. just before occurrence of breakaway. Metallographic studies on this sample of M+Zr showed a chromia scale formed underneath the alumina. SIMS depth profiles of this M+Zr specimen revealed enrichment of Y and Zr at the scale/gas interface (Fig. 12). This enrichment was similar to that of Fe and Cr, thus indicating that it probably originated from the early stages of oxidation. None of the SIMS profiles indicated any segregation of Zr at the scale/alloy interface. Hence, it is assumed that

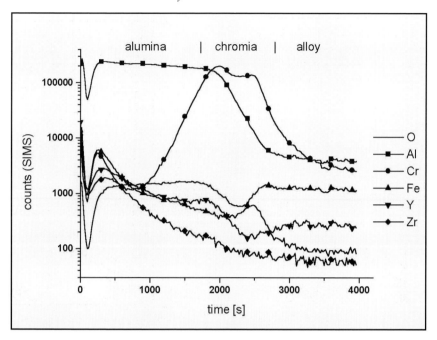

Fig. 12 *SIMS depth profiles of the oxide scale formed on the model alloy M+Zr (90 μm thick foil) after 60 h isothermal oxidation at 1300 °C in air.*

zirconium and yttrium play a major role at the beginning of oxidation and migrate to the surface of the oxide during long term exposure. By electron microprobe analyses indications were found that the positive effect of Zr on the oxidation behaviour as in the case of Ti (described in Section 3.4) might be partly related to a gettering of detrimental impurities such as carbon, sulfur and/or phosphorus.

An additional important finding for the Zr-containing alloy was that during oxidation at 1200°C the formed alumina scale exhibited a porous structure (Fig. 13). This porosity can in fact explain the higher scale growth rate observed on M+Zr after short time oxidation at 1200°C in Fig. 2. Due to its high affinity for C and N and based on literature data [11], Zr is expected to prevail in the alloy matrix in the form of tiny carbo nitrides. The finely distributed Zr(C,N)-particles become embedded into the inwardly growing alumina scale. Oxidation of these particles within the scale leads to scale microcracking and formation of pores. Consequently, the enhanced molecular oxygen transport through the defective scale causes accelerated oxidation. This mechanism was described in detail in ref. [11] for a nitrogen-contaminated batch of the commercial FeCrAl ODS alloy PM2000. An interesting observation made during cyclic exposure of alloy M+Zr was that in spite of the oxide defectiveness on this material, the scale seems to possess an excellent resistance against thermally induced stresses (Figs 1 and 2). Probably, the microcrack formation imparts to the alumina scale a certain extend of pseudo-plasticity, as was also proposed in references [12,13].

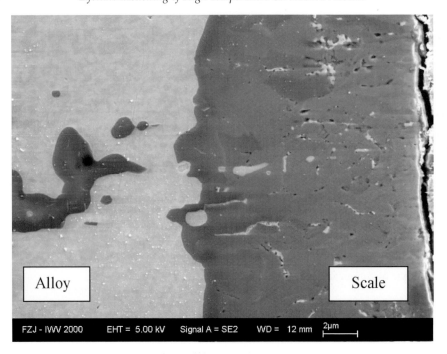

Fig. 13 *SEM cross-sectional image of oxide scale formed on model alloy M+Zr after 100 h isothermal oxidation at 1200°C in air, showing formation of porosity and microcracking.*

4. Concluding remarks

In this study one of the first attempts has been made to investigate effects of several minor alloying elements on the lifetime cyclic oxidation of Y-containing FeCrAl wrought materials. The oxidation tests and the analytical studies of the formed oxide scales and alloy matrices revealed that the alloy oxidation performance is significantly influenced by P, Ti and Zr, whereas the Ca and V additions in a concentration of 5 and 80 ppm respectively did not exhibit a clear effect.

Phosphorus-additions showed a similar benign effect at 1100 and 1200°C, but at 1300°C these P-containing alloys exhibited a dramatically decreased lifetime caused by a substantial increase in oxide growth rate. AES studies indicated that this is related to P segregation at the oxide grain boundaries and/or the oxide/metal interface.

Zirconium additions led to a significant lifetime extension. The positive effect was related to improved oxide adherence in spite of an initially enhanced oxidation rate of the Zr-containing alloy at temperatures up to 1200°C.

Titanium additions also led to improved oxide adherence. According to RBS studies this is related to an enhanced incorporation of yttrium into the oxide scale during the first hours of oxidation. Especially in the foil specimens, however, the Ti additions lead to formation of internal mixed Y/Ti/Al-oxides. The resulting fast Al-consumption can therefore cause shorter lifetimes of thin walled components.

Both Zr and Ti additions were found to prevent internal oxidation of the alloy grain boundaries (the broccoli effect), which was observed on most of the other model alloys during long term cyclic exposures. This positive effect of Zr and Ti can be related to their ability to tie up the carbon and/or nitrogen impurities in the tiny carbonitride precipitates, thus preventing formation of chromium carbides on the alloy grain boundaries. Depending on the amount and distribution of the carbonitrides the Zr- and Ti-containing alloys can show enhanced oxidation kinetics, however, with respect to the lifetime cyclic oxidation behaviour the latter effect is compensated by improved scale adherence.

The results obtained in this study strongly indicate that optimum oxidation resistance of FeCrAl-alloys can only be achieved by a combination of several reactive elements, e.g. Y+Zr, Y+Ti etc. When selecting the reactive elements, it must be considered that their effect is often very sensitive to the exposure conditions, e.g. the oxidation temperature. The type, amounts and distribution of the above additions as well as those of the typical alloy impurities, such as C and N must be carefully controlled in order to obtain extended lives of FeCrAl-based components with a given major alloy composition.

5. Acknowledgements

The authors gratefully acknowledge the financial support of their research activities by the European Community in the frame of the BRITE/EURAM project LEAFA (Life Extension of Alumina Forming Alloys in High Temperature Corrosive Environments) Project No. BE-97-4491.

References

1. J. P. Wilber, M. J. Bennett and J. R. Nicholls, The effect of thermal cycling on the mechanical failure of alumina scales formed on commercial FeCrAl-RE alloys, in *Cyclic Oxidation of High Temperature Materials, Proc. of Int. Conf. 1999* (M. Schütze and W. J. Quadakkers, eds). Publication No. 27 in European Federation of Corrosion Series. Published by The Institute of Materials, London, 1999, 133–147.

2. W. J. Quadakkers, Growth mechanisms of oxide scales on ODS alloys in the temperature range 1000–1100°C, *Werkst. Korros.*, 1990, **41**, 659–668.

3. D. R. Sigler, Aluminium oxide adherence on Fe–Cr–Al alloys modified with Group IIIB, IVB, VB and VIB elements, *Oxid. Met.*, 1989, **32**, (5/6).

4. D. P. Whittle and J. Stringer, Improvements in high temperature oxidation resistance by additions of reactive elements or oxide dispersions, *Phil. Trans. Roy. Soc. (Lond.)*, 1980, **A295**, 309–329.

5. P. Y. Hou, K. Prüßner, D. H. Fairbrother, J. G. Roberts and K. B. Alexander, Sulphur segregation to deposited Al_2O_3 film/alloy interface at 1000°C, *Scr. Mater.*, 1999, **40**, (2), 241–247.

6. W. J. Quadakkers, A. Elschner, H. Holzbrecher, K. Schmidt, W. Speier and H. Nickel, Analysis of composition and growth mechanisms of oxide scales on high temperature alloys by SNMS, SIMS and RBS, *Michrochim. Acta*, 1992, **107**, 197–206.

7. A. Rahmel and V. Kolarik, Metallography, electron microprobe and X-ray structure analysis, in *A Working Party Report on Guidelines for Methods and Testing and Research and in High*

Temperature Corrosion. Publication No. 14 in European Federation of Corrosion Series. Published by The Institute of Materials, 1995, 147–157.

8. W. J. Quadakkers and H. Viefhaus, The application of surface analysis techniques in high temperature corrosion research, in *A Working Party Report on Guidelines for Methods and Testing and Research in High Temperature Corrosion*. Publication No. 14 in European Federation of Corrosion Series. Published by The Institute of Materials, 1995, 189–217.

9. A. S. Khanna, W. J. Quadakkers, P. Kofstad and C. Wasserfuhr, The Effect of Trace Impurities of S, P and B on the High Temp. Oxidation of NiCrAl Based Alloys; *Proc. 9th EUROCORR '89*, Utrecht, NL, 1989, Proc. EG-007, Vol. II.

10. J. Quadakkers and L. Singheiser, Practical aspects of the reactive element effect, in *Proc. Int. Conf. on High Temperature Corrosion*, 22–26 May 2000, Les Embiez, France, in press.

11. W. J. Quadakkers, D. Naumenko, L. Singheiser, H. J. Penkalla, A. K. Tyagi and A. Czyrska-Filemonowicz, Batch to batch variations in the oxidation behaviour of alumina forming Fe-based alloys, *Mater. Corros.*, 2000, **51**, 350–357.

12. W. D. Kingery, *Introduction to Ceramics*. J.Wiley and Sons, New York, 1960.

13. M. Schütze, *Protective Oxide Scales and their Breakdown*. John Wiley & Sons Ltd, 1997.

6

Development of a Life Prediction Model for the Chemical Failure of FeCrAl(RE) Alloys in Oxidising Environments

J. R. NICHOLLS, R. NEWTON, M. J. BENNETT, H. E. EVANS *,
H. AL-BADAIRY[†], G. J. TATLOCK[†], D. NAUMENKO[§], W. J. QUADAKKERS[§],
G. STREHL[¶] and G. BORCHARDT[¶]

Cranfield University, MK43 0AL Cranfield, UK
*School of Metallurgy and Materials, Faculty of Engineering, The University of Birmingham, Edgbaston, Birmingham B15 2TT, UK
[†]Materials Science and Engineering, Department of Engineering, University of Liverpool L69 3GH, UK
[§]Forschungszentrum Jülich, IWV 2, 52425 Jülich, Germany
[¶]Institut für Metallurgie, TU Clausthal, 38678 Clausthal-Zellerfeld, Germany

ABSTRACT

The accurate prediction of component lifetime is essential to prevent catastrophic failure, which can have major consequences both in terms of economic issues and its impact on safety. In high-temperature applications, component life may be limited by the reaction of the material with its environment. Thus, FeCrAl(RE) Alloys, often employed in these environments, rely on an alumina scale to provide protection. The protection afforded by this scale is limited. Its failure leads to greatly enhanced corrosion rates and the ultimate failure of the component. This paper proposes a model for the chemical failure of the oxide scales following long term, high temperature exposure in oxidising environments. A number of factors contribute towards the onset of chemical failure, which incorporate both mechanical and chemical scale failure modes. These modes relate in turn to the consumption of aluminium in the alloy for the formation of the scale. It is possible to predict the onset of chemical failure by modelling the rate of aluminium depletion in the alloy. In such a model the effect of mechanical scale failure in accelerating the rate of aluminium depletion is also accounted for. In effect these modes are fundamental to the model which comprises two terms. Oxide spallation is described by the first term, whilst oxide growth is accounted for in the second. The mechanical properties of the alloy will determine which of these modes will have the dominant affect on the aluminium depletion in the alloy. Where spallation governs the mechanism, as in the case of creep-resistant alloys, the process may be expressed by a linear relationship with the aluminium consumption. However, with alloys that are less creep-resistant, oxide spallation is reduced and, therefore, the aluminium consumption will be dictated more by the oxide growth term. This establishes a situation where the time to breakaway oxidation for less creep-resistant alloys is linked to the initial aluminium content of the alloy by a power law relationship. Generally, there is good agreement with experiment and this model offers the potential for the development of a means of life prediction for the chemical failure of FeCrAl(RE) alloys in oxidising environments.

1. Introduction

FeCrAl(RE) alloys (where RE is a reactive element) are well known for their superior oxidation resistance at ultra high temperatures, above 1100°C. These excellent high temperature properties are associated with the formation of a thermodynamically stable, slow growing, alumina surface scale which protects the material against rapid corrosion [1–5]. Optimum oxidation performance is a balance, involving the aluminium content of the alloy, its chromium content, the addition of reactive elements such as yttrium (in metallic form or as an oxide dispersion) and the removal of potentially damaging tramp elements. These additions not only alter the oxidation performance of the alloys, but also their mechanical properties. For example FeCrAl-based materials with metallic yttrium additions, such as Kanthal AF* or Aluchrom YHf†, are relatively weak mechanically but have excellent oxidation resistance, being used, for example, for heater element wires and automotive catalytic converter bodies. Conversely, alloys such as PM2000§ which contains yttrium as an oxide dispersion have excellent high temperature creep strength as well as good oxidation resistance, finding wide use in heat exchangers and ultra-high temperature structural components.

In spite of these superior oxidation properties, the lifetime of components manufactured from the FeCrAl-based alloys can be limited by oxidation during long term service at temperatures around or above 1100°C [3,4,6]. The reason for this is that during high temperature exposure aluminium from the alloy matrix is consumed in the formation of the protective alumina scale. Spallation and rehealing exacerbates this loss due to cyclic plant operation, system shutdowns and restarts. Finally, when the remaining aluminium content falls below some critical concentration, the alloy can no longer reform a protective alumina scale. This results in catastrophic breakaway oxidation, associated with the rapid growth of base metal, iron oxides. The time at which this breakaway oxidation occurs is not easily predicted, it cannot solely be related to the growth rate of the alumina scale at a given time, but depends on when spallation starts, the extent of spallation, the severity of the thermal cycles and the mechanical properties of the underlying alloy as well as on the shape, chemistry and metallurgy of the particular FeCrAl alloy.

The LEAFA project has focused on the life-limiting corrosion processes, associated with the chemical failure of commercial FeCrAl(RE) alloys and many of the aspects of such failure are reported and discussed in other papers in this volume [7–12]. This paper concentrates on the development of a life prediction model for the chemical failure of a range of commercial FeCrAl(RE) alloys of European manufacture. The model aims to cover such critical variables as temperature, cyclic conditions, alloy component geometry and alloy chemistry. Particularly, the level of the alloy reservoir and how it is affected by component geometry and alloy strength has been introduced in a companion paper [11]. This life-prediction model provides one of the major deliverables of the LEAFA project.

---

* Trademark of Kanthal-Sandvik AB.
† Trademark of Krupp-VDM.
§ Trademark of Plansee.

2. Evaluation of Existing Models for Oxidation Limited Component Life

The most comprehensive modelling work to date on the chemical failure of FeCrAl based alloys was undertaken as part of a CEC COST 501 initiative aimed at designing advanced, ultra-high temperature, heat exchangers. The work undertaken jointly by Quadakkers, Bennett *et al.* [4,6,13–15] focused on the prediction of chemical failure for oxide dispersion strengthened FeCrAl based alloys. The time to breakaway oxidation for this class of alumina-forming alloy was modelled as a function of temperature, component thickness and alloy composition [6,13,15]. The model was formulated, based on the following understanding: during high temperature exposure any alloy initially gains mass due to the uptake of oxygen in forming a surface scale. For the case of FeCrAl-based alloys, this initial mass increase is associated with the formation of an alumina scale. Thus, scale growth leads to the loss of aluminium from the bulk alloy. After some long oxidation time ($t_{B/O}$), the aluminium content of the alloy may decrease below some critical level (C_B) such that an alumina scale can no longer form. This is then the life limit for the component as less protective oxides and ultimate base metal oxides must now form.

Thus, the time required for the initial aluminium content (C_o) to drop to the critical concentration (C_B) can be calculated. This is given by [13,15]:

$$t_{B/O} = \left[4.4 \times 10^{-3} \frac{\rho d (C_o - C_B)}{k} \right]^{1/n} \tag{1}$$

where n, k are oxidation parameters defining a generalised rate law, $\Delta m = kt^n$ where Δm is the mass gain per unit area (if n were equal to $1/2$ the oxidation would be parabolic), ρ is the alloy density and d is the specimen thickness, or component wall thickness. This model is illustrated in Fig. 1 for ODM751 (an ODS FeCrAl alloy) oxidised at 1100 and 1200°C [13].

In formulating this model, the following assumptions were made:

1. Oxide scale growth, without spalling, obeys a power law time dependence: $\Delta m = kt^n$, where Δm is the mass change in mgcm^{-2}, t is the time in hours and k is the oxidation rate constant.

2. Due to the high diffusion coefficients, the aluminium concentration profile in the alloy remains extensibly even over the whole plate thickness. The plate material was assumed to be of infinite length and width.

3. Catastrophic breakaway oxidation occurs when the initial aluminium concentration (C_o) drops below a critical level (C_B); which is insufficient to maintain a stable, protective alumina scale.

It was further recognised that in practice, thermal stresses and growth induced stresses will eventually cause the oxide scale to spall. Taking this spallation into account, the lifetime was predicted by a second eqn [13] given by:

$$t_{B/O} = 4.4 \times 10^{-3}\, \rho d\, (C_o - C_B)\, k^{-1/n}\, (\Delta m^*)^{1/n-1} \qquad (2)$$

This equation defines a second life-limiting condition in the oxidation lifetime diagram depicted in Fig. 1, for the case where the life of the component is controlled by spallation and rehealing. In deriving eqn (2), it was assumed that:

4. The onset of oxide spalling occurs at a critical mass gain Δm^* (effectively a critical oxide thickness), measured in mg cm^{-2}. The critical mass gain varies with alloy composition, but is believed to be independent of temperature.

5. After spallation, the reformation of the alumina scale proceeds according to the above power law time dependence, i.e. scale growth occurs like growth on a fresh metal surface.

The above equations (eqns 1 and 2), allow the time to breakaway oxidation to be estimated as a function of component thickness and alloy aluminium content, respectively, for the two cases of; (a) no oxide spallation and (b) complete loss of oxide at some critical thickness, plus repair of the spalled oxide, assuming no edge or corner effects, i.e. the component is of infinite width and length. In reality, components have finite dimensions and this was acknowledged in a subsequent modification to these equations [15]. The component thickness d in the above equation is replaced by

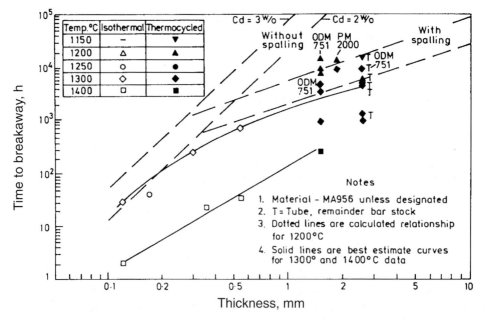

Fig. 1 *Calculated oxidation lifetime diagram from ODM751 at 1100°C and 1200°C, compared with experimental results at 1200°C [6,13].*

$$d \Big/ \left(1 + \frac{d}{b} + \frac{d}{e} \right)$$

for a rectangular specimen of length b, width e and thickness d. This correction, in fact, acknowledges that it is the volume to surface area ratio (V/A) that defines the rate of aluminium depletion, a point emphasised in ref. [11] when considering corner and edge effects.

Other approaches have taken a more empirical route, by curve fitting experimental data. For example, as part of a previous Brite/Euram programme (Improve) the breakaway oxidation of ODS alloy MA956 has been extensively studied. Following the concepts first introduced by Quadakkers *et al.* [6,13–15] sample thickness was considered an important parameter in controlling life and was extensively studied as part of this research. Samples were oxidised over a wide range of temperatures, from 1050–1400°C, for exposure times up to 22 000 h at 1050°C. The data were then fitted to a power law so providing an empirical lifetime prediction model based on experimental measurements, but with little supporting mechanistic understanding. Figure 2 presents data collected as part of the 'Improve' project [16] at temperatures of 1200 and 1300°C. The data are plotted against sample thickness and again illustrate the importance of sample thickness in defining the available aluminium reservoir. Superimposed on these data are the power law models for component lifetime at 1200 and 1300°C, determined by curve fitting the data for MA956 to a power law.

At 1200°C the empirical model for the breakaway lifetime of MA956 is:

$$t_{B/O} = 3111 \, d^{1.30} \tag{3}$$

This model fits experimental data with a regression coefficient of $R^2 = 0.972$. At 1300°C the equivalent model for MA956 is:

$$t_{B/O} = 725 \, d^{1.05} \tag{4}$$

and for this case the regression coefficient is $R^2 = 0.811$. In both of these examples d is the specimen thickness and the assumption is that the plates are of infinite dimension in the other two axes such that there are no corner of edge effects.

Clearly, both of these approaches should be consistent giving similar life predictions. Thus, the phenomenological modelling approach, based on understanding oxide growth and spallation, should be capable of predicting the empirically modelled alloy behaviour.

From a comparison of eqns (1) to (4), it becomes apparent that if the lifetime is only dependent on oxide growth, spallation and repair then the measured lifetimes should be linearly dependent on thickness. However, if no spallation occurs (for example thin foils, where the oxide thickness may never exceed a critical thickness to spall) the dependence on thickness will follow a power law with an exponent of 2 for parabolic kinetics and generally greater than this if sub-parabolic kinetics predominate. For MA9556, an ODS strengthened alloy, one would expect spallation plus repair to be the predominant mechanism. Measured exponents of 1.30 (1200°C) and 1.06 (1300°C) would suggest a combination of the above two mechanisms [16].

Thus over the temperature range 1200–1300°C for MA956, the alloy partially spalls and repairs with some areas of the surface continuing to grow by parabolic, or sub-parabolic kinetics, and other areas approximating to linear kinetics (associated with the spall and repair process).

3. Assessment of the Time to Breakaway Oxidation

Prior to the start of the LEAFA project, the main body of lifetime data for FeCrAl based materials available in the open literature was derived from two CEC research programmes, firstly COST 501 [3,4,6,13] and latterly the Brite/Euram programme 'Improve' [16]. Across these two studies six commercial FeCrAl-based alloys have been evaluated over the temperature range 1050–1400°C. The six alloys were:

(a) Three ODS alloys, produced by mechanical alloying, namely: MA956, ODM751 and PM2000.

(b) A powder metallurgical fabricated alloy, Kanthal APM, and

(c) Two wrought FeCrAlloy steels, Alloy JA13 and Aluchrom YHf.

The data presented in Fig. 2 compare the relative performance of these alloys. In addition the LEAFA project [7,12] has extensively studied the breakaway oxidation of the following alloys, following discontinuous oxidation at 1100, 1200 and 1300°C.

(d) The ODS alloy: PM2000 125 μm, 0.5 mm, 1.0 mm and 2.0 mm thick.

(e) The powder metallurgy alloy: Kanthal APM 70μm, 0.5 mm, 1.0 mm and 2.0 mm thick.

(f) Two wrought alloys: Kanthal AF at 70 μm, 0.5 mm, 1.0 mm and 2.0 mm thick and Aluchrom YHf at 50 μm, 80 μm, 0.5 mm, 1.0 mm and 2.0 mm thickness.

For the majority of this combined set of results, the long exposures to achieve breakaway oxidation conditions have necessitated the use of available resources across all of the LEAFA/ Improve partnership. The tests have generally involved discontinuous oxidation in laboratory air, using normal furnace heating and cooling rates (i.e. approx. 10^2 °C h^{-1}) with cool down cycles between 20–1000 h, depending on the length of exposure (20, 100 and 300 h were the preferred cycle intervals). Figure 3 summarises the experimentally determined times to breakaway for all seven alloys, with variable thicknesses from 50 μm to 2.7 mm, as evaluated over the temperature range 1050–1400°C. The maximum time to breakaway is in excess of 20 000 h with the minimum less than 24 h. It is this set of data that has to be modelled if a comprehensive, reliable lifetime model is to be developed.

The chemical analysis of the major additions in these seven FeCrAl(RE) alloy systems is tabulated in Table 1. Differences in composition for a given alloy/material thickness reflects batch to batch variations during alloy manufacture.

Table 1. *Analysis of FeCrAlRE alloy compositions examined (mass%)*

Alloy	Test programme/ source of data	Al	Cr	Ti	Zr	Y (Y$_2$O$_3$)
ODS Alloys						
ODM751	COST 501	4.6	16.5	0.55	–	(0.4)
MA956	COST 501					
PM2000	COST 501	5.50	20.0	0.50	–	(0.5)
MA956	Improve	4.62	19.7	0.41	–	(0.5)
PM2000	Improve	5.83	18.8	0.49	–	(0.5)
PM2000 (0.5mm)	LEAFA	4.18	19.0	0.55	–	(0.37)
PM2000 (1.0mm)	LEAFA	5.31	18.4	0.55		(0.37)
PM2000 (2.0mm)	LEAFA	5.11	19.5	0.55		(0.37)
PM Alloys						
Kanthal APM	Improve	5.86	21.1	0.21	–	0.20
Kanthal APM	LEAFA	5.84	21.0	0.026	0.11	0.19
Wrought Alloys						
FeCrAlloy JA13	Improve	5.00	16.3	0.01	–	0.32
Aluchrom YHf	Improve	5.60	20.3	<0.05	0.04	0.05
Aluchrom YHf	LEAFA	5.53	19.7	0.01	0.054	0.046
Kanthal AF	LEAFA	5.23	20.8	0.094	0.058	0.034

Note: Values for yttrium in parenthesis are Y$_2$O$_3$ additions.

4. An Updated Lifetime Prediction Model

From the foregoing, and also the better mechanistic understanding as a result of the LEAFA project [7–11], the following factors are known to influence the lifetime to breakaway oxidation.

- *Breakaway oxidation occurs when the 'aluminium reservoir' can no longer support the growth, re-formation of a protective alumina scale.*

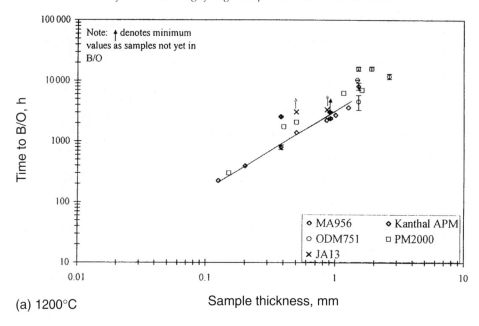

(a) 1200°C

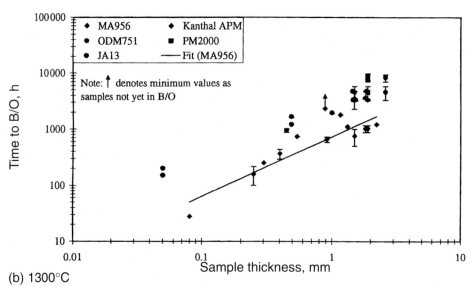

(b) 1300°C

Fig. 2 *Effect of sample thickness on times to breakaway oxidation for alloys evaluated within the 'Improve' project (data taken from ref. [16]).*

If no spallation/mechanical failure of the alumina scale occurs during service (often the case for thin foil samples, such as in automotive catalytic converters, or for weak alloys at ultra high temperatures) then the alumina scale fails due to *Intrinsic*

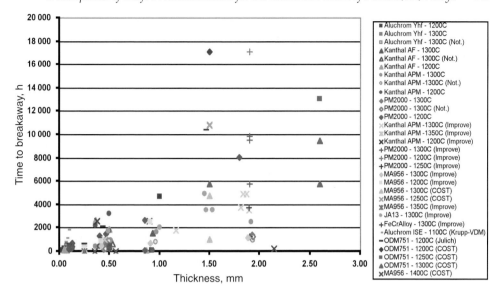

Fig. 3 *Time to breakaway oxidation data at temperatures between 1100–1400°C for FeCrAlRE alloys of various thicknesses (open symbols, marked (Not.) in the legend have not gone into breakaway at the plotted time. They went into breakaway one increment later).*

Chemical Failure (InCF)*, where less protective oxides form because the aluminium activity at the surface has been reduced to essentially to zero (It will never be zero but will be sufficiently low that chromia can form under the alumina scale and, if not chromia, then base metal oxides).

When spallation occurs, the aluminium consumption results from alumina scale loss due to the spallation processes and then from more rapid consumption during the regrowth of the protective alumina scale. If the site of spall initiates at the metal/oxide interface, then this regrowth will be at a rate comparable with the rate of early scale formation. If the site of spall lies within the scale, then regrowth at a slower rate would be expected. Thus for alloys that fail by spallation and regrowth, this is one area of uncertainty, i.e. 'Where is the site at which spallation occurs?', 'How large are the local spall areas? 'How fast does the oxide regrow in this area?'.

Ultimately, this spallation and repair process will deplete the aluminium level to a concentration where it is no longer possible to reform a protective alumina scale and at this point the alloy enters breakaway oxidation due to *Mechanically Induced Chemical Failure* (MICF). This then raises a further question of uncertainty, 'What is the critical aluminium concentration at which Mechanically Induced Chemical Failure occurs?', 'Does it change with alloy composition, thickness or alloy mechanical properties?'. All of these questions have been addressed as part of the LEAFA project [7].

---

*The concepts/terms of 'Intrinsic Chemical Failure (InCF) and 'Mechanically Induced Chemical Failure' (MICF) were first proposed by H. E. Evans for the exhaustion failure of chromia forming alloys, used within the nuclear industry [17].

• *Local spalling will occur at a critical oxide thickness,* dependent on the local mechanical properties/geometry, defect density and strain generated within the metal/oxide composite system.

• Thus the *critical oxide thickness* to spall depends on the cooling rate and creep strength of the alloy [7,9] and on the fracture toughness (strain energy release rate) of the scale/alloy interface [9]. It is therefore expected that the critical oxide thickness to spall should increase when components are cooled slowly and that it should be generally larger for the weaker alloys and should also increase if other stress relief mechanisms can relieve the driving force for fracture. These observations account for why even thin foil, strong alloys rarely show Mechanically Induced Chemical Failure but fail by Intrinsic Chemical Failure (evident by green or gold colourations under the scale, prior to classic breakaway oxidation) [7,8,10].

• The *critical oxide thickness* to spall must therefore be an important, alloy-dependent, parameter in any lifetime prediction model.

• *Prior to the onset of spallation, oxide growth will follow some diffusion controlled, protective oxidation kinetics.*

• *In areas that have spalled, oxide repairs following similar kinetics to that when first formed.*

For these FeCrAl(RE) alloys, when protected by a slow growing alumina scale, oxidation kinetics have been observed to be sub-parabolic [3,6,7,10,13–15]. Thus early oxide growth and regrowth during repair are assumed to follow sub-parabolic kinetics.

The mass gain per unit surface area (Δm) is given by:

$$\Delta m = (kt)^{1/n} \tag{5}$$

where Dm is measured in g m^{-2}, t = time in h and the oxidation rate constant (k) follows an Arrhenius temperature dependence such that:

$$k = k_o \exp\left(-\frac{Q}{RT}\right) \tag{6}$$

Hence for the early stages of oxidation, prior to the onset of spalling, or during the oxide repair processes the aluminium consumption can be calculated from the oxygen uptake, assuming that an α-Al$_2$O$_3$ scale forms.

The aluminium consumption in gm^{-2} = $\dfrac{M_{Al}}{M_{O_2}} \Delta m$ $\hspace{2cm}$ (7)

where M_{Al} is the molar mass of aluminium in α-Al$_2$O$_3$ and M_{O_2} is the molar mass of oxygen in α-Al$_2$O$_3$. Thus the mass of aluminium consumed, which equates with mass of aluminium incorporated within the scale, is 1.124 Δm.

Table 2. *Calculated critical oxide thicknesses to spall for Kanthal APM, at 0.38 mm thickness and a cooling rate of 10^4 °C h^{-1}*

Temperature, °C	Critical oxide thickness to spall, μm, at 10^4 °C h^{-1}	Estimated oxide thickness to spall, μm, at 10^2 °C h^{-1}
1000	2.3	7.3
1100	2.5	7.8
1200	2.6	8.0

At this stage, one may ask *'Can k be used to normalise any temperature dependence when modelling the lifetimes to breakaway oxidation?'*

Given that the onset of both mechanically induced chemical failure and intrinsic chemical failure require that the aluminium levels must drop to some critical, lower value before breakaway oxidation occurs (albeit different levels for the two failure mechanisms) and that the only route for this aluminium depletion is through the growth of the alumina scale, then the temperature dependence of the oxidation kinetics could well be used to model the temperature dependence of lifetimes to breakaway.

To use the temperature dependence on the oxidation kinetics in this way would require that

(a) the critical aluminium level at onset of breakaway oxidation itself is not temperature dependent (or is only weakly temperature dependent), and that

(b) the critical oxide thickness to spall does not depend on temperature.

Figure 4 (taken from ref. [7]) shows the variation of residual aluminium at the onset of breakaway oxidation for a range of alloys of different strengths and thicknesses tested at 1200 and 1300°C. Although there is much scatter, it is clear that the residual aluminium levels do not depend on temperature. They do depend, however, on both alloy mechanical properties and material thickness.

The second condition requires that the critical oxide thickness to spall is not temperature dependent. This has been addressed by Evans and Nicholls [9] in modelling the critical temperature drop necessary to cause spalling. Here it is found that the critical oxide thickness at which spalling occurs, for a given temperature drop, depends only weakly on temperature. Table 2 reproduces some predicted critical oxide thickness from this paper for temperatures of 1000, 1100 and 1200°C, at a cooling rate of 10^4°C h^{-1}, plus estimated thickness at a slower cooling rate of 10^2 °C h^{-1} (These estimates were based on the understanding that the critical temperature drop is 750°C, at a cooling rate of 10^4 °C h^{-1}, while at 10^2 °C h^{-1} it is 1030°C [9]).

Measured values of the critical mass gain (a function of oxide thickness) for the onset of spallation [7] would similarly confirm that within measurement accuracy the critical oxide thickness to spall does not vary with temperature, although it does vary with alloy composition and also sample thickness. Figure 5 reproduces

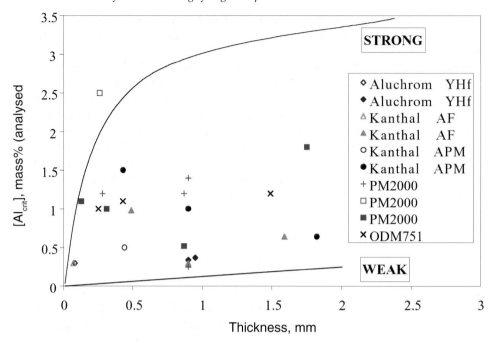

Fig. 4 *Dependance of $[Al_{crit}]$ (calculated) with thickness for four commercial FeCrAl(RE) alloys oxidised in air at 1100–1300°C. (The upper and lower boundary lines are drawn solely to delineate the data set and have no physical significance.)*

experimental data on the critical oxide thickness to spall calculated from the mass gain data published in reference [7]. Values range from 7.1 to 21.4 µm depending on

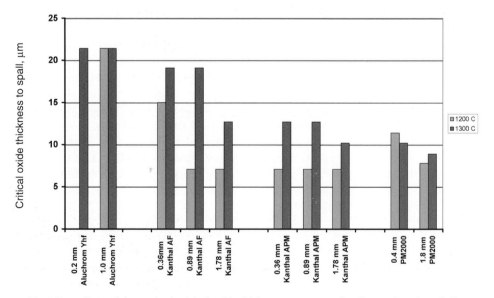

Fig. 5 *Experimental determined critical oxide thicknesses at onset of spall, as a function of alloy composition, thickness and temperature.*

alloy composition and sample thickness. For Kanthal APM at 0.36 mm thick a value of 7.1 µm is measured at 1200°C agreeing well with the predictions of Evans and Nicholls [9].

From the above discussion, it would appear that the Arrhenius temperature dependence of the oxidation rate law should be sufficient to model the temperature dependence of the lifetime to breakaway oxidation. Thus, as k is a measure of the aluminium consumption rate, and for a given alloy system (alloy composition, plus thickness) there is a fixed, usable reservoir, the parameter $kt_{B/O}$ should be a temperature independent measure of the available aluminium for alumina scale formation.

The available aluminium can be calculated from the alloy composition. A detailed analysis is given in ref. [11]. At onset of breakaway oxidation this alloy aluminium level falls below some critical level (C_B). Thus the available aluminium is given by:

$$\text{the aluminium reservoir} = \rho_m \frac{V}{A} \frac{(C_o - C_B)}{(1 - C_B)} \tag{8}$$

where ρ_m is the density of the alloy, C_o is the initial aluminium concentration, C_B is the residual aluminium concentration at onset of breakaway (i.e. the critical aluminium level necessary to ensure alumina growth/regrowth will occur). Both C_o and C_B are measured as a mass fraction (see ref. [11]). V is the volume of the material and A the surface area. The factor $(1-C_B)$ is a correction factor that must be applied to correct for the change in aluminium level within the alloy at onset of breakaway oxidation.

Equating equations (5+7) with (8) permits a life model for thin foil materials or weak substrate alloys to be developed — the implication being that under these conditions no spallation occurs and the life is purely controlled by the rate of consumption of available aluminium due to isothermal oxidation.

This model takes the form:

$$kt_{B/O} = \left[\frac{M_{O_2}}{M_{Al}} \rho_m \frac{V}{A} \frac{(C_o - C_B)}{(1 - C_B)} \right]^n \tag{9}$$

Substituting the molar mass fractions of aluminium and oxygen in alumina gives

$$kt_{B/O} = \left[0.89 \rho_m \frac{(C_o - C_B)}{(1 - C_B)} \frac{V}{A} \right]^n \tag{10}$$

This is equivalent to eqn (1), the model for non-spall conditions, as proposed by Quadakkers, Bennett *et al.* [6,13,14] and then later modified by Strehl *et al.* [11] by incorporation of the correction factors (1-C_B), except the sub-parabolic oxidation kinetics have been fitted to an equation of the form:

$$(\Delta m)^n = kt$$

with the mass gain measured in g m^{-2}. The correction factor $(1-C_B)$ has been included to account for the small change in either volume, or density, associated with the loss of aluminium [11].

Equation (10) becomes one limiting case on the life of a FeCrAl(RE) alloy component and applies when no spallation is observed throughout the component life. As such, it models behaviour under intrinsic chemical failure (InCF) and should be applicable to foil components, thin wires and weak alloys where creep within the alloy can relax all stresses such that no mechanical failure of the protective alumina scale occurs. For the conditions of Intrinsic Chemical Failure, C_B is expected to be close to zero.

5. Modelling Aluminium Consumption for Thicker Oxide Scales

For thicker oxide scales, greater than the critical oxide thickness to spall, aluminium consumption is due to oxide loss by spallation plus continued parabolic or sub-parabolic growth (protective behaviour) of the remaining, unspalled oxide. Areas that have recently spalled will also be oxidising by parabolic or sub-parabolic kinetics. It is most probable that significant spalling will only occur during each thermal cycle or system shutdown for thicker oxides, as demonstrated in the calculations of Evans and Nicholls [9]. Thus, depending on when and where the last spall occurred, a thick oxide surface will be terraced, with the most recent spalls leaving bare metal exposed, assuming spallation occurs at the metal/oxide interface.

Thus if β is the fraction of surface area with a thickness greater than x_c, the critical oxide thickness to spall, then this fraction of oxide will be lost (spalled) on the next shutdown cycle. The area will then regrow following parabolic, or sub-parabolic, kinetics and will not spall in further cycles until it again exceeds the critical oxide thickness x_c. The aluminium consumption for this spall and regrowth model will

Table 3. Statistical functions to describe the uncertainty in breakaway oxidation data; discontinuous oxidation 100 h cycles at 1300°C; measured parameter $(k_p t_{B/O})$ corrected to a V/A ratio of 0.25 mm

Alloy	Statistical distribution	Mean $k_p t_{B/O}$ (μm)^2	Standard deviation $k_p t_{B/O}$
Kanthal AF	Log-normal	1531.6	$\times \atop +$ 1.29
Kanthal APM	Log-normal	2385.4	$\times \atop +$ 1.24
Aluchrom YHf	Log-normal	880.3	$\times \atop +$ 1.23
PM2000	Bimodal 60%	1304.2	$\times \atop +$ 1.16
	Log-normal 40%	761.3	$\times \atop +$ 7.33

Note: the statistical distribution for PM2000 was bi-modal with 40% of the sample (or surface area) showing a wide variation in the susceptibility to spall. This may well reflect this alloy's increased strength and its unique microstructure.

consist of two terms, a term related to the accumulated spalled oxide and a term related to the continued growth in areas that have not spalled recently; hence:

$$\text{Al consumption} = \frac{M_{Al}}{M_{O_2}}\left[\sum_i^j \beta_i \rho_{ox} \Delta x_i + \sum_i^j (1-\beta_i)\rho_{ox} x_i\right] \quad (11)$$

where j is the total number of cycles/shut down in the life of the sample/component, β_i is the fraction spalled in any shutdown/cycle (i), ρ_{ox} is the density of the alumina scale, Δx_i is the average thickness that spalled (Δx_i will always be greater than x_c) and x_i is the oxide thickness remote from spalled area, in cycle i (i.e. the sum of all incremental growth over the period from onset of spall to cycle i, at this location).

5.1. Spallation Under Cyclic Conditions

Equation (11) provides a general equation for the aluminium consumption under any shutdown condition. This equation can be simplified when periodic shutdown cycles occur, i.e during discontinuous oxidation or during thermal cycling.

Assuming that spallation will occur locally, during the next shut down cycle, when $\Delta x_i \geq x_c$, then the mean time between spalls $\Delta t = n_c/f$, where n_c is the number of cycles required to grow an oxide thicker than x_c and f is the cycle frequency.

Thus eqn (10) becomes:

$$\text{Al consumption} = \frac{M_{Al}}{M_{O_2}}\left[\rho_{ox}(k\Delta t)^{1/n}\sum_i^j \beta_i + \rho_{ox}(kt)^{1/n}\frac{\sum_i^j (1-\beta_1)}{j}\right] \quad (12)$$

Hence the terms $\sum_i^j \beta_i$ and $\sum_i^j (1-\beta_i)$ define distributions for the fractional area that has spalled and the fractional area that has not spalled respectively up to a cycle count of j. If the cycle count is large, then these discrete distributions can be approximated by continuous statistical functions, such as the Gaussian or log-normal distributions (these distributions will be discussed in section 5.3 on handling uncertainty).

The term β is a continuous statistical function that defines the distribution for the spalled area fraction and which will take a minimum value of zero and a maximum value of unity ($\beta = \frac{1}{j}\sum_i^j \beta_i$). Furthermore, the mean number of spall cycles (j) equals $t/\Delta t$. Thus substituting for β and j in eqn (12), yields:

$$\text{Al consumption} = \frac{M_{Al}}{M_{O_2}}\left[\frac{t}{\Delta t}\beta\rho_{ox.}(k\Delta t)^{1/n} + (1-\beta)\rho_{ox}(kt)^{1/n}\right] \quad (13)$$

Further substituting the spalled oxide thickness, $\Delta x = (k\Delta t)^{1/n}$. Then

$$\text{Al consumption} = \frac{M_{Al}}{M_{O_2}}\left[\beta\rho_{ox.}\frac{k.t}{\Delta\bar{x}^{(n-1)}} + (1-\beta)\rho_{ox}(kt)^{1/n}\right] \tag{14}$$

where $\bar{x}$ is the mean oxide thickness that spalls, a value expected to be a little greater than the critical oxide thickness x_c, and β is a statistical function that defines the likely area fraction that will spall. Finally, equating this aluminium consumption to the reservoir available (eqn (8)) and rearranging gives:

$$\beta\frac{kt_{B/O}}{\Delta\bar{x}^{(n-1)}} + (1-\beta)(kt_{B/O})^{1/n} = \frac{M_{O_2}\rho_m}{M_{Al}\rho_{ox}}\frac{(C_o-C_B)}{(1-C_B)}\frac{V}{A} \tag{15}$$

Substituting for M_{Al} and M_{O_2}, this simplifies to:

$$\beta\frac{kt_{B/O}}{\Delta\bar{x}^{(n-1)}} + (1-\beta)(kt_{B/O})^{1/n} = 0.89\frac{\rho_m}{\rho_{ox}}\frac{(C_o-C_B)}{(1-C_B)}\frac{V}{A} \tag{16}$$

where β is a statistical function that defines the likely area fraction to spall, $\Delta m = (kt)^{1/n}$ defines the sub-parabolic (or parabolic) scale growth kinetics in g m^{-2}, $\Delta\bar{x}$ is the average spalled oxide thickness in m (a value just greater than x_c, the critical oxide thickness), ρ_m is the density of the alloy, ρ_{ox} the density of α-alumina, C_o the aluminium mass fraction in the alloy, C_B the critical aluminium mass fraction at the onset of breakaway and V/A defines the local volume to surface area ratio in m.

Thus the first term in eqn (16) accounts for oxide spallation and the second term for oxide growth. If no spallation occurs, i.e. for weak alloys or thin foil materials, then $\beta = 0$ and eqn (16) reduces to eqn (10). If spallation dominates, then β approaches 1 and equation (16) suggests a linear relationship between the time to breakaway ($t_{B/O}$) and V/A, the volume/surface area ratio. This limiting case is similar to eqn (2) as originally proposed by Quadakkers *et al.* [6,13].

Thus a critical parameter, which defines the shape and models predictive performance is β, the statistical function that defines the likely surface area to spall.

5.2. Handling Uncertainty in a Life Prediction Model

Ideally, a life prediction model should provide a Risk of Failure for any given lifetime. The new model aims to do this by recognising the stochastic nature of the spallation events and therefore the uncertainty in predicting a 'unique' component life. This is achieved by incorporating a statistical function for specimen to specimen variation in oxidation behaviour and for the β term in the above equation the area fraction distribution that may spall. The statistical function was evaluated experimentally by undertaking repeat oxidation studies under identical test conditions for each of the alloy systems investigated in LEAFA. Up to 8 repeat samples for some exposure conditions were undertaken, but more generally 3 repeats. Figure 6 illustrates the scatter in data observed for the repeat experiments undertaken on Kanthal AF at 1300°C. Data were measured on 0.5, 1.0 and 2.0 mm thick sample and then corrected for samples thickness (volume/surface area) using eqn (16). The data indicate that specimen to specimen variation follows log-normal statistics with a mean of 1531.6 $(\mu m)^2$ for the $k_p\ t_{B/O}$ product (53.5 m g^2 cm^4 in mass gain terms: for this statistical evaluation oxidation kinetics were assumed parabolic thus $k_p \equiv k_r$ when $n = 2$) and a standard deviation of $\overset{\times}{\div}1.29$. Thus the ratio of minimum life to maximum life at a 95% confidence would vary by a factor of 2.58 for Kanthal AF.

Similarly, the repeat oxidation data for Kanthal AF, Aluchrom YHf and PM2000 all followed log-normal statistics. Table 3 summarises the continuous statistical distribution that were most appropriate for each of these materials.

6. Comparison of the Life Prediction Model with Experimental Data

This model development has permitted the entire set of oxidation breakaway data, for a family of alloys tested under COST 501, Improve and LEAFA, to be incorporated in a single model. The model concept is mechanistically based and relies on some critical experimental parameters which are:

1.   The oxidation rate equation: $\Delta m = (k.t)^{1/n}$ where Δm is measured in gm^{-2}. When $n = 2$ the oxidation follows parabolic kinetics. The oxidation rate constant

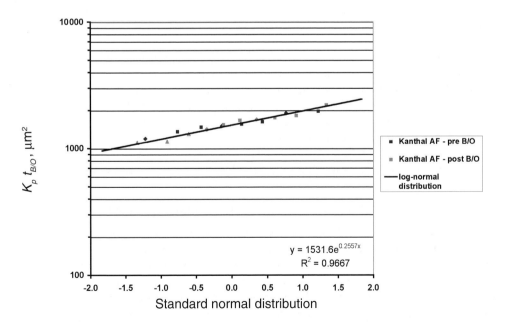

Fig. 6 Probability plot for the scatter in repeat experiments for the oxidation of Kanthal AF at 1300°C, 100 h cycles oxidised to the onset of breakaway corrosion.

k, provides a method for correcting for oxidation temperature in the model, provided the rate law exponent does not change significantly from one temperature to another. The oxidation rate constant can be accurately measured from short term, continuous or discontinuous data, provided the samples do not spall significantly.

2. The mean thickness of oxide lost due to spallation, each thermal cycle, $\Delta \bar{x}$ measured in m.

    $\Delta \bar{x}$ is expected to be larger than x_c, the critical oxide thickness to spall, depending on the cycle duration. For rapid cycles, say 1 h cycles, $\Delta \bar{x}$ will approach x_c but for long duration cycles, $\Delta \bar{x}$ will be larger than x_c, taking a value typically $\Delta \bar{x} = x_c + 2.59 \times 10^{-6} \, (k \Delta t)^{1/n}$ metres, where Δt is the cycle period in hours.

The critical oxide thickness to spall (x_c) can be estimated from short term, cyclic oxidation data by plotting the mass of spall collected against gross mass gain (see ref. [7]). Then $x_c = 0.535 \, \Delta m$, with x_c measured in µm and Δm in g m^{-2}, for an alumina scale formed on FeCrAl(RE) steel.

3. A Statistical Function that define the uncertainty in the oxidation kinetics from sample to sample and batch to batch, and incorporates β the area fraction distribution of oxide that spalls.

This function can be determined from repeat measurements taken randomly across samples and alloy batches. Care must be taken not to censor the data by limiting the sample selection to one batch of material. The calculation of these statistical functions follows the methodology outlined in Section 5.2 on dealing with uncertainty. Note for the case of β: if no spallation occurs then β by definition must be equal to zero and the materials will fail by Intrinsic Chemical Failure [InCF]. If $\beta > 0$ then the samples fail by Mechanically Induced Chemical Failure [MICF].

4. The volume/surface area ratio, measured in metres.

This parameter is related to the sample geometry and specimen dimensions and takes a value of $d/2$ for an infinite plate of thickness (d). Similarly for a wire of infinite length the ratio is $d/4$, where now d is the wire diameter.

Strictly, if the *local* volume/surface area ratio could be defined (see the approach proposed in ref. [11]), this parameter may well be able to account for *local* breakaway conditions at edges and corners.

5. The aluminium concentration in the alloy (C_o), measured as a mass fraction.

6. The critical aluminium concentration at which the alumina scale can no longer self-repair (C_B), measured as a mass fraction.

These last two parameters, in conjunction with the volume/surface area ratio define the aluminium reservoir available for protection/scale repair.

C_o can be measured from alloy batch to batch, and if the manufacturing tolerance are large can also be incorporated as a statistical function. C_B has to be determined by experiment, although it can be assumed zero for Intrinsic Chemical Failure (InCF). A number of methods have been proposed in the literature, among the most elegant is that proposed by Al-Badairy *et al.* [18], based on the re-oxidation of a previously oxidised wedge shaped sample. This approach would provide a rapid experimental method for assessing C_B, provided the samples fail by Mechanically Induced Chemical Failure (MICF).

Otherwise, there is no other alternative than to oxidise specimens through to breakaway oxidation. Foil samples cannot be used as these would fail by Intrinsic Chemical Failure (with C_B close to zero). Furthermore, if C_B is found to vary with specimen thickness (see ref. [7]) then full thickness test sections must be used, invariably involving long test durations. Equally, a further statistical function could be included for C_B if experience shows that there is a wide scatter in CB values at which breakaway corrosion is observed.

6.1. Validation of the Model

Figure 7 provides a simplified validation of the model. In this Figure, data for Aluchrom YHf, Kanthal AF, Kanthal APM and PM2000 are plotted at 1300°C. The filled symbols represent samples that have gone into breakaway oxidation, while the open symbols have not (these were generally measured at one cycle increment prior to the onset of breakaway). Superimposed on this figure are median predictions for the onset of breakaway corrosion.

In modelling the chemical failure process, oxidation is assumed to follow parabolic kinetics, thus eqn (16) simplifies to

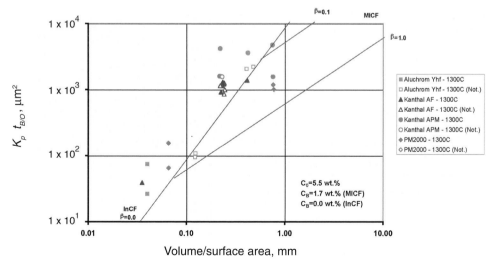

Fig. 7 *Model of Breakaway Oxidation, based on parabolic rate law oxidation, the lines super-imposed on this figure correspond to median predictions.*

$$\beta\frac{k_p t_{B/O}}{\Delta\bar{x}} + (1-\beta)(k_p t_{B/O})^{1/2} = 0.89\frac{\rho_m}{\rho_{ox}}\frac{(C_o - C_B)}{(1-C_B)}\frac{V}{A} \tag{17}$$

The assumption of parabolic kinetics provides a conservative model, as aluminium consumption would occur more quickly than for sub-parabolic kinetics as often observed in practice. Thus, in Fig. 7 the experimental data points all lie above (longer times to breakaway) the predicted trend lines. Two sets of trend lines are plotted on this Figure, those corresponding to Intrinsic Chemical Failure (with $\beta = 0$) and those corresponding to Mechanically Induced Chemical Failure (with β of 0.1 and 1.0). These model prediction curves have been calculated from eqn (17) using $C_o = 0.0$mass% for intrinsic chemical failure. For mechanically induced chemical failure curves are plotted for $\beta = 0.1$ or 1.0, with $C_B = 1.7$mass% and the critical oxide thickness to spall (Δx) taken as 9 μm (mid-way through the range of measured values, see Fig. 5).

The results and trends agree well with the experimental data. By assuming a parabolic rate constant, rather than a sub-parabolic one, the model provides a conservative estimate of component life.

Figure 8 extends this analysis and plots the total available data (from COST 501, Improve and LEAFA — the data previously plotted in Fig. 3) for exposure times up to 20 000 h at temperatures in the range 1050–1400°C. There data are again plotted on log axis as $k_p t_{B/O}$ vs V/A ratio. For thin samples ($V/A < 0.2$ mm) the slope of the graph is parabolic, with $(k_p t_{B/O})^{1/2}$ proportional to V/A, while for thick samples and stronger materials the slope is linear, i.e. $(k_p t_{B/O})$ is proportional to V/A. This behaviour is consistent with the proposed model (as can be seen from eqn (17), the simplified case that applies to parabolic kinetics).

Superimposed on Fig. 8 is a set of predicted trend lines for the onset of breakaway

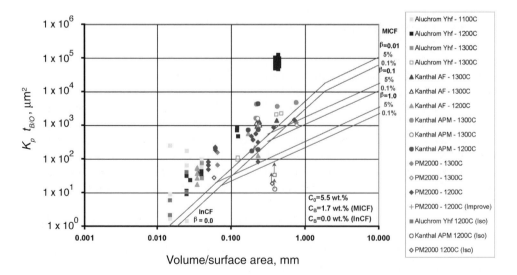

Fig. 8 *Life prediction model for chemical failure for FeCrAlRE, alumina forming alloys (InCF-Intrinsic Chemical Failure: MICF – Mechanical Induced Chemical Failure). The solid lines present model predictions at either a 5% or 0.1% risk of failure within the predicted lifetime.*

oxidation, based on the above model. Because of some uncertainty in the extent of spallation during any shutdown cycle, together with possible alloy to alloy and batch to batch variation, these must necessarily be probabilistic. Assuming that sufficient cycles occur such that a distribution in residual oxide thicknesses and the extent of spall can exist then trend lines based on the risk of localised breakaway oxidation can be superimposed on the $k_p t_{B/O}$ vs V/A diagram. Predictions for a 0.1% and 5% risk of failure are presented in Fig. 8.

This stochastic model takes the form:

$$\beta e^{-\lambda s_M} \frac{k_p t_{B/O}}{\Delta \bar{x}} + (1-\beta) e^{-\lambda s_I} (k_p t_{B/O})^{1/2} = 0.89 \frac{\rho_m}{\rho_{ox}} \frac{(C_o - C_B)}{(1 - C_B)} \frac{V}{A} \tag{18}$$

where the new terms $e^{-\lambda s_M}$ and $e^{-\lambda s_I}$ provide measures of the dispersion expected as a result of the log-normal statistical behaviour observed for the $k_p t_{B/O}$ product (see Fig. 6). These terms present multipliers that reduce the predicted lifetime to breakaway under intrinsic chemical failure ($e^{-\lambda s_I}$) and mechanical induced chemical failure ($e^{-\lambda s_M}$) conditions, with β now defining the area fraction (or the probability) of spallation.

The terms $e^{-\lambda s_M}$ and $e^{-\lambda s_I}$ can be evaluated using experiment and standard normal (Gaussian) statistics tables. The terms s_M and s_I are the standard deviations, measured on a logarithmic scale, for samples that would undergo mechanical induced chemical failure (s_M) or intrinsic chemical failure (s_I). These values can be calculated as outlined in Section 5.2, while λ is a constant that defines the level of Risk of Failure and can be derived from normal statistic tables.

For a 5% risk of failure: $\lambda = 1.645$.
For a 0.1% risk of failure: $\lambda = 3.090$.

Figure 8 presents all the measured data, not just the data for one batch of alloy. When combining such data sets the expected scatter will increase. As any predictive model must always be conservative the model must be capable of taking this into account if component failure is to be avoided. The stochastic nature of this model, plus the assumption of parabolic kinetics provides the necessary conservatism.

Combining the data presented in Table 3 (to make predictions more applicable to any FeCrAl(RE) alloy) then the expected distribution in $k_p t_{B/O}$ life values will be log-normal with a standard deviation of $\overset{\times}{\div} 1.44$. To provide a conservative prediction, for a 5%, and 0.1% risk of failure, the median prediction (50% chance of failure as plotted in Fig. 7) must be displaced to shorter lives.

For the model predictions, it is assumed that $s_M = s_I = 0.3649$ where $s = \ln(1.44) = 0.3649$.

The above trend lines are included in Fig. 8 for three levels of β: $\beta = 0$ (intrinsic chemical failure) and $\beta = 0.1$ or 1.0 (mechanical induced chemical failure) which demonstrate both the good fit and conservatism of this model. All of the breakaway data points are above the 0.1% risk of failure, while a small fraction of the measured data falls between the 0.1% and 5% risks of failure. The trend lines also demonstrate the importance of the spalls probability (β) in determining, the lifetime of a component under mechanically induced chemical failure conditions.

Thus, this modelling approach development has allowed the entire set of oxidation breakaway data, for all alloys tested and evaluated in the LEAFA project plus other relevant published studies, to be presented on a single diagram that shows the variation in time to breakaway (chemical failure) as a function of volume/surface area ratio (equivalent to half the component thickness for large sheet materials). The model can accurately predict component life for FeCrAl(RE) based materials that oxidise to form α-alumina scales over the temperature range 1050–1400°C for exposure times up to 22 000 h at 1050°C, and can provide an associated risk of failure with each predicted lifetime.

The model requires a number of critical input parameters to provide reliable prediction which, in the most part, can be estimated from short term oxidation data.

7. Conclusions

A model has been developed that allows the lifetime (to chemical failure) to be predicted under oxidising conditions for a wide range of FeCrAl based materials, both weak and strong. The model predicts the component life and provides a risk of failure within this lifetime.

The performance of the model has been demonstrated assuming parabolic kinetics and this shows a good fit between the model and experiment. This parabolic approximation provides a conservative estimate of component life, as it is widely recognised that these FeCrAl(RE) based materials oxidise following sub-parabolic kinetics.

The generalised model takes the form:

$$\beta e^{-\lambda s_M} \frac{kt_{B/O}}{\Delta \bar{x}^{(n-1)}} + (1-\beta)e^{-\lambda s_I}(kt_{B/O})^{1/n} = 0.89 \frac{\rho_m}{\rho_{ox}} \frac{(C_o - C_B)}{(1-C_B)} \frac{V}{A}$$

where $t_{B/O}$ = the time to breakaway oxidation, $\Delta m = (kt)^{1/n}$ defines the oxidation kinetics, Δx is the average thickness of oxide when it spalls. Δx takes value just greater than the critical oxide thickness to spall (x_c). β is a statistical term that accounts for the variable area fraction that spalls, during each shutdown event. (probability of spallation), $e^{-\lambda s_M}, e^{-\lambda s_I}$ are multipliers that define the risk of failure, under mechanical induced chemical failure and intrinsic chemical failure conditions respectively. λ is a constant that defines the level of risk (λ = 1.645 for a 5% risk of failure and λ = 3.090 for a 0.1% risk of failure), s_M and s_I are the standard deviations of the log-normal fit to breakaway oxidation data for an alloy under mechanical induced chemical failure conditions and intrinsic chemical failure conditions respectively. V/A defines the volume/surface area ratio, which is a characteristic parameter related to component geometry. ρ_m is the alloy density, ρ_{ox} the density of an α-Al$_2$O$_3$ scale, and C_o and C_B define the mass fraction of aluminium in the alloy at alloy manufacture (C_o) and just prior to breakaway oxidation (C_B).

A further strength of this model is that a number of the necessary model parameters can be estimated from short term oxidation studies on a batch to batch basis. These include the parabolic (or sub-parabolic) oxidation kinetics, the critical oxide thickness to spall and the statistical function β, that defines the variable surface area that may spall in any shutdown cycle. Estimation of $e^{-\lambda s_M}$ and $e^{-\lambda s_I}$ requires full lifetime testing on multiple samples, but can be undertaken under short time/high temperature conditions.

The model assumes:

(i) aluminium consumption results from alumina scale formation and possible spallation of this scale from the alloy surface during cyclic exposure;

(ii) that local spallation will occur when a critical oxide thickness is achieved;

(iii) that the critical oxide thickness to spall depends on the cooling rate, creep strength of the alloy and fracture resistance of the scale/interface;

(iv) prior to the onset spalling, oxide growth follows a generalised rate law. This allowing parabolic and sub-parabolic oxidation kinetics to be modelled; and

(v) in areas that have spalled, the oxide repairs following either parabolic or sub-parabolic kinetics.

8. Acknowledgements

This study formed part of the BRITE/EURAM LEAFA Project funded by the European Community (Project No. BE-97-4491). The authors are grateful to their Partners for the supply of the alloys tested, for the chemical analyses of the alloys and for their contribution to scientific input in discussing these results.

References

1. F. H. Stott and G. C. Wood, *Mat. Sci. Eng.*, 1987, **87**, 267.
2. G. Korb and A. Schwager, *High Temp. — High Pres.*, 1989, **21**, 475.
3. W. J. Quadakkers, *Werkst. Korros.*, 1990, **41**, 659.
4. M. J. Bennett, H. Romary and J. B. Price, *Heat Resisting Materials*. ASM International, 1991, p.95.
5. B. A. Pint, *Oxid. Met.*, 1996, **45**, 1.
6. W J. Quadakkers and M. J. Bennett, *Mater. Sci. Technol.*, 1994, **10**, 126.
7. R. Newton, *et al.*, *Lifetime Modelling of High Temperature Corrosion Processes* (M. Schütze, W. J. Quadakkers and J. R. Nicholls, eds). EFC Publication No. 34. This Volume, pp.15–36.
8. H. Al-Badairy *et al.*, *Lifetime Modelling of High Temperature Corrosion Processes* (M. Schütze, W. J. Quadakkers and J. R. Nicholls, eds). EFC Publication No. 34. This Volume pp.50–65.
9. H. E. Evans and J. R. Nicholls, *Lifetime Modelling of High Temperature Corrosion Processes* (M. Schütze, W. J. Quadakkers and J. R. Nicholls, eds). EFC Publication No. 34. This Volume pp.37–49.

10. D. Naumenko, *et al.*, *Lifetime Modelling of High Temperature Corrosion Processes* (M. Schütze, W. J. Quadakkers and J. R. Nicholls, eds). EFC Publication No. 34. This Volume pp.66–82.

11. G. Strehl *et al.*, *Lifetime Modelling of High Temperature Corrosion Processes* (M. Schütze, W. J. Quadakkers and J. R. Nicholls, eds). EFC Publication No. 34. This Volume pp.107–122.

12. A. Kolb-Telieps *et al.*, *Lifetime Modelling of High Temperature Corrosion Processes* (M. Schütze, W. J. Quadakkers and J. R. Nicholls, eds). EFC Publication No. 34. This Volume pp.123–134.

13. W. J. Quadakkers and K. Bongartz, *Mater. Corros.*, 1994, **45**, 232–241.

14. W. J. Quadakkers, D. Clemens and M. J. Bennett, in *Microscopy of Oxidation — 3* (S. B. Newcomb and J. A. Little, eds), p.195–206. Published by The Institute of Materials, London, 1997.

15. I. Guruppa, S. Weinbruch, D. Naumenko and W. J. Quadakkers, *Mater. Corros.*, 2000, **51**, 224–235.

16. Final report on Contract no. BRE-2-CT94-0605 "How to Improve the Failure Resistance of Alumina scales on High Temperature Materials". European Community, Brussels, May 1997.

17. H. E. Evans, A. T. Donaldson and T. C. Gilmour, *Oxid. Met.*, 1999, **52**, 379–402.

18. H. Al-Badairy, G. J. Tatlock and M. J. Bennett, *Mater. High Temp.*, 2000, **17**, 101–108.

7

The Influence of Sample Geometry on the Oxidation and Chemical Failure of FeCrAl(RE) Alloys

G. STREHL, V. GUTTMANN*, D. NAUMENKO[†], A. KOLB-TELIEPS[§],
G. BORCHARDT, W. J. QUADAKKERS[†], J. KLÖWER[§], P. A. BEAVEN*
and J. R. NICHOLLS[¶]

Institut für Metallurgie, TU Clausthal, 38678 Clausthal-Zellerfeld, Germany
*Institute for Advanced Materials, Joint Research Centre, 1755 ZG Petten, The Netherlands
[†]Forschungszentrum Jülich GmbH, IWV 2, 52425 Jülich, Germany
[§]Krupp VDM, 58778 Werdohl, Germany
[¶]Cranfield University, Cranfield, Bedford, MK43 0AL, UK

ABSTRACT

The influence of the geometry factor on the protection imparted by alumina scales on various FeCrAl(RE) alloys at high temperatures has been investigated. The most critical parameter involved in scale formation and the maintenance of protection is the volume to surface ratio of the component. An essential role in protection is played by the mechanical strength of the substrate because of its marked influence on the sensitivity to scale spallation. In this context the material thickness, the sample shape and the constraints resulting from the component geometry are of importance. In addition, the oxygen partial pressure in the surrounding atmosphere, which can become reduced by unfavourable geometries such as thin bores or crevices has been identified as a relevant parameter.

1. Introduction

The attractive high temperature oxidation properties of FeCrAl(RE) alloys are based on a protective alumina scale, the formation and maintenance of which relies on the availability of an aluminum reservoir in the underlying alloy [1]. The formation of a protective scale and the related aluminium consumption up to the onset of chemical failure (breakaway oxidation), as indicated by the rapid oxidation of Cr and Fe, strongly depend on the sample geometry both global and local. In particular, a low volume to surface ratio adversely affects lifetime, simply because in this case the available Al is subject to a relatively high rate of consumption.

Apart from intrinsic chemical failure (ICF), which relates to the ideal case of a fully adherent scale, mechanically induced chemical failure (MICF), which is associated with scale spallation and restoration [2], may occur. Spallation takes place if the relief of stresses in the scale arising from scale growth and cooling processes is hindered as for instance in high strength alloys, thick samples or for complex geometries that impose constraints on sample deformation. Descaling and healing, respectively, result in accelerated aluminium consumption and premature failure must be expected.

In the present work a number of examples are given of the influence of technically relevant geometrical parameters on the oxidation and the chemical failure of FeCrAl(RE) alloys. The work also includes investigations on the effect of holes and crevices, which is related to scale formation under specific design conditions.

2. Experimental

The compositions of the alloys investigated are given in Table 1. Samples with a thickness of 0.5 mm and higher were given a 1200 grit surface finish, thinner materials were used in the as-received condition. Sample dimensions, exposure temperatures and times are given together with the experimental results.

3. Results and Discussion

3.1. Volume to Surface Ratio

For a given alloy the volume to surface ratio is the most important geometric quantity determining the component lifetime during oxidation. The available aluminium is proportional to the component volume, whereas the consumption of aluminium by scale formation is proportional to the total surface area. In general, the oxidation kinetics are described in terms of mass gain (Δm) or scale thickness vs time (t)

$$\frac{\Delta m}{A} = kt^n \tag{1}$$

where A is the surface area and k and n represent the oxidation rate constant and the time exponent, respectively.

The amount of aluminium released from the metal during oxidation ($m_{Al}^{oxide}(t)$) is fully contained in the oxide (including the spalled oxide) because unlike chromia,

Table 1. Alloy compositions (values taken from [3])

Alloy	Al, mass%	Cr, mass%	Fe, mass%	Y, ppm	Zr, ppm	Ti, ppm	Hf, ppm	Mn, ppm	Si, ppm
Aluchrom YHf	5.5	19.7	Bal.	460	540	98	310	1800	2900
Kanthal APM	5.8	21.0	Bal.	0.2	1100	260	1.1	800	4000
PM 2000	5.4	19.5	Bal.	3700	14	4500	0.05	310	240

alumina does not evaporate in the temperature range investigated [4]. Thus, eqn (2) holds at all times.

$$m_{Al}^{alloy}(0) = m_{Al}^{alloy}(t) + m_{Al}^{oxide}(t) \tag{2}$$

Due to the faster diffusion of aluminium in the alloy compared to the diffusion of oxygen in the scale, a uniform aluminium content in the alloy is a reasonable approximation during the alumina-forming stage at elevated temperatures [1,5]. However, it should be noted that at times close to breakaway this assumption becomes questionable. In theory, the initial aluminium mass fraction C_0 will decrease to a critical value C_B at the onset of breakaway oxidation at the time t_B.

$$C_0 = \frac{m_{Al}^{alloy}(0)}{m^{alloy}(0)}, \quad C_B = \frac{m_{Al}^{alloy}(t_B)}{m^{alloy}(t_B)} \tag{3}$$

Calculation of the aluminium content in the oxide in terms of C_0 and C_B leads to:

$$m_{Al}^{oxide}(t_B) = m_{Al}^{alloy}(0) - m_{Al}^{alloy}(t_B) = C_0 m^{alloy}(0) - C_B m^{alloy}(t_B) \quad \Leftrightarrow$$

$$m_{Al}^{oxide}(t_B) = C_0 m^{alloy}(0) - C_B \left[m^{alloy}(0) - m_{Al}^{oxide}(t_B) \right] \quad \Leftrightarrow$$

$$m_{Al}^{oxide}(t_B)\left[1 - C_B\right] = m^{alloy}(0)\left[C_0 - C_B\right] \quad \Leftrightarrow$$

$$m_{Al}^{oxide}(t_B) = m^{alloy}(0)\frac{C_0 - C_B}{1 - C_B} \tag{4}$$

In eqn (4) the left hand side describes a surface related quantity, whereas the right hand side is volume dependent. The aluminium content in the oxide can also be calculated from the mass gain:

$$m_{Al}^{oxide}(t) = 1.12427\,\Delta m(t) \tag{5}$$

Expressing the mass of the alloy using its density (ρ_{alloy}) and including eqs (1) and (4) gives:

$$t_B = \left(\frac{V}{A} \frac{\rho_{alloy}}{1.12427} \frac{C_0 - C_B}{1 - C_B} \frac{1}{k} \right)^{\frac{1}{n}} \tag{6}$$

Volume to surface ratios for a number of typical technical geometries are given in [1]. In order to take into account quantitatively the processes of descaling and scale restoration, more complex equations than (6) are required. This increased complexity is associated with the uncertainty associated with the time dependence of the degree of scale spallation/reformation after each exposure cycle. Reference [6] provides a stochastic approach to addressing this problem. Thus eqn (6) represents the scenario when no spallation occurs (usually observed for thin section components).

Equation (6) shows that not only the geometrical term, i.e. the volume to surface ratio, which governs the Al consumption, but also the chemical terms, especially the total Al content C_0 initially available in the alloy and the critical Al level C_B at the onset of chemical failure, play a decisive role in component lifetime. Since the lifetime is related to all these values by a power law function they have to be carefully assessed and optimised for technical applications.

With regard to the initial Al level, C_0 is solely alloy dependent. It should be noted that the possibility to increase C_0 is very restricted since exceeding the 5–6 mass% Al typical for commercial FeCrAl (RE) alloys leads to difficulties in hot working of the materials [7].

The critical Al level C_B at the onset of chemical failure depends partially on the alloy composition but in addition, the sample geometry becomes important in connection with its role in inducing scale spallation, which can trigger component failure via MICF. In order to form a protective alumina scale on a blank surface after descaling, higher Al levels are required than for maintaining protection by a pre-existing, adherent scale. This is because during scale formation on a fresh surface, Al has to compete with other species such as Cr and Fe. For a relatively low Al content, oxidation of these base metals can become the dominating process. In this case, the Al content lies below C_{NOSH} (NO Self Healing [2]), i.e. self healing of the scale is no longer possible.

In [1], the term C_B in the denominator in the equation corresponding to (6) has been neglected. For alloys such as PM 2000, critical aluminium contents at time to breakaway in excess of 1.7 mass% have been measured [3]. For the same material the exponent of the growth law is often evaluated as $n = 1/3$ [8]. Using these values, lifetime predictions from the two formulae differ by a factor of 1.05, i.e. a lifetime extension of 5%. Compared to the effect of variations in the exponent n or the rate constant k, this is very small. For technical lifetime estimates the denominator can be approximated by $1 - C_B \approx 1$.

Concerning the geometrical term in (6), i.e. the volume to surface ratio, it should be noted that the latter does not only play a role for the sample as a whole, but often has a specific significance with respect to the local sample geometry. In this case, not only the locally available aluminium reservoir and the aluminium consumption at nearby surfaces but also the supply of additional aluminium by diffusion from the bulk of the component have to be taken into account.

The expected local differences in Al consumption and supply, for instance in the case of sheet samples can be illustrated by calculating the ratio for a cubic body with an edge length a ($a <$ sheet thickness) (Fig. 1). This gives V/A values of a, $a/2$ and $a/3$ for the bulk, the edges and the corners, respectively. Thus, corners clearly suffer the most. Moreover, as the volume to surface ratio decreases in this order, the number of planes through which consumed Al can be replaced in the cubic element decreases from 5 to 3. According to this ranking, breakaway oxidation of sheet samples is generally initiated at the corners and edges. In particular, sharp, i.e. rectangular corners cause a deleterious effect. In 0.5 mm thick PM 2000 sheet it was found that a corner radius of 0.5 mm was sufficient to reduce the corner effect and that with a radius of 1.5 mm the corners behaved like the edges.

Failure sensitivity at corners and edges can also be enhanced by possible non-uniformities in sample thickness. If the test pieces have to be ground and polished, even very careful preparation can sometimes lead to variations in the sample thickness and lower thickness values especially in the corner regions are often unavoidable.

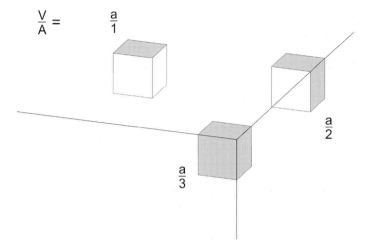

Fig. 1 *Local volume to surface ratio in the plane, at the edge and at the corner of a sheet sample, illustrated by a cube with edge length a.*

A specific aspect related to corners and edges concerns enhanced oxide cracking and spallation that results from the development of relatively high cooling stresses [9,10]. It should be noted, however, that such damage strongly depends on the degree

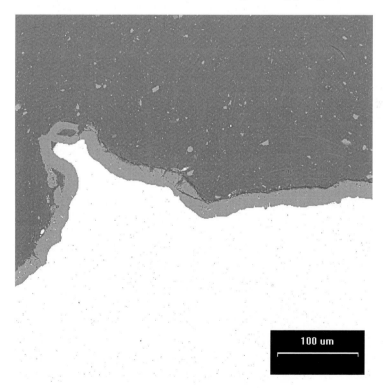

Fig. 2 *Scale adherence at the edge of a coupon of 2 mm thick Kanthal APM after 2250 h of oxidation at 1200°C.*

of stress relaxation via substrate deformation. This is demonstrated in Fig. 2 for the relatively weak alloy Kanthal APM which undergoes scale-substrate lifting as a result of scale growth stresses. Despite the development of a pronounced convex geometry, which is expected to promote spallation [9–11], the scale exhibits excellent adherence. At the crest, rehealing after prior scale lifting is documented but no spallation had taken place.

The local volume to surface ratio can also become reduced by complex sample geometries. For instance, holes drilled into sheet materials very close to each other produce locally thin bridges of material and enhanced breakaway oxidation must be expected in these regions. However, in the present work it was found that rather thin bridges are needed to cause a deleterious effect. For instance in the case of 0.5 mm thick PM 2000 sheet, sets of holes of 3 mm dia. caused premature failure when the distance between the holes was less than 1 mm. Comparing in a somewhat arbitrary way the surface to volume ratio for such a bridge with that related to a corner of the same volume it is found, that only below a distance of about 1 mm, this ratio becomes clearly higher for the bridge.

In the following, examples are given which demonstrate quantitatively the importance of the geometry effect on oxidation and breakaway oxidation. A first example which is related to the consequences of mechanical constraints during oxidation is depicted in Fig. 3. Aluchrom YHf samples of identical thickness (1 mm) and surface area (314 mm²), but exhibiting square and elongated shapes are compared. For the more rigid square sample, higher stresses have to be taken up by the oxide scale with the consequence that scale cracking together with internal oxidation are facilitated (Fig. 4). Thus compared to the ribbon sample (4 mm × 78.5 mm) which exhibits high flexibility, the oxidation rate is increased and thus a lifetime reduction should be expected.

A further example, given in Fig. 5, illustrates the oxidation behaviour of a free hanging sample, a ring type specimen and an automotive catalyst body. According to constraints in stress relaxation resulting from the specific design in the case of the

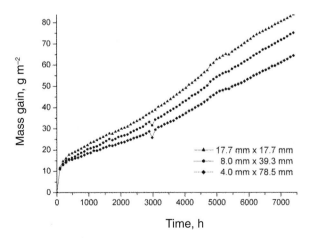

Fig. 3 *Net mass change of Aluchrom YHf oxidised in air at 1200°C. Comparison of different sheet geometries with same thickness and surface area.*

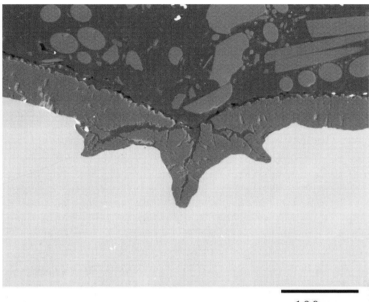

100μm

Fig. 4 *Cracked oxide and oxide intrusion in the square-shaped sample of Aluchrom YHf after 7392 h oxidation (see Fig. 3).*

catalytic converter, high cooling stresses have to be taken up in the scale and hence cracking and internal oxidation occur (Fig. 6). In a similar way, this holds for the ring-shaped sample. In contrast, stress relaxation in the free hanging specimen can easily take place by substrate deformation [2].

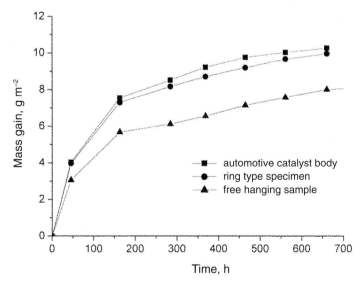

Fig. 5 *Differences in the oxidation behaviour of different geometries of Aluchrom YHf 58 μm foil oxidised at 1100°C.*

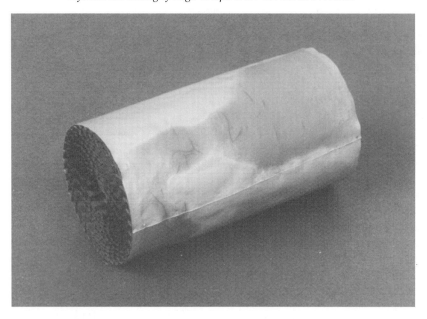

Fig. 6 *Model catalytic converter body after 660 h oxidation in air at 1100°C showing cracks in the otherwise adherent oxide (see also Fig. 5).*

Investigations on the influence of sample thickness on oxidation and chemical failure were carried out for Kanthal APM. Coupons 20 mm × 10 mm with nominal thicknesses of 0.5, 1 and 2 mm were employed. Exposures were carried out in air at 1200°C with heating rates of 6°C/min and furnace cooling with 300 h cycles. The oxidation kinetics for the various thicknesses are documented in terms of net mass change in Fig. 7. The curve for the 0.5 mm material may be considered as the standard

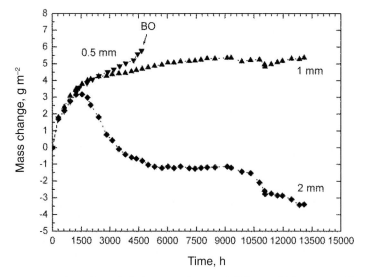

Fig. 7 *Net mass gains of Kanthal APM specimens of different thicknesses oxidised in air at 1200°C (BO = Breakaway Oxidation).*

function since at least up to 4000 h no marked spallation was observed. Moreover, the results also obeyed a power law (linear log mass – log time plot) which is consistent with the usually applied oxidation law (1).

The values of k and n were found to be 0.21 mg cm^{-2} h^{-n} and 0.39, respectively. The mass change–time curves of the 1 and the 2 mm material tended to fall below the standard curve after about 3000 and 1500 h, respectively. Moreover, negative mass changes occurred for the 2 mm material beyond about 1500 h, and after about 3000 h the mass decreased to below that of the virgin material.

Surface examination of the 0.5 mm material revealed a high density of small individual or coagulated light spots, which after time to failure of about 4600 h reached sizes of about 100–150 µm (Fig. 8). These spots correspond to local scale decohesion which took place in connection with the development of hollow-like defects at the substrate/scale interface (Fig. 9). At such places the scale became undergrown by a new scale.

In the 1 mm samples, significant descaling started after about 3000 h compared to about 1500 h in the 2 mm samples. For the latter, about 40% of the original scale had disappeared after 3000 h exposure. Scale loss was characterised by blank areas with maximum sizes of about 0.4 mm × 0.7 mm in the 1 mm and 0.6 mm × 1.5 mm in the 2 mm material, together with marked scale flaking which was more pronounced for the thicker coupons. Scale deterioration became initiated by thin intersecting cracks which, after longer exposure times, developed into V-shaped grooves. In the 0.5 mm thick material only a few thin grooves developed (Fig. 8). As illustrated in Fig. 10, the grooves separated the scale clearly into small brick-like pieces with a final size of about 0.8 mm × 1.4 mm in the thicker coupons (note: the values indicate the extensions in the short and long sample directions, respectively).

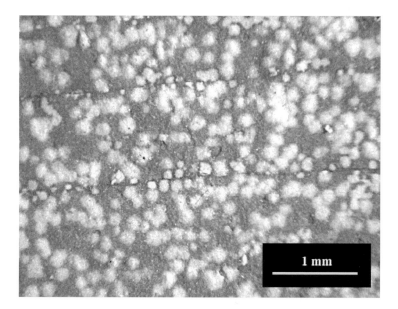

Fig. 8 *Surface view of 0.5 mm Kanthal APM after 4600 h of oxidation at 1200°C. Light spots indicate local scale lifting and interfacial pores.*

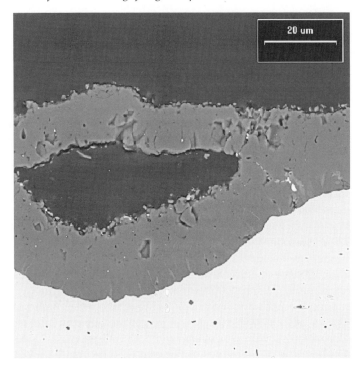

Fig. 9 *Hollow-like defect beneath the original scale undergrown by a new oxide layer on 0.5 mm Kanthal APM after 4600 h of oxidation at 1200°C.*

Fig. 10 *Surface view of scale damage on 2 mm thick Kanthal APM after 3540 h of oxidation at 1200°C, showing broad grooves and a brick-like oxide structure.*

The grooves were oriented fairly parallel to the sample axis, although severe deviations from this orientation sometimes occurred. Such scale damage appeared at the beginning of oxidation but became more pronounced with increasing exposure time as a result of preferential spallation and scale flaking taking place at the corners and edges of the bricks. In these regions, a number of defects such as pores and cracks developed in the scale and enhanced inward scale growth took place, as illustrated in Fig. 11, whereas the scale formed elsewhere was effectively defect-free.

The scale cracking mechanism in both the 1 and the 2 mm thick materials was chiefly characterised by wedge-type cracking (Fig. 12) [12]. In the 0.5 mm coupons scale buckling was also sometimes detected in connection with the interfacial defects observed (Fig. 9).

In the long term regime beyond about 10000 h (Note: the 0.5 mm samples had already failed after about 4600 h) the scale morphology became rather irregular since broad grooves developed (Fig. 10). The 1 mm samples revealed a few flakes and blank spots and about 70% of the surface was still covered by the primary scale. In the 2 mm sample a larger number of small blank areas and more scale flaking appeared and only about 40% of the original scale was still present. A rough estimate showed that in this case about 30% of the scale corresponded to the grooves. The latter developed a maximum opening at the scale–air interface of about 200 μm.

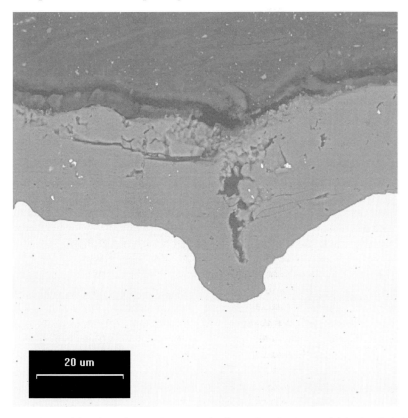

Fig. 11 *Enhanced local oxidation due to scale distortion in 2 mm thick Kanthal APM after 2250 h of oxidation at 1200°C.*

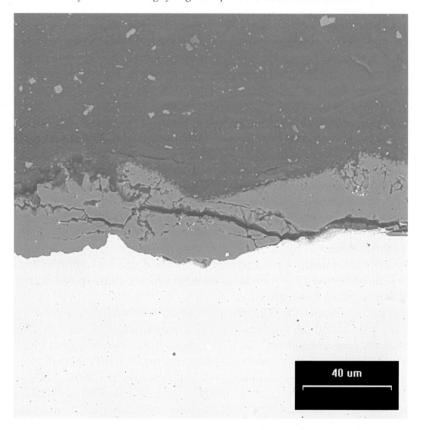

Fig. 12 *Scale distortion in the form of wedge-type cracking on 2 mm thick Kanthal APM after 2250 h of oxidation at 1200°C.*

Already at an early stage of exposure, scale growth induced substrate lifting at the edges and corners of the samples (Fig. 2), which was clearly more important in the thicker coupons. Additional deformation of the test pieces occurred by bending, twisting and elongation. These deformations were pronounced in the 0.5 mm thick samples but did not appear in the 2 mm material.

The variations in scale damage observed in Kanthal APM samples of different thicknesses can be interpreted as follows. The low creep resistance of this alloy (1000 h creep rupture strength of only 1.6 MPa at 1200°C and 11.3 MPa at 800°C [13]) means that scale growth as well as cooling stresses, which both subject the substrate to tension and the scale to compression, can be partially relieved by substrate deformation, depending on the sample thickness. The 0.5 mm samples with a low strength became severely deformed as evidenced by shape changes and the appearance of pores in the substrate. The interfacial pores (see the hollow-like defect in Fig. 9) result from the thermal stresses generated perpendicular to the scale–substrate interface during oxidation, followed by cooling. Thus, in thin samples the response of the coupon to oxidation induced stresses is essentially governed by the scale behaviour. Stresses are mainly taken up by the weak substrate and therefore only small stresses develop in the oxide.

In contrast, the stresses are not fully relieved by substrate deformation in the 1 and 2 mm thick coupons. Strong scale substrate lifting at the corners and edges indicates that deformation is limited to a thin sub-scale layer, i.e. the scale growth stresses are too low to deform the whole sample. The cooling stresses also have to be taken up preferentially by the scale in the case of a thick, i.e. high strength substrate. Thus for thicker samples the response to oxidation stresses is mainly dictated by the strong substrate. The scale is therefore subject to high stresses with the consequences of descaling. Similar conclusions have been reported for testing Kanthal APM at 1300°C [14].

Scale damage occurs preferentially by wedge-type cracking. In this case, shear cracks are generated in the scale particularly during cooling, followed by decohesion along the scale–substrate interface or within the scale [11]. The pronounced scale flaking observed especially after long term testing, seems to be an important consequence of this mechanism. The alternative process of scale buckling, which was sometimes found in the 0.5 mm samples, did not in general cause complete descaling since scale healing proceeded beneath the detached scale (Fig. 9). Thus, if the outer scale underwent fracture, as occurred after long term exposure, small flakes were lost but no blank areas appeared.

With regard to chemical failure, the present investigations have shown that the 0.5 mm samples failed after about 4600 h. Microstructural details of the failure mechanism are presented elsewhere [15]. Provided that no scale spallation occurs, the lifetime for rectangular test pieces should obey eqn (6). Lifetime predictions based on the assumption of scale integrity correspond to about 30 000 and 115 000 h for the 1 and 2 mm thick samples respectively. The marked scale loss observed particularly in the 2 mm thick coupons should substantially reduce these times. Up to 13 000 h exposure no chemical failure has been observed.

Another geometry effect concerns the fact that in the case of complex geometries within the substrate with a small volume to surface ratio, such as long thin holes or incidentally generated narrow cracks, crevices etc., the oxygen supply may become a crucial quantity, as has been shown already for chromia formers [16]. Such a situation has been simulated by a 80 mm deep hole with a diameter of only 1 mm. The oxygen supply for the formation of alumina was only sufficient down to approximately 65 mm. In the lower part of the hole a non-protective mixed Al/Cr/Fe-oxide formed and nitrogen was able to diffuse into the material. Following the Gibbs energies of formation [17] at 1200°C, titanium nitride formed first followed by aluminium nitride. Calculations made for the actual Ti and Al activities in the alloy confirmed this ranking of nitrogen affinity.

The reaction zone of nitrogen with Ti was about 1 mm deep (Fig. 13). The concentration and size of the nitrides decreased with increasing distance from the surface of the hole. Thus a reaction mechanism controlled by the diffusion of nitrogen into the material can be assumed. As soon as the available Ti is transformed into nitrides, the formation of aluminium nitride starts, using the TiN particles as nucleation points. Similar observations were made in [18,19]. After 100 h exposure, a reaction front between nitrogen diffusing into the material and aluminium diffusing in the opposite direction is marked by a high density of fairly large aluminium nitride precipitates at a depth of 100 µm.

Table 2. Gibbs energies of formation of relevant metal nitrides at 1200°C

Metal nitride	ΔG_f, kJ/mol N$_2$
TiN	–398.063
AlN	–312.458
Cr$_2$N	–39.963
CrN	–9.872
Fe$_4$N	not stable

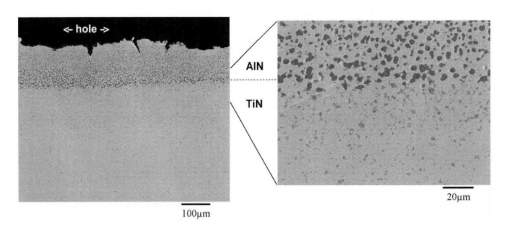

Fig. 13 *Nitride formation in the surroundings of a deep hole in PM 2000.*

From these results it can be concluded that somewhat extreme geometrical arrangements are needed to produce conditions where no protective alumina layer forms and breakdown of the metal by nitridation occurs.

4. Summary

The effect of geometry on the oxidation kinetics and chemical failure of Fe–20Cr–5Al–(RE) alloys has been investigated. Essential factors influencing the lifetime have been identified.

Because protective scale formation depends on the available aluminium reservoir and the transport of aluminium to the metal oxide interface, the volume to surface ratio plays an essential role for the lifetime. In this context the importance of both the global and the local volume to surface ratio has to be stressed. Although the material thickness is often mentioned as the main geometry factor, the overall component design, which controls the stiffness, as well as the design of local parts like corners, edges or holes have to be taken into account. Each of these can dramatically reduce the lifetime, if simple rules are ignored.

Geometries leading to an increase in the component stiffness, cause constraints for the relief of thermal stresses and thus accelerate scale damage, which can become an important factor in lifetime prediction. For Kanthal APM sheet material with a thickness of less than 0.5 mm no problems emerged with respect to scale damage due to growth or cooling stresses, because of easy stress relaxation according to the low substrate strength. In this case, however, the possibility of creep damage in the substrate has to be noted. Under service conditions, sharp corners and edges should be avoided. Already a small rounding off seems to be helpful. Closely adjacent, small holes in sheet material appear to be harmless, because the hole geometry allows sufficient Al supply to the thin-walled bridges between the holes. In contrast to this, deep thin bores into the material can cause problems even in thick components, because at greater depth the lack of oxygen prevents the formation of a protective alumina layer. Thus the propensity to nitridation induced premature failure is increased.

5. Acknowledgement

The authors gratefully acknowledge the financial support of the European Commision via the Brite Euram Project "Life extension of alumina forming alloys in high temperature corrosion environments (LEAFA)" BE-97-4491.

References

1. I. Gurappa, S. Weinbruch, D. Naumenko and W. J. Quadakkers, Factors governing breakaway oxidation of FeCrAl-based alloys, *Mater. Corros.*, 2000, **51**, 224–235.
2. G. Strehl, *et al.*, The effect of aluminium depletion on the oxidation behaviour of FeCrAl foils, in *Proc. 4th Int. Conf. on the Microscopy of Oxidation*, held at Trinity Hall, Cambridge 20–22 September, 1999 (G. Tatlock and S. Newcomb, eds). *Mater. High Temp.*, 2000, **17/1**, 87–92.
3. Brite Euram Project, "Life extension of alumina forming alloys in high temperature corrosion environment (LEAFA)" BE-97-4491, Final Report.
4. P. Kofstad, *High Temperature Oxidation of Metals*. John Wiley & Sons, Inc., New York, London, Sydney, 1966.
5. B. Lesage, L. Maréchal, A.-M. Huntz and R. Molins, Aluminium depletion in FeCrAl alloys during oxidation, *Dep. Diff. Forum*, 2000, **94–99**, 1707–1712.
6. J. R. Nicholls *et al.*, Development of a life prediction model for the chemical failure of FeCrAl(RE) alloys in oxidising environments, This volume, pp.83–106.
7. A. Kolb-Telieps, J. Klöwer, A. Heesemann and F. Faupel, High Temperature corrosion resistant Fe–Cr–Al Foils, in *HTCP 2000* (T. Narida, T. Maruyama and S. Taniguchi, eds), pp.305–308.
8. W. J. Quadakkers, Growth mechanisms of oxide scales on ODS alloys in the temperature range 1000-1100°C, *Mater. Corros.*, **41**, 659–668.
9. D. Renusch *et al.*, Effect of edges and corners on stresses in thermally grown alumina scales, *Oxid. Met.*, 2000, **53**, (1/2), 171–191.
10. A. G. Evans, G. B. Crumley and R. E. Demaray, On the mechanical behaviour of brittle coatings and layers, *Oxid. Met.*, 1983, **20**, 193.
11. H. E. Evans, Stress effects in high temperature oxidation of metals, *Int. Mater. Rev.*, 1995, **40**, 1–40.

12. H. E. Evans, Interfacial crack growth during temperature changes in *Cyclic Oxidation of High Temperature Materials* (M. Schütze and W.J. Quadakkers, eds). Publication No. 27, in the European Federation of Corrosion Series. Published by The Institute of Materials, London, 1999.

13. Kanthal Data Sheet.

14. J. P. Wilber, M. J. Bennett and J. R. Nicholls, Life-time extension of alumina forming FeCrAl–RE alloys: influence of alloy thickness, in *Proc 4th Int. Conf. on the Microscopy of Oxidation*, held at Trinity Hall, Cambridge, 20–22 September, 1999 (G. Tatlock and S. Newcomb, eds). *Mater. High Temp.* 2000, **17/1**, 125–132.

15. P. A. Beaven and V. Guttmann, Structural observations of breakaway oxidation in alumina forming Fe–Cr–Al–RE alloys (to be published).

16. M. J. Bennett, J. A. Desport, C. F. Knights, J. B. Price and L. W. Graham, Crevice corrosion of Nimonic 86 and Hastelloy X in a mixed nitrogen–oxygen environment, *Corros. Sci.*, 1993, **35**, (5–8) 1159–1165.

17. I. Barin, *Thermochemical Data of Pure Substances*. VCH, New York, 1989.

18. N. Wood and F. Starr, Oxidation of Incoloy MA 956 after long-term exposure to nitrogen-containing Atmospheres at 1200°C, in *Microscopy of Oxidation — 2, Proc. 2nd Int. Conf. on the Microscopy of Oxidation*, held at Selwyn College, Cambridge, 29–31 March (S. B. Newcomb and M. J. Bennett, eds). Published by The Institute of Materials, London 1993, p.298–309.

19. M. Turker and T. A. Hughes, Nitridation of Ferritic ODS Alloys, in *Microscopy of Oxidation — 2, Proc. 2nd Int. Conf. on the Microscopy of Oxidation*, held at Selwyn College, Cambridge, 29–31 March, 1993 (S. B. Newcomb and M. J. Bennett, eds). Published by The Institute of Materials, London 1993, p.310–320.

8

The Role of Bioxidant Corrodents on the Lifetime Behaviour of FeCrAl(RE) Alloys

A. KOLB-TELIEPS, U. MILLER*, H. AL-BADAIRY†, G. J. TATLOCK†,
D. NAUMENKO§, W. J. QUADAKKERS§, G. STREHL¶, G. BORCHARDT¶,
R. NEWTON**, J. R. NICHOLLS**, M. MAIER†† and D. BAXTER††

Krupp VDM GmbH, Plettenberger Str. 2, 58791 Werdohl, Germany
*PM ODS-Materials Plansee GmbH, 6600 Reutte, Tyrol, Austria
†University of Liverpool, Liverpool, L69 3GH, UK, §Forschungszentrum Jülich, IWV-2,
52425 Jülich, Germany,
¶TU Clausthal, Institut für Allgemeine Metallurgie, Robert-Koch-Str. 42, 38678
Clausthal-Zellerfeld, Germany
**Cranfield University, SIMS, Cranfield, Bedfordshire, MK43 0AL, UK
††Joint Research Center/IAM 1755 ZG Petten, The Netherlands

ABSTRACT

Studies have been undertaken to investigate the oxidation behaviour of industrial iron–chromium–aluminium alloys with reactive element additions in exhaust contaminant simulating atmospheres at temperatures between 1100 and 1300°C. The model atmospheres have been air + 3.2 vol.% H_2O (moist air), N_2 + 5 vol.% H_2 + 630 vppm H_2O, N_2 + 5000 vppm NO, simulated exhaust gas, simulated combustion gas, air + 0.3 vol.% SO_2 and air + 50 vppm HCl. The temperature was cycled between room temperature and test temperature with periods of 20 h or 100 h. The test results were compared to those gained in parallel tests performed in laboratory air.

The nitrogen-containing environments act as a shielding atmosphere and the retardation of oxide growth appears to be dependent on the partial oxygen pressures. Earlier breakaway is induced by moist air and by additions of 0.3% SO_2 to air and completely different oxide growth behaviour is found in air + 500 ppm HCl. This can be attributed to the formation of volatile specimens, which deteriorate the alumina scale.

1. Introduction

Due to the formation of a protective Al_2O_3 scale, FeCrAl alloys with additions of rare earth metals are widely used at temperatures above 1000°C. Many research studies deal with oxygen atmospheres, which are most important for many applications of these alloys. However, during the production of the material or, in some cases, of components and also during their application other gases are present. Therefore these studies, which are part of the BRITE-EURAM:LEAFA project [1], aim for a greater understanding of the influence of bioxidants.

2. Experimental

Six industrial FeCrAl(RE) alloys, which contained different reactive elements but similar aluminium contents of about 5 mass%, have been tested: PM 2000 (3), Kanthal APM, Kanthal AF and Aluchrom YHf. Their chemical compositions are given in

Table 1. *Chemical compositions (mass%)*

Ref. LEAFA	Aluchrom YHf	Kanthal AF	Kanthal APM	PM 2000 2 mm	PM 2000 1 mm	PM 2000 0.5 mm
Cr, mass%	19.7	20.8	21.0	19.5	18.1	18.9
Al, mass%	5.5	5.2	5.8	5.4	5.3	4.0
Y, ppm	460	340	≤ 1	3700	3700	3700
C, ppm	2100	280	290	100	300 - 600	260
S, ppm	1.3	1.5	0.82	28	30	20
O, ppm	< 10	< 10	450	2800	3100	3000
N, ppm	40	150	160	60	300	150
P, ppm	130	140	170	24	*	*
Zr, ppm	540	580	1100	14	< 10	< 10
V, ppm	860	200	360	77	*	*
Ti, ppm	98	940	260	4500	4800	4500
Cu, ppm	110	330	240	70	*	*
Ca, ppm	12	0.95	1.1	31	15	16
Hf, ppm	310	3.1	1.1	0.05	< 5	10
Mn, ppm	1800	610	800	310	430	380
Si, ppm	2900	1900	4000	240	*	*
Nb, ppm	< 50	< 50	< 50	< 50	*	*
Mg, ppm	78	17	22	30	*	*
Mo, ppm	100	58	41	30	*	*

* Not verified.

Table 1. Kanthal APM and Aluchrom YHf samples of 1 mm thickness were deposited with more than 10 vol.% of Al. From all alloys 20 mm × 10 mm coupons were cut, which were exposed to multicomponent corrodants.

Seven series of experiments have been undertaken, as follows:

1. 500 μm thick samples were discontinuously exposed in N_2 + 5 vol.% H_2 + 630 vppm H_2O, which is often used as a shield gas environment, with 100 h cycles at 1200°C. The water vapour level was achieved by passage of the N_2 + 5 vol.% H_2 gas bottle supply through a refrigerated water bath. The gas flow was continuous at about 100 mL/min.

2. 50–1000 μm thick samples were discontinuously exposed in N_2 + 5000 vppm NO, which simulated the major components in exhaust gas, with 20 h cycles at 1200°C and 1300°C. The only important N compound NO was chosen, since engine temperatures reach values from 900 to 1300°C, where the equilibrium partial pressure of NO is most relevant at levels between 500–3000 ppm, as can be seen in Fig. 1.

3. 50–100 μm thick foils were discontinuously exposed in a simulated fuel-rich exhaust gas (N_2 + 12%CO_2 + 2%CO + 10%H_2O), with 20 h cycles at 1200°C. The oxygen partial pressure at this temperature was calculated to be about 10–15 Pa.

4. To look into the general effect of humidity 70–125 μm thick foils were discontinuously exposed in air + 3.2 vol.% H_2O (moist air) and in dry cylinder air, which contained 2 ppm of water vapour, with 20 h cycles at 1300°C.

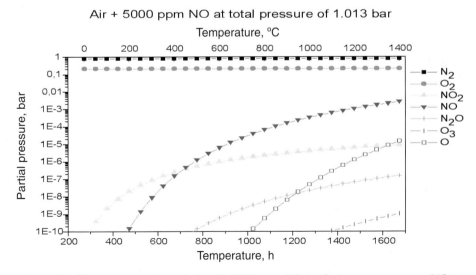

Fig. 1 *Equilibrium composition of air with 5000 ppm NO in the temperature range 0°C to 1400°C, calculated with Chemsage by TU Clausthal.*

5.　50–1000 µm thick specimens were discontinuously exposed at 1200°C in a simulated combustion gas used in power plants, N_2 + 3.4% CO_2 + 15%O_2 + 15%H_2O. The cycle time was 100 h.

6.　50–125 µm thick foils were discontinuously exposed in air + 0.3 vol.% SO_2, a major contaminant in coal fired power station environments, with 20 h cycles at 1100°C.

7.　50–1000 µm thick samples were discontinuously exposed in air + 50 vppm HCl, with 100 h cycles at 1100 and 1200°C. The gas flow was continuous at about 150 mL/min. This environment is a major contaminant in waste incinerators and biogas combustion exhaust gases.

After every cycle the furnace was cooled down to room temperature and the gross and net mass of the specimens were measured and the surface colouration was evaluated. A grey colour was associated with α-alumina, a green colour with chromia and red and black with iron oxides. The microstructure has been characterised by optical and scanning electron microscopy. The test results were compared to those gained in parallel tests performed in laboratory air.

3. Results

3.1. N_2 + 5 vol. % H_2 + 630 vppm H_2O

The initial mass gains of 0.5 mm thick Kanthal AF, Kanthal APM and PM 2000 at 1200°C are compared with previous results obtained on similar samples in laboratory air (Figs 2 and 3). The scale growth of all samples is slower in the shield gas environment.

3.2. N_2 + 5000 vppm NO

As Figs 4 and 5 show, the lifetimes of the foil specimens are in the same range as under oxidation in air and show the same development through to a greenish colouration after aluminium depletion before breakaway. The oxidation in the N_2–NO mixture is always slower but shows a similar development with respect to the colouration and the breakaway process of the samples. No evidence of internal nitridation has emerged.

The samples of Kanthal APM and Aluchrom YHf which have been CVD coated with about 10 vol.% aluminium exhibit a very poor scale adhesion, leading to spallation already after the first cycle.

3.3. Fuel-rich Exhaust Gas (N_2 + 12%CO_2 + 2%CO + 10%H_2O)

Gross mass gain of 70 µm thick Kanthal AF, 50 µm thick Aluchrom YHf and 100 µm thick PM 2000 in exhaust gas at 1200°C is shown in Fig. 6. The mass increase of all samples was slower than in air.

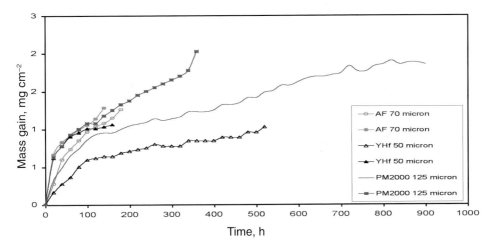

Fig. 2 *Comparison of the performance of Kanthal AF, Aluchrom YHf and PM 2000 in N_2 + 5 vol.% H_2 + 630 vppm H_2O (open symbols) and in laboratory air (filled symbols) at 1200°C, gross mass gain measured at Cranfield University.*

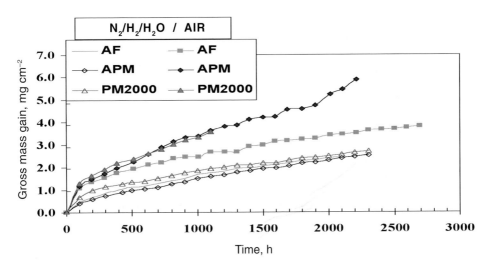

Fig. 3 *Comparison of the performance of 0.5 mm thick Kanthal AF, Kanthal APM and PM 2000 in N_2 + 5%H_2 + 630ppm H_2O and in air at 1200°C, measured by Cranfield University.*

3.4. Combustion Gas (N_2 + 3.4%CO_2 + 15%O_2 + 15%H_2O)

In combustion gas, in which PM 2000, Kanthal AF, Kanthal APM and Aluchrom YHf have been exposed at 1100 and 1200°C, only PM 2000 showed an increased oxidation rate. For the other alloys this atmosphere may be considered benign. The results for 1200°C are shown in Fig. 7.

Foils

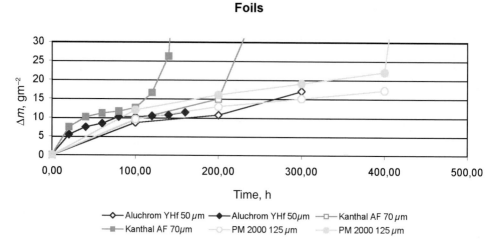

Fig. 4 *Comparison of the mass gain of the foil materials after cyclic oxidation in N_2 with 5000 ppm NO (open symbols) and in laboratory air (filled symbols) at 1200°C, measured at TU Clausthal.*

Sheet Materials

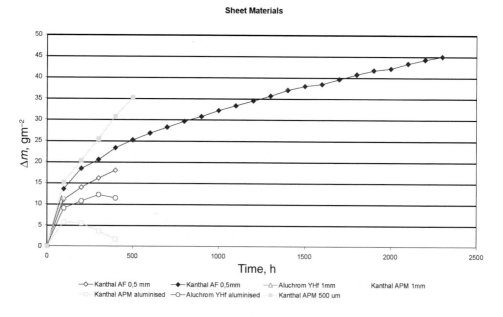

Fig. 5 *Net mass gain of thicker sheet materials after oxidation in the N_2–NO mixture (open symbols) and in air (filled symbols) at 1200°C, measured at TU Clausthal.*

3.5. Air + 3.2 vol.% H_2O (moist air)

The gross mass gain as a function of oxidation time at 1300°C is shown in Fig. 8. The onset of breakaway oxidation occurred earlier when the sample was oxidised in moist air. However, no effect of cylinder air with 2 ppm water vapour was detected. Breakaway oxidation always occurred at corners and edges of the samples where

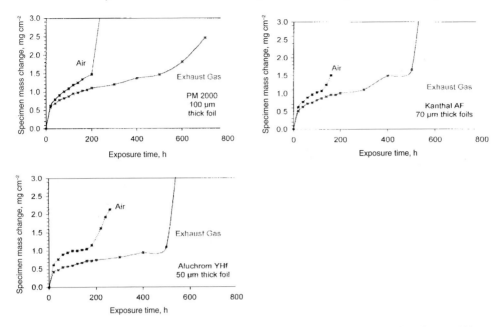

Fig. 6 *Gross mass change of Kanthal AF (70 μm), Aluchrom YHf (50 μm) and PM 2000 (100 μm) in artificial exhaust gas at 1200°C, measured at Forschungszentrum Jülich.*

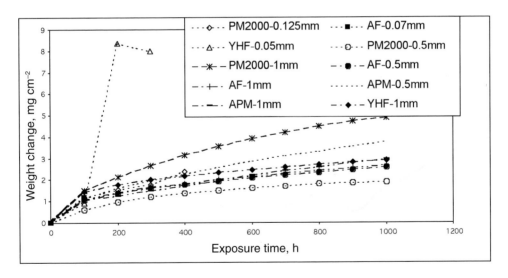

Fig. 7 *Gross mass change of several alloys in combustion gas at 1200°C, measured at JRC Petten.*

iron oxide nodules formed. Chromia formed before total failure of the protective scale. The greenish colour only lasted 1 cycle when Kanthal AF and Aluchrom YHf were oxidised at 1300°C. Details of the corrosion mechanisms at 1200°C will be given in a separate presentation in this Volume [2].

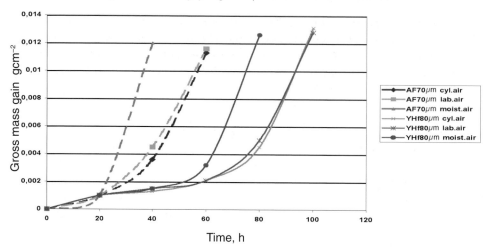

Fig. 8 *Kanthal AF (70 μm) and Aluchrom YHf (80 μm) oxidised at 1300 °C in air, cylinder air and moist air, 20 h cycles, measured at the University of Liverpool.*

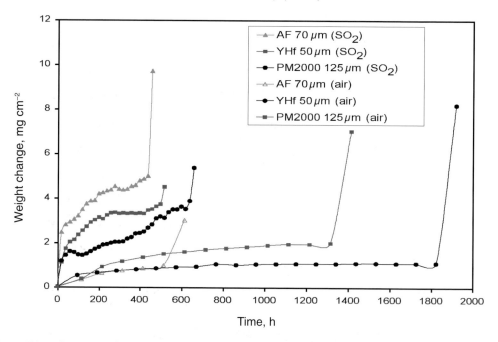

Fig. 9 *Comparison between the gross mass gain on Kanthal AF (70 μm), Aluchrom YHf (80 μm) and PM 2000 (125 μm) foils in air + 3%SO$_2$ with those in laboratory air at 1100 °C, measured at Cranfield University.*

3.6. Air + 0.3 vol.% SO$_2$

Figure 9 indicates little effect on growth kinetics when adding 0.3 vol.% SO$_2$ to the air environment. For the three alloys examined (Kanthal AF, Aluchrom YHf and PM 2000) the onset of breakaway at 1100°C was significantly reduced by the presence of sulfur dioxide in the environment. The lifetime for Kanthal AF was reduced from 550 to 430 h for a 70 mm thick foil, Aluchrom YHf was reduced from 1350 to 530 h and PM 2000 from 1830 to 630 h. Breakaway occurred with the formation of chromia underlying the alumina scale, and later non-protective iron oxide formation (evident by a rapid increase of mass gain).

3.7. Air + 50 vppm HCl

As can be seen from Fig. 10, the net mass gain is lower in the HCl-containing gas at 1200°C. In contrast to the tests performed in laboratory air, HCl induces spallation in all samples (see Figs 11 and 13). At this temperature and also at 1100°C small red and black spots can be recognised on the surfaces after relatively short exposure in the HCl-containing gas. Especially at 1100°C a cycling behaviour of the mass change with time can be observed, as can be seen from Fig. 12. At 1100°C and at 1200°C the Al coated samples showed the fastest gross mass gains and also the most severe spallation (cf. Fig. 12).

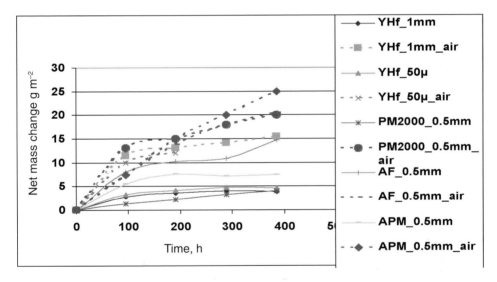

Fig. 10 *Net mass gain of several alloys in air and air + 50 ppm HCl at 1200°C, measured by Krupp VDM.*

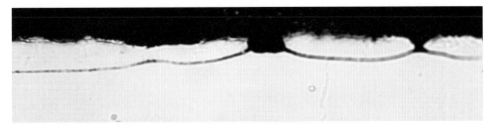

Fig. 11 *Cross-section of 1 mm thick Aluchrom YHf, locally badly adherent scales after 100 h in air + 50 ppm HCl at 1200°C.*

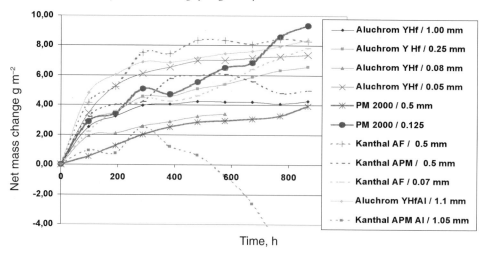

Fig. 12 *Net mass change of different samples in air + 50 ppm HCl at 1100°C, measured by Krupp VDM.*

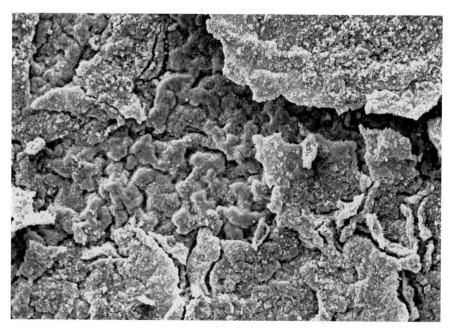

Fig. 13 *SEM micrograph of the surface of 0.5 mm thick Kanthal APM, spallation after 768 h in air + 50 ppm HCl at 1100°C.*

4. Discussion

Three of the environments, the nitrogen-oxygen-bioxidants and the synthetic exhaust, retard the breakaway compared with exposures in air. In the case of air + N_2 + 5%H_2 + 630 ppm H_2O the alumina scale, which could grow due to a sufficiently high oxygen level, protected the alloys against nitridation [3]. So this environment actually can be used as a shielding gas for FeCrAl(RE) alloys. Similar results in exhaust gas have been found by Sigler [4] and have been attributed to different oxide morphologies. The oxidation mechanisms are the same in these three environments and in air, the formation and almost parabolic growth of alumina is followed by that of chromia and iron oxides, which then leads to chemical failure breakaway.

The experiments show that the growth rates of the aluminium oxide in the three low-pO_2 test gases (N_2/NO_x, shield gas and exhaust gas) are substantially slower than in air. Therefore it is assumed that the oxygen partial pressures in these environments led to the retardation of the oxidation rate.

A similar oxidation mechanism is found in moist air and in air + 0.3%SO_2, but these environments induce earlier breakaway. The SO_2-bearing environment affects the oxidation behaviour dramatically by internal sulfidation. The H_2O atmosphere seems to enhance the oxygen transport to the metal surface. The behaviour of Fe20Cr5Al foils in water-containing atmospheres, i.e. similar oxidation rates, but reduced lifetime compared to air, may be explained by the following mechanism. The first serious crack of the alumina scale gives water access to the metallic surface. Presumable volatile hydroxides ($Al(OH)_3$, $Cr(OH)_3$, and $CrOOH$), form and accelerate the breakaway process.

In the above mentioned environments the specimens fail due to chemical breakaway when most of the aluminium is consumed and other oxides form. Therefore for the specimens with 5.5 mass% Al, the foil thickness is critical. Spallation was only noticed on thicker coupons. However, the growth of alumina in the HCl-containing atmosphere is not parabolic. At 1100°C a moderate mass gain in the early stages and occasional negative slopes occur. Even for the foils the alumina scale spalled. The mechanistic understanding for the behaviour of specimens in HCl-containing atmospheres is based on the observation, that an influence is visible before the first cracking or spalling of the alumina scale. Furthermore very fine spall was found early on and also the scale adherence in some specimens seems to be weaker. Red and black spots on top of the alumina indicate that iron is involved in the process, but not chromium, because green spots are missing.

With oxygen HCl can react at 1100 and 1200°C to form H_2O and Cl_2. Similar reactions have already been found at 600°C [5]. This might promote the formation of volatile $FeCl_2$ and $AlCl_3$. These compounds deteriorate the alumina scale, which in turn gives access of the gaseous species O_2, HCl, H_2O and Cl_2 to the metal/oxide interface. The spots on the surface could be iron oxides formed from $FeCl_2$ and O_2. Obviously it is not possible to check if a similar process also occurs for aluminium, because the scale already consists of alumina. These ideas are backed by the observation that the gross mass increases at a lower rate than the net mass and that the processes accelerate as soon as the first cracks appear in the alumina.

For the coated specimens it can be concluded that at least the process used in this work, coating with 10 mass% Al by CVD, does not guarantee sufficient protection. On the contrary these specimens failed first in all environments investigated.

5. Conclusions

1. For Fe–20Cr–5.5Al–(RE) alloys produced as wrought alloys or by powder metallurgy the following atmospheres act as shielding gases: air + N_2 + 5 vol.%

H_2 + 630 vppm H_2O, N_2 + 5000 vppm NO, N_2 + 12%CO_2 + 2%CO + 10%H_2O. N_2 + 3.4%CO_2 + 15%O_2 + 15%H_2O is benign for the weaker alloys but detrimental for PM 2000. An explanation for this behaviour is thought to be the ratio of the oxygen partial pressures in air to that in the bioxidant gas.

2. Air + 3.2 vol.% H_2O and air + 0.3 vol.% SO_2 enhance the oxidation process. Air + 50 vppm HCl seems to lead to the formation of volatile specimens and active oxidation.

6. Acknowledgements

We are grateful to the European Commission for financial support under the LEAFA project no. BRPR-CT97-0562 and to our partners for the supply of the alloys tested, for the chemical analysis of alloys and for their contribution to scientific input in discussing these results.

References

1. BRITE-EURAM project, Contract no. BRPR-CT97-0562.
2. H. Al-Badairy *et al.*, This volume, pp.50–65.
3. M. J. Bennett, R. Newton and J. R. Nicholls, *EUROCORR 2000*, available on CD ROM from The Institute of Materials, London, UK.
4. D. R. Sigler, *Oxid. Met.*, 1991, **40**, (5/6), 555–583.
5. A. Zahs, M. Spiegel and H. J. Grabke, *Mater. Corros.*, 1999, **50**, 561–578.

9

The Role of the Production Route on the Early Stage of Oxide Scale Formation on FeCrAl-Alloys

H. HATTENDORF, A. KOLB-TELIEPS, TH. STRUNSKUS*,
V. ZAPOROJCHENKO* and F. FAUPEL*

Krupp VDM, P.O. Box 1251, 58742 Altena, Germany
*Lehrstuhl für Materialverbunde, Technische Fakultät der Universität Kiel, Kaiserstraße 2,
24143 Kiel, Germany

ABSTRACT

In an oxidation test at 1100°C in air clad and hot dipped FeCrAl alloys showed a superior oxidation resistance compared to that of homogeneous FeCrAl alloys with a similar total Al concentration. The heats with a higher total silicon concentration gave the best oxidation resistance.

XPS depth profiles showed a high Cr and Fe concentration in the oxide scale of the homogeneous material after 3, 30 or 60 min at temperatures between 500 and 900°C and up to 30 min at 1100°C and lower Cr and Fe concentrations for the coated material. This indicates that formation of α-Al_2O_3 takes place earlier in the oxide scale of the homogeneous material that in the coated material, where the formation of the α-Al_2O_3 starts later. The coated material has a more perfect scale, larger grains, and a lower oxidation rate than the homogeneous material.

1. Introduction

Foils of FeCrAl alloys are used as a substrate in metal-supported automotive catalytic converters. In Europe, new exhaust emission standards, such as EURO LEVEL IV, require new designs of automotive catalytic converters and measures to reduce exhaust emissions during the cold-start phase. This can be achieved by using thinner metal support foils, which heat up faster, additionally aided by a larger reaction surface resulting from a higher cell density.

The market share of metal-supported converters is increasing, since they offer more opportunities for further developments [1]. Conventionally they are produced from 50 µm or 30 µm thin foils of Fe–20Cr–5Al (mass%) with additions of reactive elements of which Y and Hf previously proved to be superior to other elements [2]. FeCrAl alloys owe their excellent oxidation resistance at high temperature to a slowly forming layer of α-Al_2O_3. However, in thin foils this process gradually consumes the aluminium content of the alloy so that after extended service periods the protective aluminium oxide layer will fail. To achieve a sufficient or at least similar oxidation resistance for foils thinner than 30 µm to that of 50 µm thick foil, it is necessary to increase the Al content to 7 mass% or more.

Such foils can be produced by coating an FeCrAl alloy containing a reduced aluminium content with pure aluminium or an Al–Si alloy, for example, by hot dipping or cladding at intermediate thickness followed by cold rolling and diffusion annealing. The compound is easier to work than the homogeneous alloy and additionally has a superior oxidation resistance during long term exposure of up to 500 h at 1100°C [3].

The purpose of this work has been to examine the differences in scale formation in the early stages of oxidation between FeCrAl alloys produced by the conventional route and by coating. The first oxides are formed during heating to final temperature. So the chemical composition of the scale was examined for different temperatures and short times.

2. Experimental Details

2.1. Production

Fe–20Cr–5Al (mass%) (Aluchrom Y Hf) was produced by a conventional route, by ingot melting, hot rolling and cold rolling with intermediate heat treatments. This production technique limits the aluminium content to less than approximate 7 mass%, because Fe–20Cr–5Al (mass%) tends to embrittle during hot rolling. Since the hot forming stage is easier for alloys with a lower Al content, the idea for a new production techniques was to start with FeCrAl strip with reduced Al content and additions of Y and Hf produced on the conventional route. At an intermediate thickness of 0.1–1.5 mm the strip was hot dipped with Al-10mass% Si or clad with Al and then further reduced to the final thickness. Several industrial heats were produced by hot dipping or cladding.

The coated samples of Table 1 were produced by melting the FeCrAl heats with a lower Al content and additions of Y, Hf and Zr, hot rolling followed by cold rolling to 1.4 – 0.6 mm, coating with Al or Al-10mass% Si and cold rolling to 50 μm thickness. The homogeneous samples of Table 1 were produced by melting the FeCrAl heats, hot rolling followed by cold rolling to 50 μm thickness with an intermediate annealing. The surface roughness, R_a of all samples was between 0.14 μm and 0.29 μm.

2.2. Long Term Oxidation Tests

The influence of the production route on oxidation resistance was compared for the heats listed in Table 1. After cleaning these foils, samples simulating catalytic converters were produced and oxidised for 500 to 1000 h in a chamber furnace in dry air. After every 100 h the furnace was cooled down to room temperature during 24 h, all samples were then removed from the furnace and their mass change determined. The samples were then reheated to the final temperature in 1 h. For the samples of the homogeneous and coated heats after 500 h (labelled M5 in Table 1) and after 200 h (labelled M2 in Table 1) the thickness of the scale was examined in a metallographic cross section and a fractured section with SEM.

2.3. Chemical Composition of the Scale

Small pieces of 10 × 10 mm² of cold rolled 50 μm thick strips were cut form the homogeneous heat sample H5-3 and from the coated sample C5-1 (Table 1). The

Table 1. *Chemical compositions in mass% of the alloys used for the oxidation tests (Fe balance)*

Heat No.	Al coated	Total Al	Cr	Mn	Total Si	Y	Hf	Zr	N	C	S	
H5-1	-	5.5	20.1	0.18	0.27	0.06	0.05	0.04	0.003	0.030	0.002	
H5-2	-	5.6	20.5	0.11	0.27	0.05	0.04	0.04	0.007	0.025	0.002	
H5-3	-	5.6	20.4	0.20	0.25	0.06	0.05	0.04	0.004	0.022	0.004	M2
H5-4	-	5.6	20.1	0.12	0.25	0.06	0.06	0.04	0.005	0.025	0.002	
C5-1	2.4	5.2	17.2	0.29	0.53	0.03	0.05	0.05	0.005	0.031	0.002	M2
H7-1	-	6.7	20.2	0.16	0.26	0.05	0.04	0.04	0.006	0.02	0.002	
H7-2	-	6.7	19.8	0.18	0.29	0.06	0.04	0.04	0.005	0.023	0.002	
H7-3	-	6.7	20.2	0.23	0.23	0.06	0.04	0.05	0.002	0.02	0.002	M5
C7-1	2.1	7.8	20.2	0.19	0.33	0.05	0.03	0.05	0.005	0.025	0.002	
C7-2	3.8	7.1	17.3	0.15	0.29	0.05	0.05	0.01	0.004	0.023	0.002	
C7-3	5.3	7.3	18.3	0.13	0.20	0.03	0.05	0.01	0.008	0.023	0.002	M5
C7-4	3.6	6.9	17.3	0.15	0.58	0.05	0.05	0.01	0.004	0.023	0.002	
C7-5	4.6	7.4	17.2	0.29	0.76	0.03	0.05	0.05	0.005	0.031	0.002	M5

H; homogenous heat; C: hot dipped or clad; M2: microsection after 200 h at 1100°C; M5: microsection after 500 h at 1100°C.

samples were annealed at temperatures between 500 and 1100°C in air for times between 3 and 60 min. They were put in the preheated furnace and removed after the annealing time from the furnace and air cooled.

The chemical composition of the oxide scale formed on these samples during the early stages of oxidation was examined by means of XPS in conjunction with ion-beam sputtering. The typical sputter rate was approximate 0.2–0.25 nm/min. The sputtered area was 6×6 mm^2.

3. Results

3.1. Long Term Oxidation Tests

Figures 1(a) and 1(b) show the results of the long term oxidation tests at 1100°C. The mass gain of the coated material is less than the mass gain of the homogeneous heats up to 500 h. A further reduction of the mass gain seems to be achieved due to the

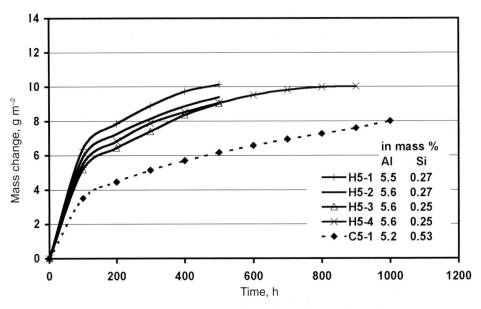

Fig. 1(a) *Long term oxidation tests at 1100 °C of the homogeneous heats and the coated heat of an approximate total concentration of 20 Cr, 5.5 mass% Al, balance Fe all listed in Table 1.*

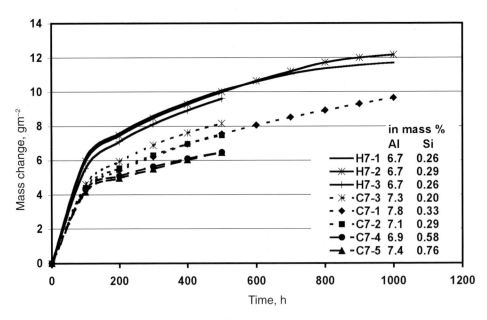

Fig. 1(b) *Long term oxidation tests at 1100 °C of the homogeneous and the coated heats of an approximate total concentration of 20 Cr, 7 mass% Al, balance Fe all listed in Table 1.*

higher silicon concentrations of the samples C5-1, C7-4 and C7-5. There is no noticeable influence of the aluminium concentration of the coated samples on the oxidation rate between 5 and 8 mass% Al up to 500 h.

The scale thickness of several samples after 200 h and after 500 h oxidation was measured on metallographic cross sections (Table 2). The scale thickness of the coated samples is thinner than that on the homogeneous alloy. No spalling was observed.

Figure 2 shows the SEM image of fractured sample sections with 7 mass% Al. The scale morphology of the coated samples is different from that of the homogeneous sample. In the homogeneous sample the outermost layer of the oxide scale consists of equiaxed grains followed by a layer of columnar grains. In the coated sample the outermost layer also consists of equiaxed grains, but the columnar grains in the layer underneath have a clearly larger column diameter.

3.2. Chemical Composition of the Scale

Figure 3 shows the XPS depth profiles of the oxide scale before an annealing treatment for the homogeneous heat H5-3 and the coated heat C5-1 with an approximate total concentration of 20 Cr, 5.5 mass% Al, balance Fe (Table 1). The Fe-, Cr- and Al-concentrations are shown as a fraction of the sum of the Fe-, Cr- and the Al-amount in mass%. For the oxygen concentration the maximum value is scaled to 1.0.

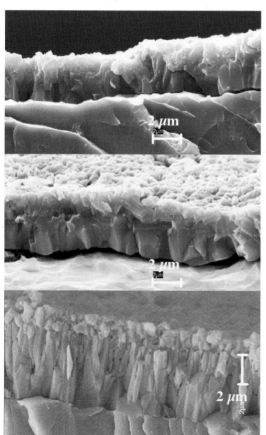

Coated sample C7-5,
7.4 Al, 0.76 Si mass%

Coated sample C7-3,
7.3 Al, 0.20 Si mass%

Homogeneous sample H7-3,
6.7 Al 0.23 Si mass%

Fig. 2 *SEM image of fractured sections of a homogeneous and two coated samples with different silicon contents.*

Table 2. Scale thickness of several samples after oxidation tests

Heat No.	Total Al, mass%	Cr, mass%	Total Si, mass%	Time, h, at 1100 °C in air	Scale thickness, μm
H5-3	5.6	20.4	0.25	200	3 - 4
C5-1	5.2	17.2	0.53	200	2 - 3
H7-3	6.7	20.2	0.23	500	6
C7-3	7.3	18.3	0.20	500	3 - 4
C7-5	7.4	17.2	0.76	500	3 - 4

H; homogeneous; C: hot dipped or clad.

Both samples are covered with a thin oxide scale, which is slightly thicker for the coated sample. Near the surface of the homogeneous sample the Fe concentration decreases slightly, while the Cr and Al concentration increases slightly. Apart from the oxygen at the outermost layer only the Al of the coating appears in the coated sample as expected.

Figures 4–7 show the XPS depth profiles of the oxide scales, formed after annealing at temperatures between 500 and 1100°C in air, for times between 3 and 60 min for the homogeneous heat H5-3 and the coated heat C5-1 with an approximate total concentration of 20 Cr and 5.5 mass% Al, balance Fe (Table 1).

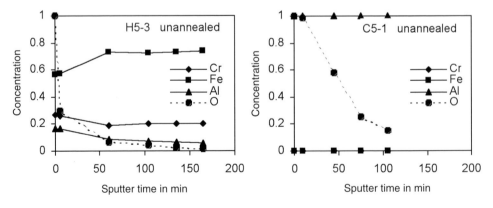

Fig. 3 XPS depth profiles of the oxide scale without annealing for the homogeneous heat H5-3 and the coated heat C5-1 with an approximate total concentration of 20 Cr, 5.5 mass% Al, balance Fe (as detailed in Table 1). The Fe-, Cr- and Al-concentrations are shown as fraction of the sum of the Fe-, Cr- and the Al-amount in mass%. For the oxygen concentration the maximum value is scaled to 1.0.

The results can be summarised as follows:

3.2.1. Homogeneous heat
Low annealing temperatures (Figs 4–6)

- The oxide scale formation on the homogeneous material starts with the formation of an oxide with mostly Fe, medium Cr and a small amount of Al.

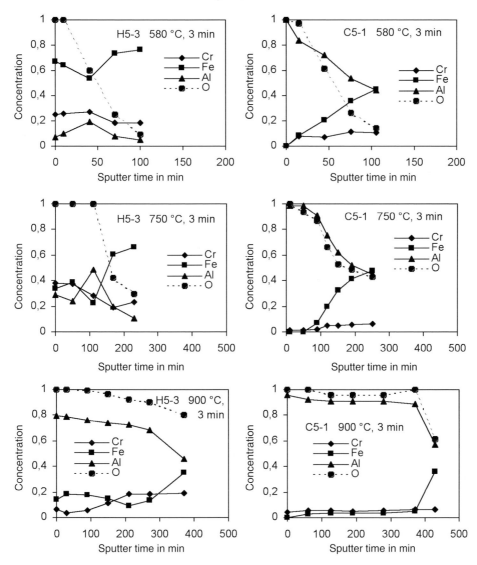

Fig. 4 *XPS depth profiles of the oxide scale after annealing at temperatures between 580°C and 900 °C in air for 3 min for the homogeneous heat H5-3 and the coated heat C5-1 with an approximate total concentration of 20 Cr, 5.5 Al mass%, balance Fe (as detailed in Table 1). The Fe-, Cr- and Al-concentrations are shown as fraction of the sum of the Fe-, Cr- and the Al-amount in mass%. For the oxygen concentration the maximum value is scaled to 1.0.*

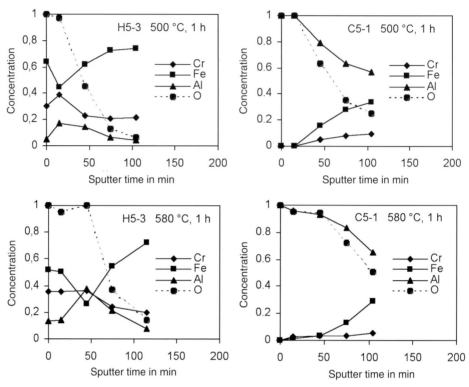

Fig. 5 *XPS depth profiles of the oxide scale after annealing at 500 and 580°C in air for 60 min for the homogeneous heat H5-3 and the coated heat C5-1 with an approximate total concentration of 20 Cr, 5.5 Al mass%, balance Fe (as detailed in Table 1). The Fe-, Cr- and Al-concentrations are shown as fraction of the sum of the Fe-, Cr- and the Al-amount in mass%. For the oxygen concentration the maximum value is scaled to 1.0.*

At 500 and 580°C the Fe concentration in the outermost part of the scale is higher than 40%, the Cr concentration between 20% and 40% and the Al concentration lower than 20%. At 750°C the Fe and Cr concentration in the outermost part of the scale is between 20 and 40% and the Al concentration increases to over 20%.

- At 500, 580 and 750°C the Al concentration shows a maximum in the scale near the scale/metal interface layer, which increases with temperature and time up to 60% after 30 min at 750°C.

- For 500, 580 and 750°C the Al concentration in the scale increases with temperature and time, and the Cr and Fe concentrations decrease.

- After 3 minutes at 900°C the Fe concentration in the outermost part of the scale is lower than the Cr concentration, but higher in the layer near the bulk. The Al concentration is higher than the Cr and the Fe concentration and higher than after annealing at 750°C.

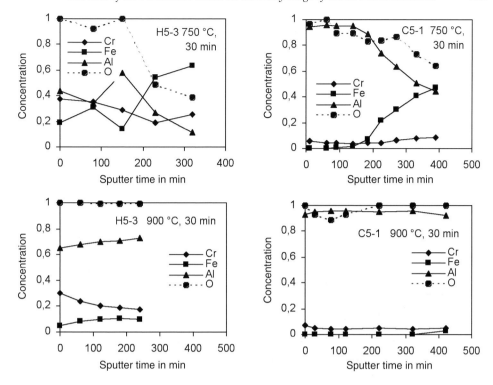

Fig. 6 *XPS depth profile of the oxide scale after annealing at 750 and 900°C in air for 30 min for the homogeneous heat H5-3 and the coated heat C5-1 with an approximate total concentration of 20 Cr, 5.5 mass% Al, balance Fe (as detailed in Table 1). The Fe-, Cr- and Al-concentrations are shown as fraction of the sum of the Fe-, Cr- and the Al-amount in mass%. For the oxygen concentration the maximum value is scaled to 1.0.*

- The scale thickness increases with temperature and annealing time.

High annealing temperatures (Figs 6 and 7)

- After 30 mins at 900°C and both annealing times at 1100°C only the outermost layer of the scale is shown. There the Cr content is higher than the Fe content and decreases continuously with depth (sputter time) starting with 30–40% at the surface. The Al concentration is higher than the Cr and the Fe concentration.

3.2.2. Coated heat
Low annealing temperatures (Figs 4–6)

- In the coated material the oxide scale formation starts with the formation of an aluminium oxide with no detectable concentrations of Cr and Fe in the outermost layer clearly to be seen at 500 and 580°C. In spite of the thickness of the coating of more than 1 μm, an amount of iron and chromium after only 1 h at 500°C or 3 min at 580°C, respectively, was found up to the scale/metal interface layer in a depth of several nm.

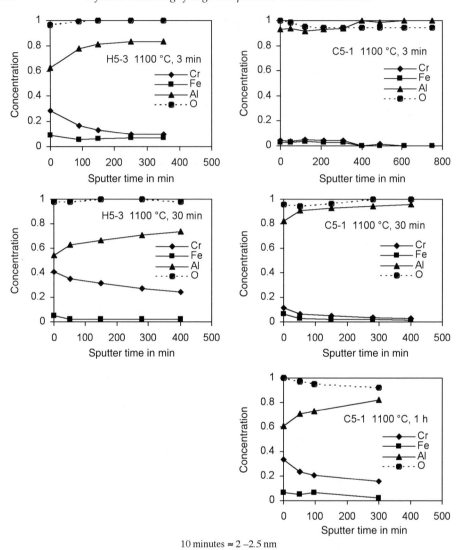

10 minutes ≈ 2 –2.5 nm

Fig. 7 *XPS depth profiles of the oxide scale after 1100°C in air for 3 to 60 min for the homogeneous heat H5-3 and the coated heat C5-1 with an approximate total concentration of 20 Cr, 5.5 mass% Al, balance Fe (as detailed in Table 1). The Fe-, Cr- and Al-concentrations are shown as fraction of the sum of the Fe-, Cr- and the Al-amount in mass%. For the oxygen concentration the maximum value is scaled to 1.0.*

- At longer times and at higher temperatures a small amount of Fe and Cr diffuses into the scale.

- Between 500 and 900°C the Al concentration in the scale is always higher than 90%. The Cr and the Fe concentrations after 3 minutes are always lower than 5%.

- The thickness increases with temperature and time.

High annealing temperatures (Figs 6 and 7)

- After 30 min at 900°C and all annealing times at 1100°C only the outermost layer of the scale is shown. After 30 min at 1100 or 900°C the Cr concentration is higher than the Fe concentration. Both the Cr and the Fe concentrations decrease with the depth (sputter time) starting from about 10 or 5% respectively at the surface, which is markedly lower than in the homogeneous sample. After one hour at 1100°C the starting Cr concentration at the surface increases to 30%, which is the value of the homogeneous material at 1100°C.

3.2.3. Summary of XPS-depth-profile measurements
- There is a high concentration of Cr and Fe in the scale of the homogeneous material at the beginning of the oxidation in contrast to the nearly pure aluminium oxide layer formed on the coated material at temperatures up to 900°C and short times at 1100°C.

- At all temperatures up to 750°C and after 3 min at 900°C, where the thickness of the scale can be seen, the scale of the coated material is of equal thickness or slightly thicker than the scale of the homogeneous material, especially at longer times.

- After 30 min at 900°C and all annealing times at 1100°C only the outermost layer of the scale is shown. Both the Cr and the Fe concentrations decrease with the depth (sputter time) starting from a markedly lower value in the coated alloy. After one hour at 1100°C the starting Cr concentration at the surface in the coated alloy is of a similar value to that of the homogeneous material.

4. Discussion

After more than 100 h at 1100°C the mass gains of the coated samples are lower than those of the homogeneous samples as already reported [2,3], as are the corresponding scale thicknesses. This is different during the first minutes of oxidation during the heating to 1100°C, when the respective scale thicknesses are comparable or slightly larger for the coated sample. Additionally, a further reduction of the mass gain for longer times seems to be achieved with a higher total silicon content, as has been described elsewhere [4].

For short times at the beginning of the oxidation there is a remarkable high concentration of Cr and Fe in the scale of the homogeneous material. Similar high Fe and Cr concentrations were reported [5] after 5 min at 800°C in dry air. The presence of Cr_2O_3 is known to enhance the transformation of metastable aluminium oxides to α-Al_2O_3 [6,7]. So, because of the high Cr and Fe concentration in the oxide scale of the homogeneous material at all temperatures between 500 and 900°C and up to 30 min at 1100°C, the formation of α-Al_2O_3 should take place earlier in the oxide scale of the homogeneous material.

The situation is quite different in the oxide scale of the coated material, where, at the beginning of the oxidation at lower temperatures, the Fe and the Cr concentrations are low or nearly zero. At higher temperatures, and longer times, Cr and Fe diffuse into the oxide scale of the coated material too, so the formation of α-Al_2O_3 starts there later. The concentrations of Cr and Fe are always lower than in the homogeneous material for temperatures up to 900°C and up to 30 min at 1100°C, so the nucleation rate should be lower and larger grains in the scale are to be expected [6], as was in fact found in the scale after 500 h at 1100°C.

It has been reported [8–10], that in FeCrAl alloys containing yttrium the Al_2O_3 scale grows mainly by inward diffusion of oxygen along the grain boundaries of the scale. So the larger grains in the scale of the coated material lead to a decreased oxidation rate and thus a thinner scale for oxidation times longer than 100 h at 1100 °C than in the case of the homogeneous material. This behaviour is observed, although the thickness of the scale on the coated material is equal or slightly thicker at the beginning of scale formation as reported above, because of the higher growth rate of the metastable Al_2O_3 [6,7].

This result is consistent with the TEM examination in [11]. There the phase formation of the oxides for samples coated by vapour deposition, as well as for homogeneous samples, both of 20Cr 5mass% Al, Fe balance and containing rare earth are described. For the homogeneous sample the complete transformation to α-Al_2O_3 is much earlier than for the coated material. Finally the scale has coarser grains in the coated case and finer (columnar) grains in the homogeneous case.

Additionally, there should be an influence of silicon, which is known to result in a further decrease of the oxidation rate, but the main effect in decreasing the oxidation rate is the high aluminium concentration in the scale at the beginning of the oxidation.

5. Conclusions

- In an oxidation test at 1100°C, in air, clad and hot dipped FeCrAl (RE) alloy heats show a superior oxidation resistance up to 500 h in comparison to the homogeneous heats with a similar total Al concentration. The heats with a higher total silicon concentration show the best oxidation resistance.

- Scale formation on the homogeneous heats starts with the formation of an oxide with a high concentration of Fe and Cr. The Al content in the scale then increases with temperature and time, and the Cr and Fe concentrations decrease.

- In the coated material the scale formation starts with the formation of an aluminium oxide at the surface layer with no detectable concentration of Cr and Fe. At longer times and at higher temperatures a small amount of Fe and Cr diffuse into the scale. The Al concentration in the scale is always higher than 90%.

- At the beginning of the oxidation the oxide scale thickness on the coated material is equal or slightly larger than that of the homogeneous material, at oxidation times longer than 100 h at 1100°C the scale of the homogeneous sample is thicker than on the coated alloy.

- Because of the high Cr and Fe concentrations in the oxide scale of the homogeneous material at all temperatures between 500 and 900°C and up to 30 min at 1100°C, the formation of α-Al_2O_3 should take place earlier in the oxide scale of the homogeneous material in contrast to the coated material, where the formation of the α-Al_2O_3 starts later. The result for the coated material is a more perfect scale with larger grains and a lower oxidation rate than in the case of the homogeneous material.

- This explains the superior oxidation resistance of the clad or the hot dipped heats in comparison to the homogeneous heats with a similar total Al concentration. For the heats with a higher total silicon concentration there should be additionally an influence of the silicon, which should result in a further decrease of the oxidation rate.

6. Acknowledgements

The authors would like thank Mr R. Hojda and Mr Peters from Thyssen Krupp Duisburg for the hot dipping, and Mr Theile from Wickeder Westfalenstahl for cladding treatments. Thanks also to Mrs A. Liebelt, Mr H. Niecke and Mr D. Siepmann for carefully carrying out the oxidation tests and Mrs A. Kalinowski and Mr H. Kossowski for metallographic and SEM investigations.

References

1. W. Maus, Metal catalytic converter substrates and their contribution to the ecological demands of the future, in *Metal-Supported Automotive Catalytic Converters* (H. Bode, ed.). Frankfurt, Germany, pp.1–13, 1997.
2. J. Klöwer, A. Kolb-Telieps, U. Heubner and M. Brede, *Corrosion '98*, Paper No. 746, NACE International, Houston, Tx, 1998.
3. A. Kolb-Telieps, J. Klöwer and R. Hojda, in *Metal-Supported Automotive Catalytic Converters*, pp. 99 – 104, (ed. H. Bode), Frankfurt, Germany, 1997.
4. J. Klöwer, *Mater. Corros.*, 2000, **51**, 373 – 385.
5. M. Göbel, *et al.*, in *Metal-Supported Automotive Catalytic Converters* (H. Bode, ed.), Frankfurt, Germany, pp. 99–104, 1997.
6. M. W. Brumm and H. J. Grabke, *Corros. Sci.*, 1992, **33**, 1677–1690.
7. W. C. Hagel, *Corrosion*, 1965, **21**, 316 – 326.
8. F. A. Golightly, F. H. Scott and G. C. Wood, *Oxid. Metals*, 1976, **10**, 163; 1980, **14**, 218.
9. W. J. Quadakkers, H. Holzbrecher, K. G. Briefs and H. Beske, *Oxid. Metals*, 1989, **32**, 67.
10. B. A. Pint and L. W. Hobs, *Oxid. Metals*, 1994, **41**, 203.
11. A. Andoh, S. Taniguchi and T. Shibata, Phase transformation and structural changes of alumina scales formed on Al-deposited Fe–Cr–Al foils, in *Proc. Int. Symp. on High-Temperature Corrosion and Protection 2000* (T. Narita, T. Maruyama and S. Taniguchi, eds), Hokkaido, Japan, 2000, pp. 297–303. Published Science Reviews, UK, 2000.

10

Modelling Internal Corrosion Processes as a Consequence of Oxide Scale Failure

U. KRUPP, S.Y. CHANG, A. SCHIMKE and H.-J. CHRIST

Institut für Werkstofftechnik, Universität Siegen, Germany

ABSTRACT

A numerical computer model has been developed which is able to describe the transition from protective oxide scale formation to internal oxidation limiting the service life by a severe degradation of the mechanical properties of high-temperature materials, e.g. by embrittlement or dissolution of strengthening phases in the sub-surface region. The model contains conditions leading to cracking and spalling of a thermally grown oxide scale and a finite difference solution of the diffusion equations which yields, in combination with a thermodynamic software module ChemApp, the time- and location-dependent concentrations of the species participating in the corrosion process. For the examples of the Ni-base alloys CMSX-4, CMSX-6, SRR99 and Nicrofer 7520 Ti (alloy 80 A), the susceptibility to internal oxidation and nitridation even under isothermal conditions was shown experimentally. Tensile tests carried out on specimens which were internally nitrided in a non-oxidative nitrogen-based atmosphere revealed the detrimental effect of internal corrosion on the mechanical properties.

1. Introduction

Many high-temperature applications, e.g. as in power plant or aeroengine components, require beside the resistance against corrosion excellent mechanical properties even at temperatures above 800°C [1]. This can be achieved by a fine-balanced chemical composition leading to the formation of slow-growing protective oxide scales (Cr_2O_3, Al_2O_3) and the precipitation of strengthening phases — in Ni-base superalloys the ordered γ' phase (Ni_3Al,Ti). In the case of gas turbine blades and vanes a combination of a ceramic thermal barrier coating (TBC) and a MCrAlY coating increases the protection against high-temperature corrosion but only if the component remains completely covered during the whole service life. For high-temperature applications corrosion is often seen to be life-determining and more or less independent of the superimposed mechanical loading. Service life is terminated when the ability to form a protective oxide scale stops, resulting in catastrophic base metal oxidation that quickly reduces the load-carrying cross-section of the component. As shown by schematic mass change vs time diagrams in Fig. 1, thermal cycling conditions do accelerate this process due to the brittleness and the lower thermal expansion of the oxides scale as compared to the base metal with the consequence of repeated spalling [2]. Because of the rather simple experimental procedure and evaluation of cyclic oxidation tests these are often used to estimate service life time [3].

However, the overall corrosion process becomes more complex if the oxide scale or the base metal undergoes microcracking due to mechanical and thermal stresses. The cyclic oxidation kinetics are then strongly influenced by a changed oxidation rate and mechanism as well as the influence of internal corrosion [4].

In several studies, higher oxidation rates, deeper depletion zones of oxide-forming elements and massive internal oxidation were found if high-temperature exposure was combined with creep or fatigue loading [5–7]. Local transition to internal oxidation can be attributed to an enhanced consumption of the oxide-forming elements due to alternating cracking/spalling and healing of the oxide scale. Then oxygen, nitrogen and other corrosive elements can penetrate into the alloy leading to a deep degradation of the mechanical properties of the material by the formation of brittle precipitates (e.g. internal Al_2O_3, TiN, and AlN) accompanied by the consumption of strengthening alloying elements like Al and Ti, which are responsible for the presence of the γ' phase in Ni-base superalloys [8,9]. It is obvious that areas weakened in such a way are preferential sites for crack initiation and enhanced propagation. Such a damage mechanism is schematically shown in Fig. 2.

The objective of the present study was the development of a model, to describe and predict the microstructural changes in the sub-surface area quantified as a consequence of mechanically or thermally induced cracking and spalling of the oxide scale. Supplemented by thermogravimetric, microstructural and mechanical studies, the applicability of such a model to life-prediction is demonstrated. In addition to this, the need to introduce parameters describing the physical interactions between the diffusion, chemical reactions and mechanical stresses determining the overall corrosion process is evaluated.

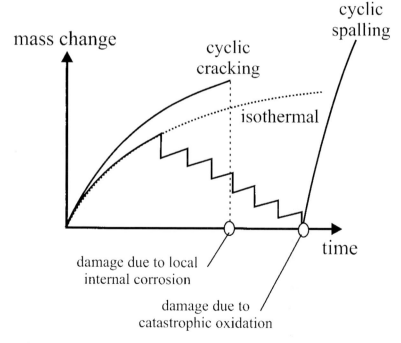

Fig. 1 *Schematic representation of the oxidation kinetics during isothermal and cyclic exposure.*

2. Materials and Experimental Methods

As test materials for the present study, three commercial single-crystalline Ni-base superalloys, CMSX-4, CMSX-6 and SRR99, and a polycrystalline Ni-base alloy Nicrofer 7520 Ti (alloy 80 A) were used. Their chemical compositions are given in Table 1. Additionally several Ni-base model alloys of the system Ni-Cr-Al-Ti were studied.

Isothermal and cyclic thermogravimetric studies were carried out on small polished (1 µm diamond suspension) discs with a diameter of approx. 12 mm and a thickness of approx. 1 mm using a self-designed vacuum-tight thermobalance equipped with a Sartorius microbalance of a sensitivity of 1 µg. An electronically-controlled (by photo diodes) specimen lift allows thermal-cycling experiments. To

Table 1. Chemical composition of the Ni-base alloys studied (in mass%)

Alloy	Ni	Cr	Al	Ti	Co	Ta	Mo	W	Fe	Re	Hf
CMSX-4	Bal.	6.0	5.6	1.0	10.0	6.0	0.6	6.0	–	3.0	0.1
CMSX-6	Bal.	9.6	4.7	4.6	4.8	1.8	2.8	0.1	–	–	–
SRR99	Bal.	8.5	5.6	2.2	5.2	2.8	0.6	9.5	–	–	–
Nicrofer 7520 Ti (alloy 80A)	Bal.	20.2	1.6	2.7	0.01	–	–	–	0.2	–	–

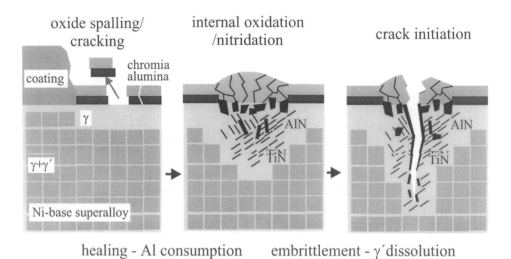

Fig. 2 Schematic representation of failure of Ni-base superalloys due to oxide spalling/cracking followed by internal oxidation and nitridation.

quantify the damage potential of internal corrosion (here the emphasis is put on internal nitridation) electropolished cylindrical tensile specimens were exposed to a nitrogen-based atmosphere consisting of N_2 50, He 45, H_2 5vol.% which had to pass a small package of Ti sponge just in front of the specimen in order to reduce the oxygen concentration and so avoid oxidation of the specimens. After the nitridation treatment tensile and fatigue tests at room temperature were carried out using a Schenck servohydraulic testing machine. The thermogravimetric and mechanical studies were supplemented by microstructural examinations, mainly using analytical scanning electron microscopy.

3. Experimental Results

The following results as examples represent the continuation of an extensive study on the selective oxidation and internal nitridation of Ni-base superalloys the results of which were published in detail in earlier papers, e.g. in [10,11]. Concerning cyclic oxidation, the superalloy CMSX-6 in particular exhibits a pronounced spalling behaviour which is probably due to its high Ti content leading to an alumina scale with limited adherence. Figure 3 shows the isothermal and thermal-cycling oxidation kinetics of CMSX-6 during exposure at 1000°C to air. In contrast to this, cyclic exposure of the chromia-forming Ni-base alloy Nicrofer 7520 Ti does not result in spalling (Fig. 3). Compared to the isothermal kinetics the cyclic mass gain is slightly increased in a similar way to that observed for the first approximately 10 h during cyclic exposure of CMSX-6.

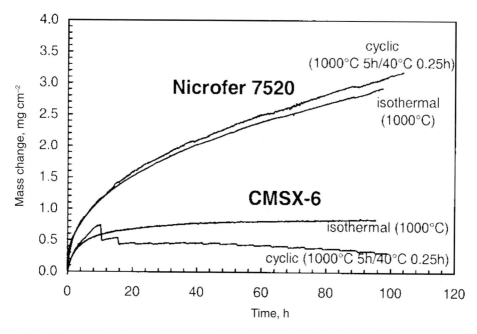

Fig. 3 *Thermogravimetrically-measured mass change vs time during isothermal and thermal-cycling exposure to air.*

The increased oxidation kinetics during thermal cycling can be attributed to microcracking of the chromia scale. The surface of the isothermally exposed specimen of Nicrofer 7520 Ti, as depicted in Fig. 4(a), reveals a crack-free and more homogeneous morphology as compared to the cyclically exposed specimen (Fig. 4b). As can be seen by corresponding cross-sections in Fig. 4(c) and 4(d) the cyclically exposed specimen suffered massive internal oxidation by formation of Al_2O_3 and occasionally TiN below a cracked but adherent chromia scale which is slightly thicker than the isothermally-grown chromia scale.

The mass change during exposure of the alumina formers CMSX-4, CMSX-6, and SRR99 is of course much lower. Even the oxidation rate of CMSX-6, which is more than twice that of the superalloys CMSX-4 and SRR99 [12], reaches only approximately

(a)   5µm

(b)   5µm

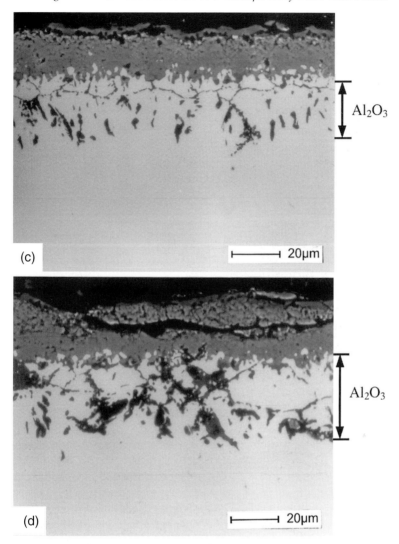

Fig. 4 (opposite and above) *Surface and cross-section of Nicrofer 7520 Ti after 200 h exposure to air under isothermal (a,c) and thermal cycling conditions (b,d).*

a third of the rate of the chromia-forming alloy Nicrofer 7520 Ti. It is worth mentioning that an intact chromia scale does not act as an effective barrier against nitrogen penetration. Preferred locations for non-protective oxide formation are edges, where in addition to stresses caused by temperature cycles geometry-induced growth stresses can result. Figure 5 gives two examples of such a situation for specimen edges of the alloys CMSX-6 and Nicrofer 7520 Ti suffering massive internal corrosion.

Although the superalloys SRR99 and CMSX-4 form well-adherent alumina scales which are not prone to cracking and spalling, local internal nitridation was always observed. Figure 6 shows internal AlN and TiN precipitates below an Al_2O_3 scale in the transition stage to internal oxidation within the Ni-base superalloy CMSX-4 after 600 h exposure at 1000°C to air.

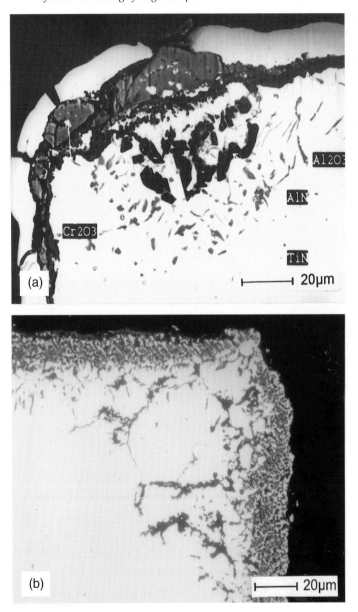

Fig. 5 *Edges revealing massive internal corrosion of the Ni-base alloys CMSX-6 after 300 h (AlN (black and blocky) and TiN (grey) and Nicrofer 7520 Ti after 100 h (Al$_2$O$_3$) exposure at 1000 °C to air.*

The local inhomogenity of the internal oxidation and nitridation is probably a consequence of dendritic segregation typically resulting from the solidification process of single-crystalline superalloys [13]. The segregation effect becomes obvious by analysing high-temperature exposed surfaces, as shown of the example of the superalloy SRR99 in Fig. 7(a). As proved by energy-dispersive X-ray spectroscopy

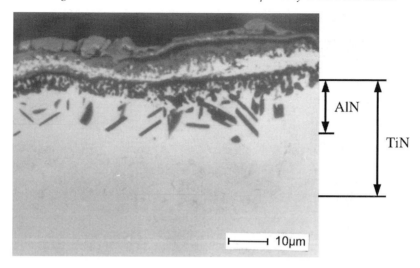

Fig. 6 *Internal precipitation of TiN and AlN below an inhomogeneous alumina scale within the superalloy CMSX-4 after 600 h exposure at 1000°C to air.*

(EDS) the coarse-grained dendritic areas are enriched in Cr and Ni while the fine-grained interdendritic areas are enriched in Al and Ti. Figures 7(b) and 7(c) represent the corrosion behaviour of SRR99 exposed for 100 h at 1000°C to air at two different locations. The cross section in Fig. 7(b) shows a homogeneous Al_2O_3 scale — probably in an interdendritic area — separated by an even interface from the base metal and followed by an approximately 4 μm thick zone in which the γ' phase is dissolved by the Al consumption due to alumina scale formation. In contrast, the cross-section depicted in Fig. 7(c) reveals an inhomogeneous alumina scale — probably formed in a dendritic area — with a wavy interface, which does not protect against internal precipitation of Ti nitrides. The consequence is an increase in the γ' dissolution zone, to a thickness of approx. 10 μm. According to the concentration profiles measured by EDS in Fig. 7(d) (corresponding to the cross-section in Fig. 7c), the dissolution can be attributed to the sub-surface depletion of both Al as well as Ti. Although the Ti concentration in solid solution is difficult to determine by EDS, when within a zone of finely dispersed Ti nitrides, the often observed coincidence of the extent of the γ' dissolution zone and the internal nitridation zone [14] allows the assumption to be made that internal nitridation deteriorates the mechanical properties of Ni-base superalloys.

Experiments to characterise interactions between internal corrosion products and mechanical damage processes were carried out by means of mechanical testing on nitrided tensile specimens of Nicrofer 7520 Ti. Comparing the stress vs strain plots of tensile tests carried out on an as-received specimen and an internally nitrided specimen, as can be seen in Fig. 8, one can observe a slightly more brittle behaviour of the nitrided specimen.

Cross sections of the gauge length of the specimens show that various microcracks initiated within the internal nitridation zone containing AlN (depth approx. 180 μm) and TiN (depth approx. 380 μm). The crack path seems to follow the grain boundaries which are nearly completely covered by TiN precipitates.

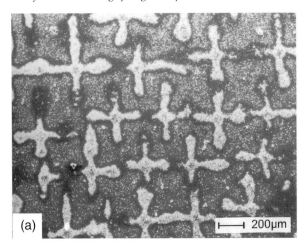

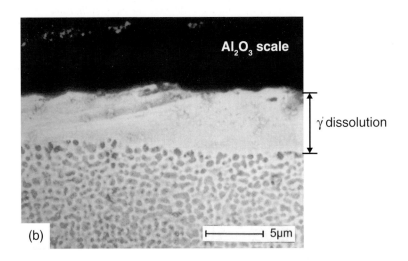

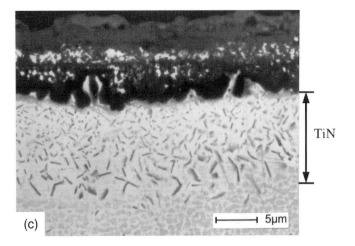

(d)

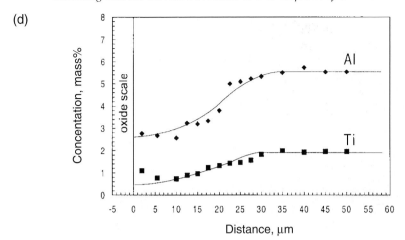

Fig. 7 (Opposite and above) Influence of dendritic segregation in the superalloy SRR99: oxidised surface (a), opposite and cross sections of two different locations of the same specimen (b,c) with corresponding concentration profiles of Al and Ti measured by EDS after 100 h exposure at 1000 °C to air (d).

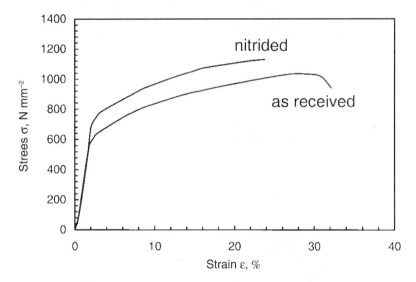

Fig. 8 Stress/strain diagrams of tensile tests on specimens of the alloy Nicrofer 7520 Ti.

4. Computer Modelling and Discussion

The computer model which has been developed consists of two parts which are mathematically separately treated but mutually interact. The first part calculates oxide growth and failure determining the time/number of thermal cycles until cracking of the oxide scale take place. According to a model proposed by Evans *et al.* [15,16] the thermal stress σ_{Ox} arising during cooling of the specimen by ΔT can be expressed by the following equation:

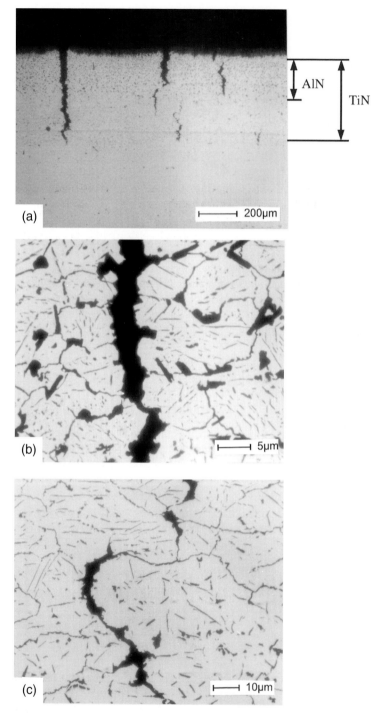

Fig. 9 *Multiple cracking in the internal nitridation zone (AlN and TiN) in the gauge length of a tensile specimen of the alloy Nicrofer 7520 Ti: Macroview (a) and details in the AlN + TiN zone (b) and in the deeper TiN zone (c).*

$$\sigma_{Ox} = \frac{E_{Ox}\Delta T\Delta\alpha}{(1-v_{Ox})} \tag{1}$$

where E_{Ox} denotes the Young's modulus of the oxide, v_{Ox} the Poisson ratio of the oxide, and $\Delta\alpha$ the difference of the coefficient of thermal expansion between oxide and alloy. The critical stress σ_c given by

$$\sigma_c = \left(\frac{E_{Ox}\gamma_0}{d(1-v_{Ox})}\right)^{\frac{1}{2}} \tag{2}$$

yields the condition for oxide spallation/cracking. That means that the energy released by cracking becomes higher than the energy γ_0 required to form the crack surfaces in the oxide scale. The thickness d of the layer is calculated by means of the parabolic rate constant k_p. Once a crack has been formed, oxygen and nitrogen (if exposure to air or nitrogen is assumed) can penetrate into the alloy. A transition to internal oxidation and nitridation will occur, when the diffusion of aluminium to the surface is not sufficient to form a protective scale. With certain restrictions, the depth ξ of internal corrosion resulting in the formation of B oxide in an alloy AB can be calculated analytically by Wagner's treatment of internal oxidation [17] using the following parabolic equation:

$$\xi = \sqrt{\frac{2c_O^s D_O t}{c_B^0}} \tag{3}$$

with the oxygen concentration c_O^s at the surface, the diffusivity of oxygen D_O in the alloy and the initial concentration c_B^0 of the alloying element B. Because complex corrosion phenomena do not fulfil the boundary conditions of eqn (3), in the present study a numerical model was applied. In many cases simultaneous internal precipitation of more than one species and internal transformations can be observed. For this reason a two-dimensional finite-difference simulation [18] was used to calculate the diffusion processes of the species participating in the internal corrosion process. According to Fig. 10 and eqn. (4) the concentration $C_{i,j}^{n+1}$ at the location $x = i\Delta x$, and $y = j\Delta y$, and at the time $t = (n+1)\Delta t$ can be calculated from four concentrations at the previous time step $t = n\Delta t$.

$$C_{i,j}^{n+1} = C_{i,j}^n + \frac{\Delta T}{\Delta X^2}\left(C_{i+1,j}^n - 2C_{i,j}^n + C_{i-1,j}^n\right) + \frac{\Delta T}{\Delta Y^2}\left(C_{i,j+1}^n - 2C_{i,j}^n + C_{i,j-1}^n\right) \tag{4}$$

Note that all the parameters used are dimensionless by relating them to the exposure time t, specimen thickness d, surface concentration c^s, and diffusion coefficient D:

$$X = x/d$$
$$T = Dt/d^2$$
$$C = c/c^s \tag{5}$$

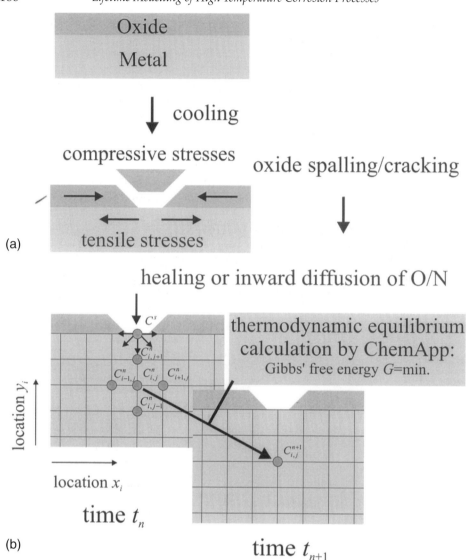

Fig. 10 *Concept of the computer-model to calculate internal corrosion (b) as a consequence of oxide scale failure (a).*

A main advantage of the numerical treatment is the possibility of incorporating thermodynamic subroutines for each diffusion step. This was done by applying the thermodynamic software library ChemApp [19]. Using data for the system Ni–Cr–Al–Ti–N the one-dimensional finite-difference simulation has already been successfully applied to the internal nitridation of Ni-base model alloys [20]. In the present study the two-dimensional simulation was used to calculate the internal nitridation as a consequence of a cracked oxide scale.

For an experimental verification of the calculated result, a crack in the oxide scale of a pre-oxidised specimen was generated using a diamond wire saw. Then the specimen was exposed for 100 h at 1000°C to the oxygen-free nitrogen atmosphere. Figure 11(a) shows the extent of the internal nitridation, which is in a good agreement with the simulated concentration profile of TiN, even though the chromia scale does not block nitrogen penetration into the alloy as assumed in the calculation. The permeability of chromia to nitrogen should be an important point to be considered for the application of chromia-forming alloys at high temperatures.

Specimens which were exposed under service conditions usually exhibit many small cracks in the oxide scale with only a small number of them resulting in massive internal oxidation and nitridation. In order to model service life of such high-temperature components reasonable criteria have to be found which define the situation where internal corrosion has grown to the extent that failure results. If one assumes that internal corrosion starts once a certain amount of oxide has spalled,

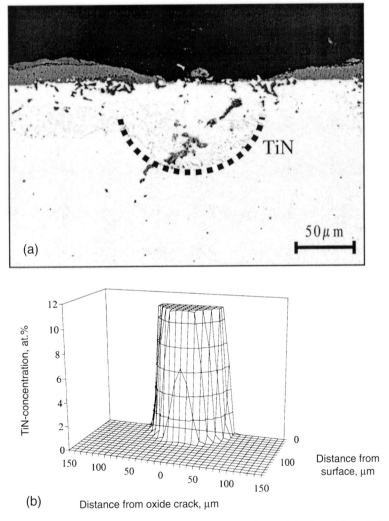

Fig. 11 *Internal nitridation of NiCr20Ti2 as a consequence of a crack in the oxide scale (a) and calculated concentration profile of the TiN precipitates (b).*

such a life-prediction can be derived by a combination of COSP (cyclic oxidation simulation program) [21], which describes the spalling process, and the kinetics of internal corrosion of the non-protected base alloy. The time to failure is then the sum of the time until spalling gives a local rise to a transition to non-protective internal oxidation and the time needed to grow the (brittle) internal oxidation/nitridation zone so that the unaffected cross section of the component is reduced to a thickness which is no longer sufficient to carry mechanical loading. Figure 12 shows for the example of the model alloy NiCr20Ti2 the thermogravimetrically-measured oxidation

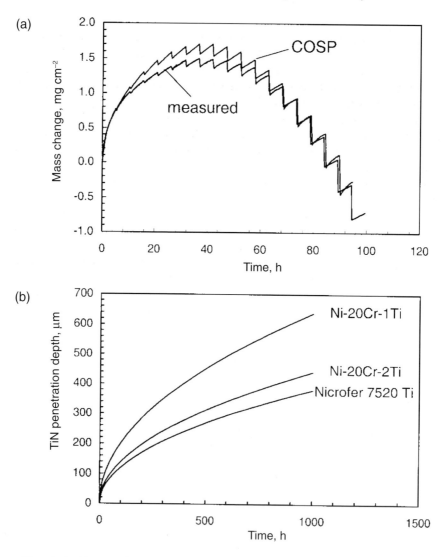

Fig. 12 *Application of the COSP model to simulate the measured thermal-cycling oxidation kinetics of the model alloy NiCr20Ti2 at 1000 °C to air (a) and internal nitridation kinetics of the same alloy compared to NiCr20Ti2 and Nicrofer 7520 Ti during exposure at 1000 °C to oxygen-free N₂ atmosphere (b).*

kinetics during thermal cycling (Fig. 12a, 1000°C, air) and the internal nitridation kinetics in the absence of a protective scale (Fig. 12b, 1000°C, N_2 atmosphere). The increased internal nitridation kinetics calculated for NiCr20Ti1 is shown in comparison with the alloys NiCr20Ti2 and Nicrofer 7520 Ti and demonstrates the detrimental effect of the reduction of the initial concentration of the nitride-forming element Ti (Fig. 12b, compare eqn., (3)).

5. Conclusions

Even during short-term exposure to air, the Ni-base superalloys CMSX-4, CMSX-6, SRR99 and Nicrofer 7520 Ti suffer locally massive internal corrosion by Al_2O_3, TiN, and AlN precipitation. This can be attributed to an inhomogeneous, cracked or spalled alumina scale, which cannot heal if the Al concentration required for superficial scale formation has fallen below a critical value. Then, nitrogen and oxygen are no longer hindered from penetration into the alloy and this leads to a degradation of the material's properties by near-surface embrittlement and dissolution of the strengthening γ' phase. This process may, under certain conditions, result in a service life of high-temperature components that is shorter than predicted by cyclic oxidation testing, where usually the transition to overall catastrophic oxidation of the base metal is used as the failure criterion.

In the present study the effects are discussed which may lead to local spalling or cracking. These processes are incorporated into a two-dimensional finite-difference simulation of internal corrosion processes. The resulting model allows the calculation of the concentration profiles of all the reacting species assuming local thermodynamic equilibrium and applying commercial thermodynamic software ChemApp. A promising concept to estimate the reduction of service life is seen in a combination of the COSP model and this internal corrosion simulation model.

References

1. D. P. Whittle, *Proc. High Temperature Alloys for Gas Turbines*. Liège, Belgium, 1978, 109–123.
2. K. N. Lee, V. K. Arya, G. R. Halford and C. A. Barrett, *Met. Mater. Trans. A*, 1996, **27A**, 3279–3291.
3. C.-O. Moon and S.-B. Lee, *Oxid. Met.*, 1993, **39**, 1–13.
4. C.T. Sims, N. S. Stoloff and W. C. Hagel, *Superalloys II*. John Wiley & Sons, 1987.
5. B. Pieraggi, *Mater. Sci. Eng.*, 1987, **88**, 199–204.
6. S. Esmeaeili, C. C. Engler-Pinto Jr., B. Ilschner and F. Rézai-Aria, *Scr. Metall. Mater.*, 1995, **32**, 1777–1781.
7. S. Osgerby and B.F. Dyson, *Mater. Sci. Technol.*, 1990, **6**, 2–8.
8. U. Krupp, S. Y. Chang and H.-J. Christ, *Proc. 5th Int. Symp. on High Temperature Corrosion and Protection of Materials*, May 2000, Les Embiez, France (*Mater. Sci. Forum*, in print).
9. M. Göbel, A. Rahmel and M. Schütze, *Oxid. Met.*, 1994, **41**, 271–300.
10. U. Krupp and H.-J. Christ, *Oxid. Met.*, 1999, **52**, 277–298.
11. U. Krupp and H.-J. Christ, *Metal. Mater. Trans. A*, 2000, **31A**, 47–56
12. S. Y. Chang, U. Krupp and H.-J. Christ, in *Cyclic Oxidation of High Temperature Materials, Mechanisms, Testing methods, Characterisation and Lifetime Estimation* (M. Schütze and W. J.

Quadakkers, eds). Publication No. 27 in European Federation of Corrosion Series. Published by The Institiute of Materials, London, 1999, p.63–81.

13. D. Goldschmidt, *Mater. wiss. Werkst. tech.*, 1994, **25**, 373–382.

14. S. Osgerby, NPL Report CMMT (A) 221, Published National Physical Laboratory, TW11 0LW, UK, 1999.

15. H. E. Evans, *Int. Mater. Rev.*, 1995, **40**, 1–40.

16. H. E. Evans, *Mater. Sci. Eng.*, 1989, **A120**, 139–146.

17. C. Wagner, *Z. Elektrochem.*, 1959, **63**, 7, 772–790.

18. J. Crank, *The Mathematics of Diffusion*. Oxford University Press, London, 1975.

19. G. Erickson and K. Hack, *Metal. Trans., B*, 1990, **21B**, 1013–1023.

20. U. Krupp and H.-J. Christ, *Oxid. Met.*, 1999, **52**, 3/4, 299–320.

21. C. E. Lowell, C. A. Barrett, R. W. Palmer, J. V. Auping and H. B. Probst, *Oxid. Met.*, 1991, **36**, 81–112.

Investigation and Modelling of Specific Degradation Processes

1.2 Iron oxide and chromia formers

11

Effects of Minor Alloying Additions on the Oxidation Behaviour of Chromia-Forming Alloys

B. GLEESON and M. A. HARPER[*]

Department of Materials Science & Engineering, Iowa State University, Ames, IA 50011, USA
[*]Special Metals Corporation, 3200 Riverside Drive, Huntington, WV 25705, USA

ABSTRACT

Minor alloying additions can significantly influence oxidation behaviour, particularly under thermal cycling conditions. This paper provides a brief overview of the oxidative effects of elements that are common minor additions to commercial, chromia-forming alloys. The principal elements of concern are Si, Ti, Al, Mn, and reactive elements such as Ce, La, and Zr. It is shown that variation of minor element contents within the specified range of a given alloy can result in markedly different cyclic oxidation behaviour. In most cases the oxidative effects of minor elements must be analysed collectively rather than independently. An important consequence of this is that accurate modelling of oxidation performance requires an understanding of this apparent interdependence between the minor alloying elements. It will be shown that although our current level of understanding is incomplete, important trends and interdependencies can be identified.

1. Introduction

Chromia-forming alloys represent a group of heat-resistant materials that contain sufficient chromium to form a continuous Cr_2O_3 scale layer when exposed to an oxygen-bearing atmosphere at elevated temperatures. For commercial Fe-, Ni- and Co-base alloys designed for service above 650°C, the chromium content is typically between 16 and 30 mass%. A maximum of 30 mass% is set to avoid α-Cr precipitation, which decreases both the workability and creep strength of the alloy.

Chromia-forming Fe–Ni–Cr, Ni–Fe–Cr, and Ni–Cr alloys are used extensively in high-temperature industrial applications owing to the stability and protective nature of a Cr_2O_3 scale. However, the oxidation performance of these alloys is highly variable and, accordingly, cannot be simply generalised. This variability in performance is apparent in Fig. 1, which shows the cyclic oxidation kinetics of various wrought, chromia-forming alloys. Each cycle consisted of holding the alloys at 1000°C in air for one day followed by cooling to room temperature. It is seen that for most of the alloys the mass change as a function of oxidation cycles changed from positive to negative, with the rates of the latter being alloy dependent. The occurrence of negative mass change is due to scale spallation during the cooling period in a thermal cycle. Continual Cr_2O_3 spallation and re-formation eventually depletes the alloy subsurface in chromium to the extent that Cr_2O_3 formation is no longer kinetically possible.

What ensues is breakaway oxidation in which spallation and reformation of the less protective base-metal oxide(s) occur [1]. It is seen in Fig. 1 that for Alloys 800HT and 800 the transitions to breakaway behaviour occur after about 80 and 110 one-day cycles, respectively. The time to breakaway is a measure of the useful service lifetime of a given alloy.

The observed variation in cyclic oxidation resistance of the different chromia-forming alloys is due largely to differences in the alloy compositions, the manner by which the scales spall (i.e. partial or complete spallation), and the extents to which element enrichment and depletion occur within the underlying alloy [2]. All of these factors are interdependent and, in order to develop accurate lifetime models, it will be necessary to understand and ultimately to quantify these interdependencies. To date, the only substantial studies to quantify interdependencies have been those by Barrett [3–5], who statistically analysed the cyclic oxidation kinetics of a number of commercial aerospace alloys to arrive at phenomenological equations relating spallation behavior to alloy composition. The equations calculate a composition-dependent oxidation parameter, K_a, which can vary from zero for the case of no scale spallation to about 100 for the case of extensive spallation. The equations highlighted the sensitivity of scale spallation to alloy composition and, in particular, the apparent interaction between alloying elements. Alloying elements such as Al, Ta, and Cr were found to be beneficial to scale spallation resistance, while Ti, Nb, Hf, Zr, Re, and C were detrimental. The equations are limited, however, because many commercial Cr_2O_3-forming alloy compositions are outside the composition database from which

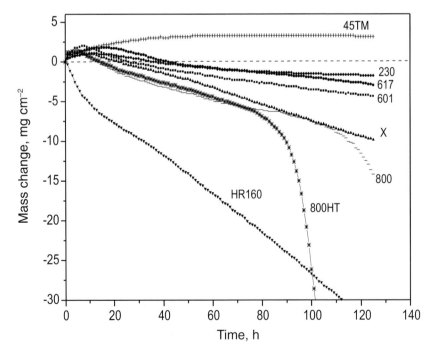

Fig. 1 *Cyclic oxidation kinetics of various Cr_2O_3-forming alloys at 1000°C in air. Each cycle consisted of one day at temperature.*

the equations were established. Notwithstanding, the statistical approach used by Barrett represents what is believed to be one of the most tractable ways to quantify the interdependency of alloy composition and scaling behaviour.

The purpose of this paper is to provide a brief overview of previous and recent results highlighting the effects of minor alloying elements on the oxidation behaviour of chromia-forming alloys. Minor elements in the context of this overview are Si, Ti, Al, and Mn in the amount 0.1–3.0 mass%, and reactive elements such as Ce, La, and Zr in the amount 0.005–0.3 mass%. All of these elements are intentional or incidental additions to the alloy and are differentiated from tramp elements such as Pb, Bi, S, Sn, P, etc., which are trace-level impurities. The effects of tramp elements will not be discussed in this paper. It is emphasised that the minority element effects discussed here are specific only to chromia formers. Similar trends may or may not pertain to alumina formers. It is further noted that the oxidation performance of chromia formers depends on the nickel content of the alloy, with nickel treated here as a major element. Figure 2 plots the time to crossover from positive to negative mass change as a function of nickel content for each of the alloys shown in Fig. 1. It is seen that the time to crossover increases with increasing nickel content in the alloy. Explanations to account for this are not conclusive and include nickel beneficially influencing scale adhesion and mechanical properties, reducing the rate of chromium diffusion in the Cr_2O_3 scale, and retarding the breakthrough transformation, Cr_2O_3 to $FeCr_2O_4 + Fe_2O_3$ [6]. The influence of nickel limits any generalised interpretations of the oxidative effects of minority elements.

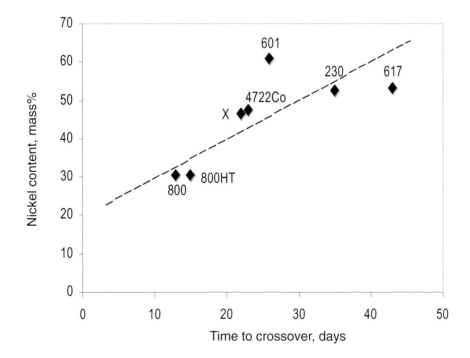

Fig. 2 *Effect of nickel content on the time to crossover from positive to negative mass change during cyclic oxidation (data from Fig. 1).*

2. Minority Elements in High-Temperature Alloys

All high-temperature alloys contain intentional or incidental levels of minor elements which, depending on level, may affect oxidation behaviour, mechanical behaviour, and weldability. Silicon is typically added during the melt processing of the alloy for the purpose of deoxidation and any excess silicon will be in solid solution. Manganese is also added during melt processing and plays the dual role of desulfurising and deoxidising, with the former having the principal role. Aluminium and titanium are added to improve high-temperature strength. The two elements react together with nickel to form $Ni_3(Al,Ti)$-based γ' precipitates that, when small in size, are potent strengtheners. Titanium also has a separate effect from aluminium through its stabilisation of carbon with respect to intergranular corrosion resistance. Reactive elements like La and Ce sometimes improve hot workability of difficult grades of alloy but are often added specifically for improving oxidation resistance, particularly under thermal cycling conditions.

The levels of minor elements in a given alloy can vary within a specified range. This is exemplified in Table 1, which shows the specified limiting compositions of the extensively used 800 and 800HT Fe–Ni–Cr alloys. Variations in element levels within the specified range can result in markedly different cyclic-oxidation behaviours. As a case in point, Fig. 3 compares the mass-change behaviour of two different 800 alloys as a function of one-day cycles at 1000°C in air. The alloys were

Table 1. *Limiting chemical compositions (mass%) specified by Special Metals Corporation for alloys 800 and 800HT*

Element	Alloy 800	Alloy 800HT
Fe	39.5 min	39.5 min
Ni	30.0–35.0	30.0–35.0
Cr	19.0–23.0	19.0–23.0
Mn	1.5 max.	1.5 max.
Si	1.0 max.	1.0 max.
Al	0.15–0.6	0.15–0.6
Ti	0.15–0.6	0.15–0.6
Al + Ti	–	0.85–1.2
C	0.1 max.	0.06–0.1
S	0.015 max.	0.015 max.
Cu	0.75 max.	0.75 max.

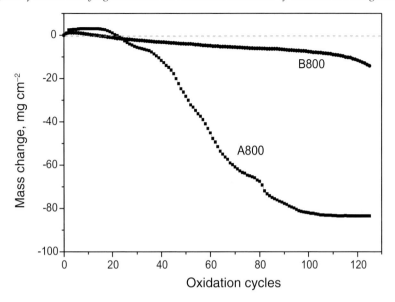

Fig. 3 *Cyclic oxidation kinetics of 800 alloys from two different suppliers. Oxidising conditions were 1000°C in air, with each cycle consisting of one day at temperature.*

from two different suppliers, with a main composition difference being that the alloy from supplier A (A800) contained 0.13 Si + 0.41 Ti and the one from supplier B (B800) contained 0.38 Si + 0.32 Ti. (Unless stated otherwise, all compositions will be given in mass percent.) The alloys were not very different microstructurally and it can be inferred that the much poorer oxidation performance of A800 is primarily attributable to its comparatively lower Si and higher Ti contents. This result effectively highlights the sensitivity of oxidation performance to minor-element contents.

From the standpoint of weldability, the desired minority element levels in iron-based high-temperature alloys are approximately 0.8–1.0 Mn, 0.2 Al, 0.2 Ti, 0.1 Si. Manganese is considered beneficial for minimising fissuring and for preventing hot cracking in welds, whereas Al, Ti and Si are considered detrimental because they increase the tendency for hot cracking. The levels of minority elements for optimising other key properties such as strength and oxidation resistance are typically incongruent and it is the goal of the alloy producer to determine target levels that optimise the combination of properties. Achieving such a goal requires a sound understanding of the effects of minor alloying elements on oxidation performance. The following discusses in greater detail the oxidative effects of specific minority elements. It will be shown that although our current level of understanding is incomplete, important interdependencies and trends can be identified.

2.1. Manganese

Manganese is generally found to be detrimental to the oxidation performance of chromia-forming alloys. For stainless steels and Fe-base superalloys the maximum specified manganese level is typically 1.5–2.0 mass%, while for Ni-base superalloys the maximum is no more than about 0.5 mass%. As a consequence, the detrimental effect of manganese is most likely to be observed when oxidising stainless steels and Fe-base superalloys.

Manganese forms a more stable oxide than chromium and thus MnO may be expected to develop at the base of the Cr_2O_3 scale layer. However, the manganese tends to diffuse through the Cr_2O_3 layer and establish an outer $MnCr_2O_4$ scale [7–9]. As shown by Lobnig *et al.* [10], the lattice diffusivity of manganese in Cr_2O_3 is about two orders of magnitude faster than both nickel and iron at 900°C, thus explaining the increased tendency for manganese to oxidise at the outer surface. The duplex $Cr_2O_3/MnCr_2O_4$ scale structure is less protective and more prone to spallation than an exclusive Cr_2O_3 scale structure.

Figure 4 compares the mass-change behaviour of eight Ni-base 890 alloys of slightly different compositions as a function of 7-day cycles at 1000°C in air + 5 vol.% H_2O. The measured compositions of the alloys are included in this Figure. The oxidation behaviour of six of the alloys containing around 0.02% Mn were comparable and significantly better than the two alloys containing around 0.85% Mn. The higher manganese contents in these last two alloys caused them to exhibit steady mass losses much earlier and at rates much greater than the other alloys.

2.2. Silicon

The silicon content in high-temperature alloys is usually limited to less than 3% because of mechanical, weldability, and thermal-stability constraints. For most wrought alloys the silicon content is kept below about 1.0%. Low silicon contents of 0.05–1.0% may or may not be sufficient to form a continuous SiO_2 layer at the alloy/scale interface, but are sufficient to have a beneficial effect on oxidation kinetics, particularly by facilitating exclusive Cr_2O_3 scale formation [11–14]. For instance, Jones and Stringer [14] showed that the oxidation mode for the Co–25Cr alloy changes from the development of a fast-growing duplex structure to the formation of a protective Cr_2O_3 scale with the presence of as little as 0.05% silicon in the alloy. Results from a recent analytical study on the transient oxidation behavior of a Ni-base chromia-forming alloy [15] suggest that the beneficial effect of silicon (and aluminium) is established at room temperature through the formation of a passive oxide film rich in silicon (and aluminium). The thickness of this passive layer was of the order of 3 nm. It is possible that upon initial heating, the presence of the passive layer causes the outward flux of chromium to be significantly greater than the inward flux of oxygen to the extent that the initial nucleation of Cr_2O_3 occurs primarily at the alloy surface, thus helping to facilitate Cr_2O_3-scale formation.

Evans *et al.* [16] reported the influence of silicon additions on the oxidation resistance of Fe–20Cr–20Ni-base alloys in a CO_2-based atmosphere at 900°C and found that the rate of scale growth (primarily Cr_2O_3) was a minimum for an alloy containing about 0.9% silicon. The minimum in the rate constant coincided with the presence of a continuous SiO_2-rich layer at the alloy/scale interface, which was inferred to act as a barrier to chromium diffusion. Localised internal protrusions of SiO_2 developed in alloys containing more than about 0.9% silicon.

A negative aspect associated with silicon addition is that the extent of oxide scale spallation under conditions of thermal cycling tends to increase with increasing silicon content above about 0.5% [17,18]. This is indicated by the results shown in Fig. 3. Higher silicon additions also tend to reduce the creep strength of most high-temperature alloys and worsen fusion-line cracking in weldments [19].

Compositions: mass%

Alloy	C	Mn	Fe	Si	Ni	Cr	Al	Ti	Mo	Nb	Ta	Zr
A	0.13	0.027	25.08	1.73	43.01	27.20	0.294	0.276	1.514	0.397	0.245	0.0074
B	0.13	0.021	26.76	1.70	43.02	27.24	0.206	0.256	0.007	0.378	0.225	0.0088
C	0.13	0.019	27.00	1.69	42.92	27.30	0.220	0.272	0.004	0.391	0.003	0.0105
D	0.12	0.020	25.20	1.77	43.98	26.24	0.214	0.278	1.490	0.398	0.222	0.0090
E	0.12	0.019	26.71	1.63	44.04	26.28	0.203	0.273	0.055	0.386	0.217	0.0115
F	0.12	0.018	26.94	1.86	43.81	26.30	0.210	0.277	0.006	0.390	0.007	0.0135
G	0.12	0.887	25.83	1.81	43.90	26.27	0.202	0.272	0.006	0.386	0.203	0.0131
H	0.12	0.850	25.29	1.54	44.19	26.32	0.758	0.274	0.006	0.388	0.206	0.0219

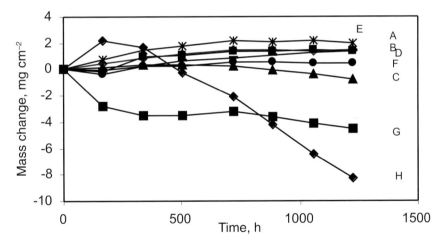

Fig. 4 *Cyclic oxidation kinetics of 890 alloys from different heats. Oxidising conditions were 1000°C in air + 5 vol.% H_2O, with each cycle consisting of one week at temperature.*

Gulbransen and Andrew [20] reported that the beneficial effects of silicon additions to a Ni–20Cr alloy are optimised if the silicon content is kept low (0.25%) and a small amount of manganese (~0.05%) is added. The reasons for the apparent synergistic effect between silicon and manganese additions are still not well understood.

2.3. Aluminium and Titanium

The creep strength of Ni-base superalloys increases with increase in the coherent γ'-$Ni_3(Al,Ti)$ precipitate volume fraction, $f_{\gamma'}$ [21]. Most wrought, precipitation-hardenable Ni-base superalloys have an $f_{\gamma'}$ between 0.2 and 0.4; any higher than about 0.45 renders the alloy difficult to deform by hot or cold rolling. A high $f_{\gamma'}$ also degrades the weldability of the alloy due to an increased susceptibility to strain-age cracking. In general, Ni-base superalloys are found to be workable if (Al + Ti) < 5 and weldable if (2Al + Ti) < 6 [22]. However, these limits are not rigorous and Al + Ti contents are often kept much lower to ensure reliability and durability. Care is also taken to keep the Ti:Al ratio in the alloy sufficiently low to avoid formation of a plate-like, hexagonal-closed packed form of η-Ni_3Ti after prolonged heating. This phase is generally detrimental to creep strength and toughness [23]. Thus, in wrought, chromia-forming Ni-base superalloys, the aluminium content is kept below about 1.7% and the titanium content below about 0.6%.

Iron decreases the high-temperature stability of γ' and favours the formation of β-NiAl, which is not an effective strengthener. For this reason, Fe–Ni–Cr-base superalloys hardened by intermediate precipitation rely mainly on the formation of η-Ni_3Ti or γ''-Ni_3Nb. However, these precipitates are unstable at higher temperatures, with their long-term strengthening effects having an upper temperature limit of about 650°C. Thus, the wrought, chromia-forming Fe-base superalloys typically contain no more than about 0.6% Al and 0.6% Ti.

From the standpoint of oxidation resistance, previous results indicate that aluminium is beneficial and titanium is detrimental [2]. Both metals form a more stable oxide than Cr_2O_3; however, aluminium oxidises essentially only in the region of the alloy/scale interface whereas titanium tends to oxidise at both the scale surface and the alloy/scale interface [24–26]. Thus, titanium behaves in a manner similar to that of manganese. The main difference is that titanium undergoes more extensive internal oxidation within the subsurface region of the alloy, particularly at grain boundaries. This difference may be attributable to titanium being less mobile than manganese in the Cr_2O_3 scale. It is interesting to note that the addition of 0.5–3.0% titanium has been shown to impart sulfidation resistance to chromia-forming alloys [27, 28].

For the aluminium to have a beneficial effect it must be at a level above 0.4%, which is sufficient to establish an Al_2O_3-rich layer at the base of the Cr_2O_3 scale. This layer, which acts to decrease the thickening kinetics of the Cr_2O_3 scale by reducing the outward migration of chromium cations, does not necessarily need to be continuous to confer a beneficial effect. As discussed by Barrett and Lowell [29] in their study of the cyclic oxidation behaviour of commercial alloys at 1150°C, the most critical factor in minimising scale spallation is to keep the scale as thin as possible.

2.4. Reactive Elements

Chromia scales can be very susceptible to spallation during thermal cycling. It has long been established that the addition of small amounts (0.005–0.3%) of reactive elements (RE), such as Y, La, Zr, Ce or Hf, can improve considerably the resistance of these scales to spallation [30]. An example of this effect is shown in Fig. 5 for the case of six different wrought, Ni-base 693 alloys oxidised in air + 5% H_2O at 1100°C with one-week thermal cycles. As can be seen in the accompanying table of alloy compositions, the best performing alloys were those that contained the highest

Compositions: mass%

Alloy	C	Mn	Fe	Cr	Al	Nb	Si	Ti	Ce
A	0.016	0.180	8.84	29.22	0.32	0.06	0.11	0.37	5E-04
B	0.160	8.50	29.93	0.31	0.02	0.25	0.25	0.37	0.021
C	0.051	0.160	7.59	30.04	0.33	0.99	0.28	0.36	5E-04
D	0.032	0.160	7.71	30.06	0.31	0.10	0.28	1.02	5E-04
E	0.027	0.160	7.48	30.05	0.32	0.99	0.27	0.40	0.018
F	0.039	0.020	8.54	30.33	0.30	0.30	0.11	0.26	0.012

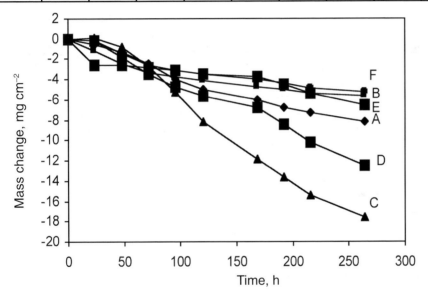

Fig. 5 *Cyclic oxidation kinetics of 693 alloys from different heats. Oxidising conditions were 1100°C in air + 5 vol.% H_2O, with each cycle consisting of one week at temperature.*

amount of cerium (i.e. 0.018% vs 0.0005%). The results in Fig. 5 also reveal the detrimental effect of increasing the titanium level from 0.37 to 1.02% (alloy A vs D) and increasing the niobium level from 0.06 to 0.99% (alloy A vs C). However, it is further revealed that the detrimental effect of niobium can be nullified by increasing the cerium level from 0.0005 to 0.018% (alloy C vs E).

When present in a small but sufficient amount, the reactive element tends to segregate to the alloy/scale interface and scale grain boundaries, and have the apparent effect of suppressing cation diffusion in the scale. According to Pint [31], the suppression of cation diffusion improves scale adhesion by reducing scale-growth stresses associated with oxide formation within grain boundaries of the scale and also inhibiting the development of voids at the alloy/scale interface. An additional, and perhaps coinciding effect is for the RE to react with impurity sulfur in the alloy to form sulfides or S–RE-containing complexes, thus preventing the interfacial segregation of free sulfur and consequential weakening of the alloy/scale interfacial bond strength [32,33]. Alloys containing high concentrations of RE (>1%) tend to form intermetallic phases which are prone to preferential oxidation, thus decreasing the oxidation resistance of the alloy [34].

3. Summary

It has been shown in this brief overview that variations in minor alloying element contents can significantly affect the oxidation behaviour of an alloy, even if the variations are within specifications. Key minor alloying elements in chromia formers are believed to be Si, Al, Ti, Mn, and reactive elements. The oxidative effects of these elements can be summarised as follows:

Mn • generally detrimental at concentrations above about 1 mass%
   • small addition (< 0.5mass%) may be beneficial as it interacts synergistically with Si to promote Cr_2O_3 scale formation at Cr concentrations below about 20mass%
Si • acts to promote Cr_2O_3 scale formation
   • tends to be detrimental to spallation resistance at concentrations above about 0.5mass%; however, this depends significantly on principal alloying element concentrations
Ti • generally detrimental at concentrations above about 0.5 mass%
   • tends to oxidise both at the scale surface and intergranularly in the subsurface region of the alloy
Al • generally beneficial at concentrations above about 0.4 mass%
   • oxides of both Al and Si tend to become quickly established in the subscale region during the early stages of oxidation
RE • small additions (~0.005–0.3mass%) of reactive elements such as La and Ce improve spallation resistance of oxide scale layer
   • high concentrations (> 1mass%) of RE tend to reduce the oxidation resistance of the alloy by forming intermetallic phases that oxidise preferentially

It is believed that a successful lifetime prediction model for chromia-forming alloys will need to take into account minor alloying addition contents, as well as the oxidising environment. The effects of minor alloying additions are perhaps best treated via the development of statistical-based equations for oxidation parameters such as the attack parameter, K_a, used by Barrett [3–5].

References

1. J. R. Nicholls and M. J. Bennett, *Mater. High Temp.*, 2000, **17**, 413–428.
2. B. Gleeson, High-temperature corrosion of metallic alloys and coatings, in *Corrosion and Environmental Degradation of Materials* (M. Schütze, ed.), Volume 19 of the Series: Materials

Science and Technology (R. W. Cahn, P. Haasen and E. J. Kramer, eds). Wiley-VCH, Germany, 2000, pp.173–228.

3. C. A. Barrett, *A Statistical Analysis of Elevated Temperature Gravimetric Cyclic Oxidation Data of 36 Ni- and Co-base Superalloys Based on an Oxidation Attack Parameter.* National Aeronautics and Space Administration, Washington, DC, Technical Memorandom 105934, NASA Lewis Research Center, 1992.

4. C. A. Barrett, *10,000-Hour Cyclic Oxidation Behavior at 982°C (1800°F) of 68 High-Temperature Co-, Fe-, and Ni-Base Alloys.* National Aeronautics and Space Administration, Washington, DC, Technical Memorandom 107394, 1997.

5. J. L. Smialek, C. A. Barrett, and J. C. Schaeffer, *Design for Oxidation Resistance*, in ASM Handbook, Volume 20: Materials Selection and Design. ASM International, Materials Park, OH, 1997, pp.589-602.

6. A.J . Sedricks, *Corrosion of Stainless Steels*. John Wiley & Sons, Inc, New York, USA, 1996, pp. 392–399.

7. F. H. Stott, F. I. Wei and C.A. Enahoro, *Werkst. Korros.*, 1989, **40**, 198–205.

8. M. D. Merz, *Metall. Trans.* A, 1980, **11A**, 71–83.

9. D. L. Douglass and F. Rizzo-Assuncao, *Oxid. Met.*, 1988, **29**, 271–287.

10. R. E. Lobnig, H. P Scmidt, K. Hennesen and H. J. Grabke, *Oxid. Met.*, 1992, **37**, 81–93.

11. F. H. Stott, G. J. Gabriel, F. I. Wei and G. C. Wood, *Werkst. Korros.*, 1987, **38**, 521–531.

12. J. F. Radavich, *Corrosion*, 1959, **15**, 613t–617t.

13. B. Gleeson and M. A. Harper, *Oxid. Met.*, 1998, **49**, 373–399.

14. D. E. Jones and J. Stringer, *Oxid. Met.*, 1975, **9**, 409–413.

15. B. Ahmad and P. Fox, *Oxid. Met.*, 1999, **52**, 113–138.

16. H. E. Evans, D. A. Hilton, R. A. Holm and S. J. Webster, *Oxid. Met.*, 1983, **19**, 1–18.

17. R. C. Lobb, J. A. Sasse and H. E. Evans, *Mater. Sci. Technol.*, 1989, **5**, 828–834.

18. D. L. Douglass and J. S. Armijo, *Oxid. Met.*, 1970, **2**, 207–231.

19. R. H. Kane, in *Heat-Resistant Materials, Proc. 1st Int. Conf.* (K. Natesan and D. J. Tillack, eds). ASM International, Materials Park, OH, 1991, pp.1–8.

20. E. A. Gulbransen and K. F. Andrew, *J. Electrochem. Soc.*, 1959, **106**, 941–948.

21. J. J. Jackson, M. J. Donachie, R. J. Henricks and M. Gell, *Metall. Trans. A*, 1977, **8A**, 1615–1620.

22. M. Prager and C. S. Shira, *Weld. Res. Counc. Bull.*, 1968, **128**.

23. G. K. Bouse in, *Superalloys 1996* (R. D. Kissinger, D. J. Deye, D. L. Anton, A. D. Cetel, M. V. Nathal, T. M. Pollock and D. A. Woodford, eds). The Minerals, Metals and Materials Society, 1996, pp.163–172.

24. D. R. Sigler, *Oxid. Met.*, 1996, **46**, 335–364.

25. F. H. Stott and F. I. Wei, *Mater. Sci. Technol.*, 1989, **5**, 1140–1147.

26. C. L. Angerman, *Oxid. Met.*, 1972, **5**, 149–167.

27. A.S. Nagelberg, *Oxid. Met.*, 1982, 17, 415-427.

28. G.Y Lai in *High Temperature Corrosion in Energy Systems* (M. F. Rothman, ed.). The Metallurgical Society of AIME, Warrendale, PA, 1985, pp. 227–236.

29. C. A. Barrett and C. E. Lowell, *Oxid. Met.*, 1975, **4**, 307–355.

30. D. P. Whittle and J. Stringer, *Phil. Trans. R. Soc. Lond. A*, 1980, **295**, 309–329.

31. B. A. Pint, *Oxid. Met.*, 1996, **45**, 1–37.

32. J. G. Smeggil, A.W. Funkenbusch and N. S. Bornstein, *Metall. Trans.*, 1986, **17A**, 923–932.

33. J. L. Smialek, *Metall. Trans.* A, 1991, **22A**, 739–752.

34. F. S. Pettit and G. H. Meier, in *Processing and Design Issues in High Temperature Materials* (N. S. Stoloff and R. H. Jones, eds). The Minerals, Metals and Materials Society, Warrendale, PA, 1997, pp.379–390.

12

Significance of Scale Spalling for the Lifetime of Ferritic 9–10%Cr Steels During Oxidation in Water Vapour at Temperatures Between 550 and 650°C

R. J. EHLERS, P. J. ENNIS, L. SINGHEISER, W. J. QUADAKKERS and T. LINK*

Forschungszentrum Jülich, Germany
*Siemens PG, Mülheim a. d. Ruhr, Germany

ABSTRACT

Three ferritic 9–10% Cr steels were tested with respect to their oxidation behaviour in Ar-50vol.%H_2O up to exposure times of 10 000 h. Scale growth and spalling characteristics were studied by discontinuous gravimetry, optical metallography, scanning electron microscopy and X-ray diffraction. Inter- and extrapolation of oxidation data were found to require very long term testing because scale spallation in most cases appears to start only after several thousand hours of exposure. The scale thickness at which scale spallation starts differs for each exposure temperature, i.e. a critical scale thickness for the onset of spallation can not be defined. The initiation of scale cracking seems to be correlated with the type and morphology of voids and pores within the scales which are shown to be affected by alloy microstructure.

1. Introduction

The demand for a reduction of CO_2 emissions requires increased thermal efficiencies of conventional, fossil fuel fired power plants [1]. Therefore, new advanced plant designs are beeing developed to achieve steam parameters in the range of 600– 650°C and 300 bar [2]. To meet these stringent stress conditions, new ferritic–martensitic 9–10%Cr steels have been developed to replace alloys such as 1CrMoV and 12CrMoV, which are commonly used construction materials for operating temperatures of approximately 525 and 565°C respectively [3]. One of the lifetime limiting factors of the new high strength steels at the high service temperatures between 600 and 650°C is the oxidation/corrosion occurring in the service environments. It is well known, that in this temperature range all 9–10%Cr steels possess excellent oxidation resistance during exposure in air or dry gases such as N_2–1%O_2 [4]. However, in water containing environments (Fig. 1), the corrosion rates can significantly be increased [5–7]. In the present study, the oxidation behaviour in a simulated steam environment of P92 as a representative of the new class of commercial high strength 9–10%Cr steels is investigated together with that of 9Cr–1Mo- and 10Cr–1Mo–1W model steels.

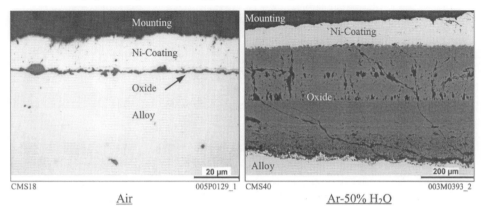

CMS18
Air
005P0129_1 CMS40
Ar-50% H₂O
003M0393_2

Fig. 1 *Oxide scales on 9% Cr-steel P92 after exposure for 10 000 h at 650 °C in air and in Ar-50vol.%H₂O.*

2. Experimental

The composition of the three studied materials is listed in Table 1.

From these materials specimens with a size of $20 \times 10 \times 2$ mm were machined. For the oxidation studies, the specimens were ground to 800 grit surface finish and exposed to an Ar-50vol.%H₂O atmosphere. Oxide scale characterisation after exposure was carried out using optical metallography, X-ray diffraction (XRD) and scanning electron microscopy (SEM).

The long term exposures were carried out in a test equipment shown schematically in Fig. 2. To measure the mass changes of the specimens during exposure, the tests were interrupted every 250 h. For this purpose, the specimens were cooled down to room temperature in dry argon. After the mass measurements, the specimens were again heated to exposure temperature in dry argon. When the test temperature was reached, the atmosphere was switched to the wet test gas. The heating rate was 10°C/min, the flow rate of the gas 10 L/h. Due to expected spallation of the rapidly growing oxide scales, the specimens were placed in crucibles so that the amount of spalled oxide could be measured in addition to the mass change of the specimen (Fig. 3). This approach allows two types of mass change curve to be derived for each specimen i.e. the net mass change, which is the mass change of the specimen, and the gross mass change, which represents the total oxygen uptake of the specimen, thus including the oxygen tied up in the spalled oxide as well as that tied up in the adhering scale.

Table 1. *Composition of the investigated steels (mass%)*

Steel	C	Cr	Mo	W	Si	Mn	V
P92	0.11	9	0.5	1.8	0.03	0.46	0.2
9Cr–1Mo	0.17	9.2	1.55	–	0.07	0.06	0.29
Fe–10Cr–1Mo–1W	0.08	10.3	1.05	1.05	0.1	0.43	0.2

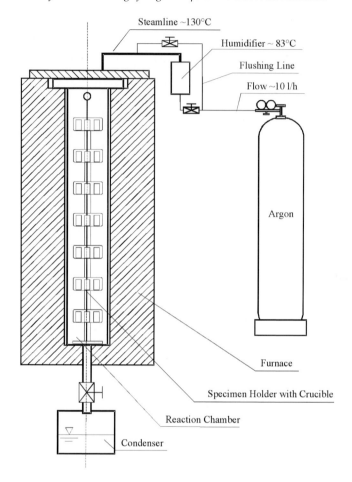

Fig. 2 *Experimental set-up for the oxidation tests.*

3. Results and Discussion

3.1. Scale Structure and Morphology

The scale structure observed after oxidation at temperatures between 550 and 650°C in Ar-50vol.%H_2O was very similar for all studied steels. Figure 4 shows a schematic illustration of a typical oxide scale. The thin outer Fe_2O_3 layer was in most cases absent, although it frequently appeared after long exposure times (>3000 h).

Figure 5 shows a typical cross section of an oxide scale on a 9% Cr steel after 1000 h oxidation at 650°C in Ar-50vol.%H_2O. The oxide mainly consists of two thick spinel layers and a smaller zone of internal oxidation. Between the outer, pure Fe_3O_4 and the inner, two-phase (Fe, Cr, Mn)$_3O_4$ layer, a straight border line exists, which in this stage of the oxidation where no spallation has yet occurred, represents the original specimen surface.

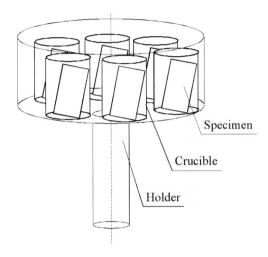

Fig. 3 *Quartz holder with specimens in crucibles.*

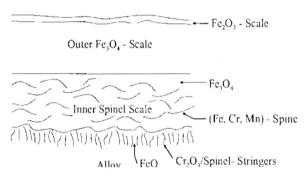

Fig. 4 *Schematic illustration of typical oxide scale on 9–10% Cr-steels at temperatures between 550 and 650°C in Ar-50vol.% H_2O.*

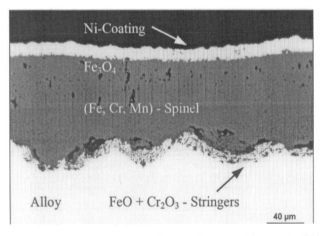

Fig. 5 *Scale on 9Cr–1Mo steel after oxidation for 1000 h at 650°C in Ar-50vol.%H_2O.*

Figure 6 shows the distribution of iron and chromium in the oxide scale on the 9Cr–1Mo steel. The interface between the inner and outer spinel layer is clearly visible. The Figure shows that no chromium is present in the outer spinel. It is only in the inner layer that chromium is detected. The presence of Cr within the inner scale should not be considered as a chromium enrichment but a result of the scale growth by outward transport of iron cations and the resulting inward movement of the scale/metal interface thereby incorporating the Cr-rich stringers existing in the internal oxidation zone, into the growing scale. The diffusivity of chromium in the temperature range 550 to 650°C is relatively slow compared to the scale thickening rate.

This is illustrated by the element distribution (Fig. 6) showing only a very small region of detectable chromium depletion in the alloy near the oxide/alloy interface and a slight chromium enrichment within the oxide in the immediate vicinity of this depletion zone.

In scales which exhibit a compact morphology as shown in Fig. 5, generally no haematite is found at the scale/gas interface although this oxide phase should exist in equilibrium with the gas atmosphere. In the compact scale, the transport of iron cations is apparently high compared to the oxygen transfer from the gas atmosphere to the oxide surface. This results in a high Fe activity at the scale/gas interface which thus does not allow magnetite to become transformed into hematite. Hematite formation is only observed after longer exposure times, i.e. after the scales have developed substantial porosity. The reason for this effect is that at the interface between the outer and inner spinel layer, the forming voids tend to coagulate into

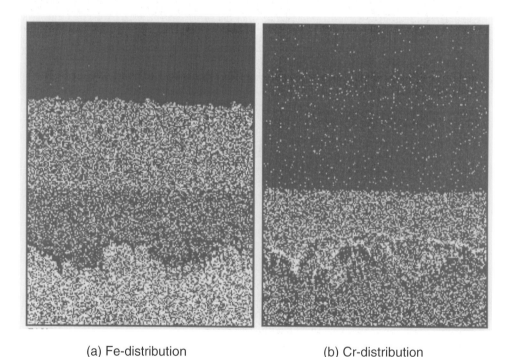

(a) Fe-distribution   (b) Cr-distribution

Fig. 6 *Element distributions in scale of Fig. 5. (a) Fe-distribution; (b) Cr-distribution.*

large pores which eventually leads to a virtual separation between inner and outer layer (Fig. 7). This results in a hampered transport of Fe cations from the scale/alloy to the scale/gas interface. Consequently, the Fe activity near the oxide surface decreases to such an extent, that in the outer part of the scale, Fe_2O_3, i.e. the oxide which is in equilibrium with the test atmosphere, can be formed.

3.2. Long Term Exposure

The mass changes of the 10Cr–1Mo–1W steel during oxidation in Ar-50vol.%H_2O at 625 and 650°C for up to 2500 h are shown in Fig. 8. The solid lines represent extrapolations of the mass change data at the two temperatures up to 10 000 h. The prediction is based on the equation:

$$\frac{\Delta m}{A} = K \cdot t^n \tag{1}$$

where Δm is the mass change of the specimen, A the surface area, K the rate constant, n the rate exponent and t the oxidation time. For $n = 0.5$ eqn (1) represents the well-known parabolic rate law for oxide scale growth.

In Fig. 8 the exponent n equals approximately 0.3 for the 650°C curve up to 2500 h, and 0.32 for the 625°C curve. In Fig. 9, the extrapolated curves are compared

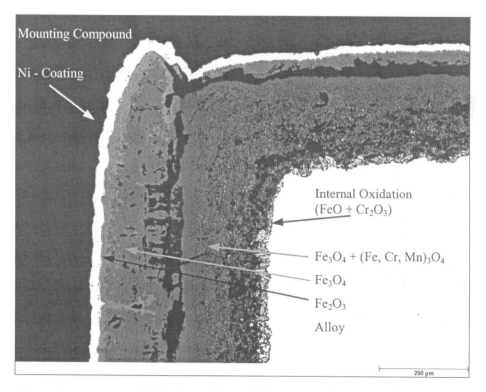

Fig. 7 *Cross-section of 10Cr–1Mo–1W steel after exposure for 10 000 h at 650°C in Ar-50vol.%H_2O.*

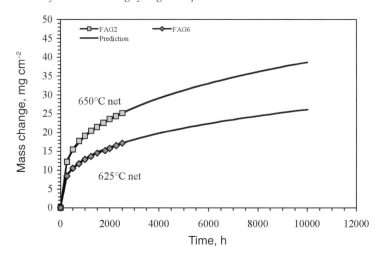

Fig. 8 *Measured mass changes up to 2500 h and the long time extrapolation for 10Cr–1Mo–1W steel during oxidation at 625 and 650°C in Ar-50vol.%H$_2$O.*

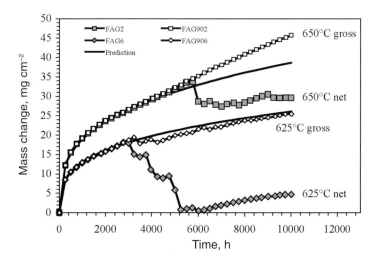

Fig. 9 *Measured mass changes as function of time during oxidation of 10Cr–1Mo–1W steel at 625 and 650°C in Ar-50vol.%H$_2$O compared with the extrapolation from Fig. 8.*

with the actually obtained long term mass change data until 10 000 h. This comparison shows a large difference between extrapolation and measured gross mass change especially at 650°C. At this temperature the overall oxidation rate strongly increases during exposure, i.e. the slope of the gross mass change curve changes from 0.3 during the first 2500 h to 0.53 for times greater than 6000 h. Thus, the latter value is even larger than the frequently assumed $n = 0.5$, representing a parabolic rate law for the oxidation rates of steels in steam. The comparison of gross and net mass changes in Fig. 9 clearly shows that this large difference between extrapolated curves and the actually measured gross mass change, which is a direct measure of the overall loss in wall thickness, is caused by occurrence of oxide spallation after approximately 6000 h.

For the specimen exposed at 625°C the extrapolation of the short term data seems to fit reasonably with the real mass changes measured for up to 10 000 h. However, Fig. 10 clearly shows that at this temperature, also due to scale spallation, the slope of the gross mass change curve increases after longer exposure times. It rises from around 0.3 during the first 2500 h to approximately 0.4 for times between 4000 and 10 000 h.

Figures 11 and 12 show the gross and net mass changes of the 10Cr–1Mo–1W steel during oxidation at 600 and 550°C. Especially the double logarithmic plot in Fig. 12 illustrates that also at these temperatures the slopes of the gross mass change curves change after longer exposure times due to oxide spallation.

These results clearly illustrate, that extrapolation of oxidation rates for 9–10%Cr steels in water vapour measured during exposures up to only a few thousand hours, can lead to a substantial underestimation of the actually occurring oxidation rates because substantial scale spalling after longer times can significantly increase the metal loss rate [8].

3.3. Scale Spallation

Figure 13 shows the net mass changes of P92 as function of time during exposure in Ar-50vol.%H_2O at 600, 625 and 650°C. This diagram illustrates that no clear correlation exists between oxidation rate and time for the onset of scale spallation. At 600°C spallation occurs after 2250 h where the net mass change is about 16 mg cm^{-2}. At 625°C spallation occurs after around 8250 h, i.e. when the mass change equals 35 mg cm^{-2}. At 650°C no significant spallation occurs up to an exposure time of 10 000 h, where the mass change exceeds 47 mg cm^{-2}. The respective metallographic cross sections (Figs 14–16) confirm that no clear correlation exists between scale thickness and the onset of spallation.

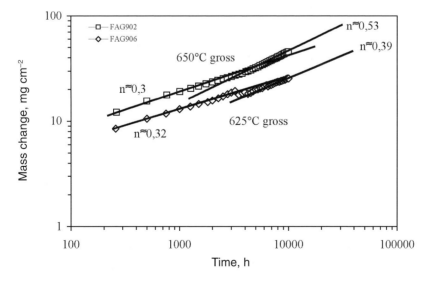

Fig. 10 *Double logarithmic plot of gross mass changes as function of time during oxidation of 10Cr–1Mo–W at 625 and 650°C in Ar-50vol.%H_2O.*

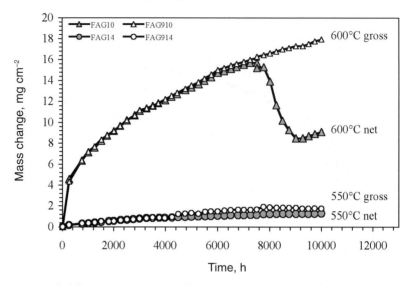

Fig. 11 *Mass changes as function of time during oxidation of 10Cr–1Mo–1W at 550 and 600°C in Ar-50vol.%H$_2$O.*

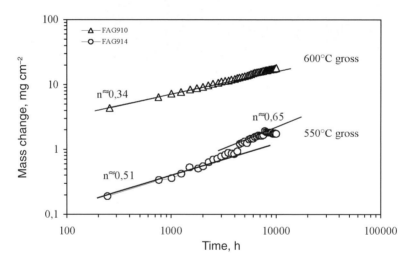

Fig. 12 *Double logarithmic plot of gross mass changes as function of time during oxidation of 10Cr–1Mo–W at 550 and 600°C in Ar-50vol.%H$_2$O.*

Figure 14 indicates that at 600°C the scale has spalled within the (Fe, Cr, Mn)$_3$O$_4$ spinel layer well before the specimen was eventually taken out of the furnace.

In Figure 15 spallation has occurred at the interface between inner and outer spinel layer. Subsequently, both (Fe, Cr, Mn)$_3$O$_4$ sub-layers were overgrown by Fe$_3$O$_4$.

The P92 exposed to the highest test temperature of 650°C shows, in spite of substantial crack formation, no substantial spallation even after reaching a scale thickness of approximately 400 μm (Fig. 16).

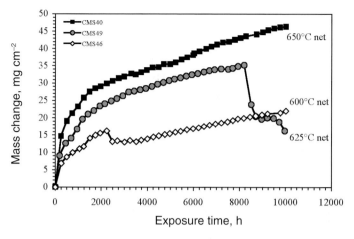

Fig. 13 *Net mass changes as function of time for P92 during oxidation at 600, 625 and 650°C in Ar-50vol.%H₂O.*

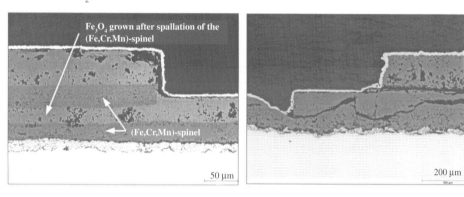

Fig. 14 *3000 h at 600°C.*   **Fig. 15** *10 000 h at 625°C.*

Metallographic cross-sections of P92 showing different types of scale spallation after oxidation in Ar-50vol.% H₂O.

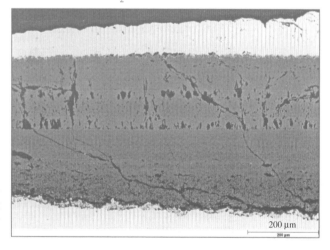

Fig. 16 *Metallographic cross-sections of P92 showing crack formation in scale after oxidation in Ar-50vol.% H₂O.*

3.3.1. Crack and void patterns

A substantial mismatch exists in the oxide scales between the thermal expansion coefficients (CTE's) of the alloy and those of the formed oxides. This mismatch reaches a maximum of about 12×10^{-6} K^{-1} at approximately 550°C between magnetite (which has the higher CTE value) and the alloy [9]. This difference in thermal expansion leads to a tensile stress in the oxide when the specimen is cooled down to room temperature for the mass measurement, assuming that the oxide grew virtually stress-free at the oxidation temperature. Due to this tensile stress, cracks are initiated in the scale. For the long term oxidation behaviour of the materials, the place in which the cracks are initiated and how they propagate within the oxide is of vital importance (Fig. 17). Generally it can be said that the occurring crack initiation/propagation modes can result in two different types of spallation. In one case spalling occurs between the two spinel sub-layers (Fig. 18), in the other case it primarily occurs at the oxide/alloy interface (Fig. 19). The reason for the two types of oxide spallation seems to correlate with the different types of pore morphology in the scale during long term exposure

The rate determining step for the oxidation of the investigated Cr steels in steam is the iron transport through the inner part of the oxide scale, which includes the (Fe, Cr, Mn)$_3$O$_4$ layer and the zone of internal oxidation. The iron transport in the outer, single-phase Fe$_3$O$_4$ scale is substantially faster than in the inner spinel, which results in pore formation at the interface between the outer and inner layer (Fig. 18). If the oxide spalls at the interface between both spinel layers (Fig. 18), the oxidation rate is still controlled by the iron transport through the inner spinel and the zone of internal oxidation. Consequently, the scale thickening rate after spalling will not substantially differ from that before occurrence of scale spalling. If, however, spallation occurs at the scale/alloy interface (Fig. 19), the oxidation rate of the steel will become significantly higher than that before occurrence of spallation.

The location where spallation occurs is obviously the 'weakest-part' of the oxide scale. This weak part is the place, where the main part of the pores is coagulating to larger voids. Thus, void and pore formation is playing a crucial role in the initiation

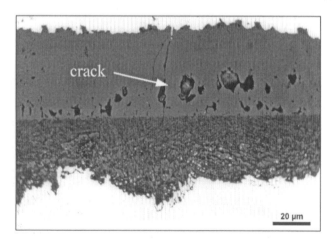

Fig. 17 *Oxidation of 9Cr-steel for 1000 h at 650°C in Ar-50vol.% H$_2$O.*

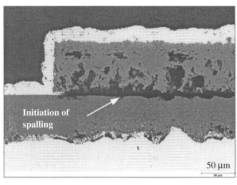 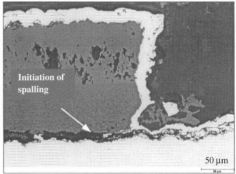

Fig. 18 *10Cr–1Mo–1W, 3000 h.*   **Fig. 19** *10Cr-steel, 1000 h.*

Different types of scale spallation during oxidation at 625°C in Ar-50vol.% H_2O.

and propagation of cracks and therefore the mode of oxide spallation and the overall metal loss rate during long term service.

3.3.2. Void formation as a function of alloy micro structure

Based on the above findings, knowledge of the mechanisms of pore formation in the scale is of great importance for evaluating the long term oxidation behaviour of the 9–10% Cr steels. The numerous observations made in the present study strongly suggest, that the eventual pore morphology and distribution in the scale depends on the way in which the vacancies within the scale coalesce to pores. The latter seems to be affected by the presence of vacancy 'condensation-sites'. Figures 20 and 21 show a clear difference in pore morphology in the outer, pure Fe_3O_4 layer, the inner two-phase layer consisting of Fe_3O_4 + (Fe, Cr, Mn)$_3O_4$ and in the Cr_2O_3/FeO internal oxidation layer adjacent to the alloy. Numerous tiny voids are present in the inner spinel layer (Fig. 21) which seem to be correlated with the (Fe,Cr,Mn)$_3O_4$ stringers embedded in the Fe_3O_4 [10, 11]. Apparently these stringers act as nucleation sites for the voids. Comparison with the alloy microstructure (Fig. 22) strongly suggests, that the morphology of the (Fe, Cr, Mn)$_3O_4$ stringers in the inner scale is a 'fingerprint' of the carbides formed on grain boundaries and martensite laths in the alloy.

If this would be the case, alloy microstructure should have a pronounced effect on scale morphology and pore distribution in scale.

In order to investigate the effect of alloy microstructure on the oxidation behaviour, a P92 steel was taken in 'as-received' condition with fully martensitic microstructure (Fig. 23). From the same heat (CQM) a second series of specimens with a ferritic microstructure was produced (Fig. 24) by suitable heat treatment. The latter was established by austenising a piece of P92 for two hours at 1050°C, cooling in the furnace to 780°C, holding that temperature for 8 h and then slowly cooling down to room temperature in the furnace. This leads to the ferritic microstructure shown in Fig. 24. For the oxidation tests, specimens with the different heat treatments were exposed for 1250 h at 650°C in Ar-50vol.%H_2O. The mass change data as function of time is shown in Fig. 25 The results indicate that no significant difference exists in the overall oxidation rates of both specimens. However, the metallographic cross

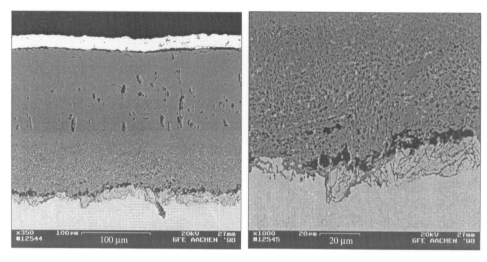

Fig. 20 *10Cr–1Mo–1W.*   **Fig. 21** *10Cr–1Mo–1W.*

Oxidation for 1000 h at 650°C in Ar-50vol.% H₂O.

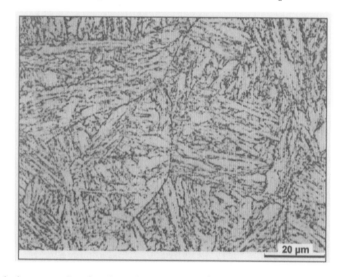

Fig. 22 *Etched cross-section showing microstructure of 10Cr–1Mo–1W steel.*

sections reveal striking differences in oxide morphology. The outer spinel phase on the martensitic steel (Fig. 26) shows similar vertical cracks as visible in the cross-section of the oxidised ferritic specimen (Fig. 27). However, the morphology of the inner spinel layers and the zones of internal oxidation appear to be fingerprints of the alloy microstructure. Especially, the morphology of the chromia stringers in the internal oxidation zone clearly show the same morphology as the chromium carbide precipitates of the martensitic (Figs 23 and 28) and ferritic materials (Figs 24 and 29) respectively.

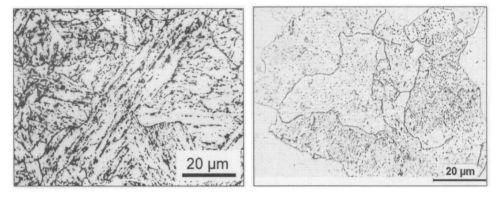

Fig. 23 *CQM (martensite).*   **Fig. 24** *HAA (ferrite).*

Etched cross-sections of P92 after different heat treatments.

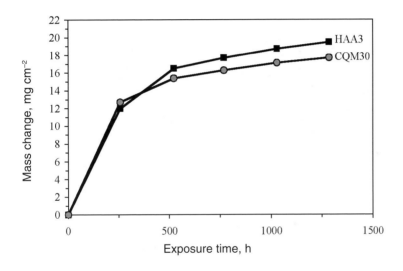

Fig. 25 *Mass changes of martensitic (CQM30) and ferritic P92 (HAA3) during exposure at 650°C in Ar-50vol.% H$_2$O.*

As already indicated in Fig. 21, the voids within the scale seem to nucleate at boundaries between different oxide phases in the scale and in the internal oxidation zone, e. g. between Fe$_3$O$_4$ and the Cr rich precipitates of (Fe, Cr, Mn)$_3$O$_4$. As the latter mainly result from oxidation of carbides on grain boundaries and/or martensite laths, the original alloy microstructure can have a substantial effect on void morphology (Figs 28, 29) and consequently on the long term spallation mode.

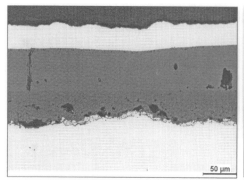 

Fig. 26 CQM30 (martensite).   *Fig. 27* HAA3 (ferrite).

Cross-section of P92 after martensitic and ferritic heat treatment and subsequent oxidation for 1250 h at 650°C in Ar-50vol.% H₂O.

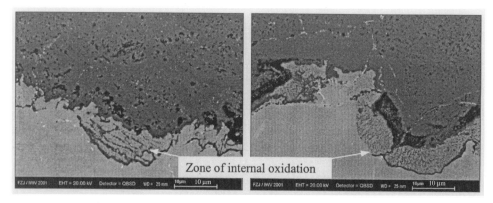

Fig. 28 CQM30 (martensite).   *Fig. 29* HAA3 (ferrite).

High magnifications of scale/alloy interfaces from specimens shown in Figs 26 and 27 respectively.

4. Conclusions

During testing of ferritic–martensitic 9–10% Cr steels in water vapour, extra- and interpolation of results obtained from relatively short term data can lead to erroneous conclusions in respect to the long term behaviour of materials. Scale spallation, which can have a substantial effect on the overall material loss rate, in some cases starts to occur only after several thousand hours of exposure. Consequently, no critical scale thickness for the onset and mode of spallation can be defined. The onset of spallation seems to depend on the oxidation temperature and the occurrence of the void morphology developing in the scale. The latter is clearly associated with interfaces between various oxide phases, e. g. between the Cr-rich precipitates of $(Fe, Cr, Mn)_3O_4$ spinel and the 'Fe_3O_4 matrix' in the inner scale. As the Cr-rich spinel precipitates result from oxidation of carbides on grain boundaries and martensite laths, alloy microstructure can have a substantial effect on void formation and crack initiation in the oxide scale.

References

1. K. Weizierl, Kohlekraftwerke der Zukunft, *VGB Kraftwerkstechnik*, 1994, **74** (2), 109.

2. R. Blum, J. Hald, W.Bendick, A. Rosselet and J. C. Vaillant, Neuentwicklungen hochwarmfester ferritisch-martensitischer Stähle aus den USA, Japan und Europa. *VGB Kraftwerkstechnik*, 1994, **74** (8), 641–652.

3. J. B. Marriott and K. Pithan, *A Review of the Properties of 9–12% Cr Steels for use as HP/IP Rotors in Advanced Steam Turbines*. Office for Official Publications of the European Communities, Catalog number, CD-NA-1187-EN-C (1989).

4. M. Thiele, H. Teichmann, W. Schwarz and W. J. Quadakkers, Corrosion behaviour of ferritic and austenitic steels in the simulated combustion gases of power plants fired with hard coal and brown coal, *VGB Kraftwerkstechnik*, 1997, **77** (2), 129–134.

5. A. Rahmel and J. Tobolski, Einfluß von Wasserdampf und Kohlendioxid auf die Oxidation von Eisen und Sauerstoff bei hohen Temperaturen, *Corros. Sci.*, 1965, **5**, p. 333.

6. Y. Ikeda and K. Nii, Mechanism of accelerated oxidation of Fe–Cr-alloys in water vapour containing atmosphere, *Trans. Jpn Inst. Met.*, 1984, **26**, 52.

7. W.J. Quadakkers, M. Thiele, P. J. Ennis, H. Teichmann and W. Schwarz, Application limits of ferritic and austenitic materials in steam and simulated combustion gas of advanced fossil fuelled power plants, in *Proc. EUROCORR '97*, Trondheim, Norway, NTNU Gløshaugen N-7034, 1997, Vol II, pp.35–40.

8. R. J. Ehlers, Oxidation von ferritischen 9–12%Cr-Stählen in wasserdampfhaltigen Atmosphären bei 550 bis 650°C, Diss. RWTH Aachen, Feb. 2001, Germany.

9. M. Schütze, *Protective Oxide Scales and their Breakdown*. Wiley, Chichester, UK, 1997.

10. M. Thiele, W. J. Quadakkers, F. Schubert and H. Nickel, Hochtemperaturkorrosion von ferritischen und austenitischen Stählen in simulierten Rauchgasen kohlebefeuerter Kraftwerke; *Berichte des Forschungszentrums Jülich*, Jül-3712, ISSN 0944-2952, Jülich, FRG, 1999.

11. H. Nickel, Y. Wouters, M. Thiele and W. J. Quadakkers, The effect of Water Vapour on the Oxidation Behaviour of 9%Cr Steels in Simulated Combustion Gases, 9. Tagung Festkörperanalytik, Chemnitz 23–26 June 1997, Proceedings in *Fresenius' J. Anal. Chem.*, 1998 (361) 540–544.

13

Understanding the Breakaway Corrosion of Ferritic Stainless Steels in Water Vapour

A. GALERIE, S. HENRY, Y. WOUTERS, J.-P. PETIT*, M. MERMOUX*,
C. CHEMARIN* and L. ANTONI†

Laboratoire de Thermodynamique et de Physicochimie Métallurgiques, UMR CNRS/UJF/INPG 5614,
Ecole Nationale Supérieure d'Electrochimie et d'Electrométallurgie de Grenoble,
BP 75 F-38402 Saint Martin d'Hères Cedex, France
*Laboratoire d'Electrochimie et de Physicochimie des Matériaux et des Interfaces,
UMR CNRS/UJF/INPG 5631
†Usinor Recherche et Développement – Centre de Recherches d'Ugine,
Avenue Paul Girod, F-73403 Ugine, France

ABSTRACT

It is known that the lifetime of stabilised ferritic stainless steels can be reduced by breakaway corrosion in presence of water vapour. After a passive period where a thin chromia scale protects the steel, massive iron-containing oxide nodules appear and rapidly cover the surface, leading to huge weight gains and reduced life.

Using photoelectrochemical and Raman spectroscopy experiments, the initial growth of iron oxides was shown to be located at the alloy-chromia scale interface, resulting from the inward transport of OH particles and reaction with the dechromised substrate. The necessary thermodynamic modifications to induce iron oxidation suggest that nodule nucleation takes place where local metal–oxide decohesion occurs. The competitive growth of iron and chromium oxides is to likely be controlled by the relative acidic properties of these oxides, as OH species should dissociate more rapidly on the most acidic oxide (Fe_2O_3).

In addition, systematic measurements of the time to breakaway as a function of temperature, H_2O pressure, chromium and silicon contents of the steels have been performed. The results are shown to obey a simple law which could allow the life time in any conditions to be predicted.

1. Problem Presentation

Chromium-containing ferritic stainless steels are commonly used for their good behaviour in oxidising environments at high temperatures, which results from the formation of a thin, adherent and impervious chromia-rich scale. They are, for example, largely used in automotive exhaust systems and their use will increase in the near future. However, they can suffer rapid degradation in pure water vapour, as iron rich oxides may develop, forming big nodules, firstly on edges and ridges of the samples, later on the flat areas. These nodules rapidly coalesce, leading to catastrophic oxidation [1,2]. It was observed that the nodules develop both inwards and outwards, forming nearly pure hematite Fe_2O_3 in their outer part and iron-chromium spinel $(Fe,Cr)_3O_4$ in their inner part (Fig. 1).

Nodule formation is associated with a rapid increase of the reaction rate (breakaway) and the initial slow parabolic rate law is followed by much more rapid quasi-linear kinetics (Fig. 2).

The purpose of this paper is to propose a mechanism for nodule nucleation and growth and to study the influence of temperature, water vapour pressure and composition on the life time of the steels.

2. Materials

Five ferritic stainless steels were prepared by Centre de Recherches d'Ugine with different chromium (12%, 15%, 18%) and silicon (~0%, 0.5%, 1.0%) contents, as presented in Table 1. All steels contained 0.2 % Ti as a ferritic stabiliser ensuring stability of the bcc lattice structure at any temperature. They were cold rolled into sheets of ~1 mm thick and cut into rectangular samples, generally 2×1.5 mm, which were SiC polished to the grade 1200 prior to oxidation. The steel containing 15%Cr and 0.5%Si was used for the determination of oxidation mechanisms, the other ones mainly for life time evolutions studies.

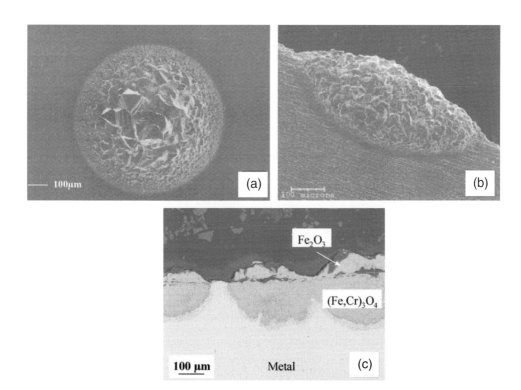

Fig. 1 *SEM observations of nodules formed on the Fe–15%Cr–0.5%Si–0.2%Ti ferritic steel after 50 h oxidation at 900°C. Atmosphere: Ar + 150 mbar H_2O. (a) nodule on a flat area; (b) nodule on a sample ridge; (c) cross-section of a flat area.*

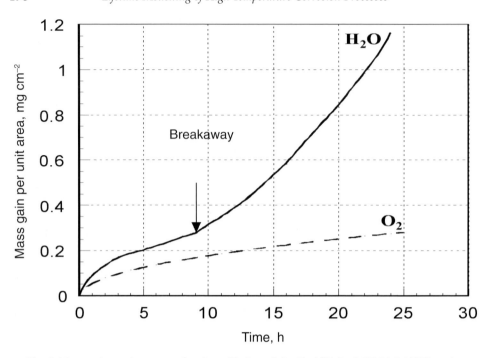

Fig. 2 *Mass gain vs time curve for the oxidation of the Fe–15%Cr–0.5%Si–0.2%Ti steel at 900°C under 150 mbar H₂O pressure showing breakaway oxidation. Comparison with oxidation in oxygen where no breakaway occurs.*

Table 1. *Chromium and silicon contents of the ferritic stainless steels prepared for the present work*

	12% Cr	15% Cr	18% Cr
~ 0% Si		X	
0.5% Si	X	X	X
1.0% Si		X	

3. Nodule Nucleation and Growth

3.1. Evidence for Nodule Nucleation at the Alloy–Scale Interface

Although the formation and growth of nodules is described in the literature, no discussion of the location of their nucleation in the corrosion scale is available. In our view, this location is one of the keypoints when trying to discuss the mechanism of nodule appearance. A series of oxidation experiments was therefore undertaken with the Fe–15%Cr–0.5%Si–0.2%Ti steel at 900°C under 150 mbar H₂O for different exposure times and the nature of the oxidation products in the scale was investigated by photoelectrochemistry and Raman spectroscopy experiments.

Typical photoelectrochemical energy spectra of thin oxide scales (thickness ~ 0.15 and ~0.30 µm), grown in water vapour are presented in Fig. 3. It can be observed that four contributions to the photocurrent appear with increasing light energy. Their thresholds are as follows : Eg_1 ~2.2 eV, Eg_2 ~2.5 eV, Eg_3 ~3 eV, Eg_4 ~3.5 eV. From bandgap data reported in the literature [3,4], it can be concluded that the first contribution corresponds to Fe_2O_3, whereas contributions 3 and 4 are signatures of Cr_2O_3. The photocurrent starting at ~2.5 eV is attributed to the presence in the scales of $(Fe,Cr)_2O_3$, solid solution of hematite and chromia, which is expected to have a bandgap value between that of Fe_2O_3 and Cr_2O_3. Figure 3 also shows that the ratio between the photocurrent due to chromia and that due to the other phases increases with increasing scale thickness. This suggests that, after the very early initial stage, the oxidation corresponds mainly to chromia growth.

In addition, the various phases observed by photoelectrochemistry were also detected by Raman spectroscopy. Raman imaging measurements were performed on several oxidised samples, in particular for low oxidation durations. Figure 4 shows the optical image of the surface of a Fe–15%Cr–0.5%Si–0.2%Ti steel exposed to H_2O at 900°C for 2 min and the Raman images corresponding to hematite, chromia and their mixed oxide. Let us first note that the various Raman images reproduce perfectly the optical image. It is also clearly seen that the various phases are not distributed in an homogeneous way on the surface of the oxide scale. One can notice in particular, that chromium oxide is not present in all areas of the scale for such a short oxidation treatment, as previously observed by other techniques [5].

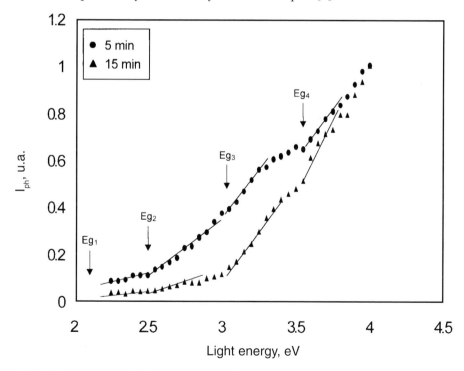

Fig. 3 *Photocurrent vs light energy spectra of oxide films grown during 5 and 15 min on the Fe–15%Cr–0.5%Si–0.2%Ti steel at 900°C under 150 mbar H_2O.*

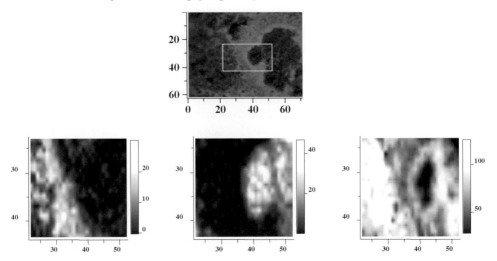

Fig. 4 *Optical image of a thin oxide scale grown at 900°C in H_2O atmosphere. Growth time: 2 min. Raman images of the spatial distribution of Cr_2O_3 (left), Fe_2O_3 (middle), $(Fe,Cr)_2O_3$ (right). X and Y dimensions are in μm. Raman images were recorded with the 514.5 nm laser line of an Argon ion laser under a ×100 microscope objective.*

Considering now thicker scales, particular attention was paid to oxides grown just before the appearance of the first nodule on the flat areas of the samples. For this study, oxidation conditions of 900°C, 150 mbar H_2O were chosen, where the chromia scale was ~3 μm. For that thickness, nodular corrosion was already observed on sample edges and ridges. The photoelectrochemical spectrum recorded on such a sample is presented in Fig. 5 and contains the two contributions of external (3.5 eV) and internal (3.0 eV) chromia, as previously reported for the oxide scale grown on pure chromium [4]. The interesting point is the significant reappearance of the two low energy photocurrent contributions, the first at ~2 eV attributed to Fe_2O_3 and the second at ~2.5 eV possibly due to the Fe_2O_3–Cr_2O_3 solid solution. As these contributions appear for thick scales after having previously disappeared, it can be postulated that they are related to one or several nodules at the beginning of their growth.

Raman spectroscopy imaging was performed on cross sections of the same type of samples. As observed in Fig. 6, small Fe_2O_3-rich domains are indeed detected near the alloy-scale interface, corresponding to nodule nucleation at this interface.

3.2. Proposed Mechanism

3.2.1. Nodule nucleation
Only nodules appearing on the flat areas of the samples are considered here. Nucleation of nodules at sample edges and ridges is clearly the result of scale crack appearance and is not discussed here.

Considering the respective oxygen partial pressures of the Fe_2O_3/Fe and Cr_2O_3/Cr systems in the temperature range studied, it is not possible for hematite to nucleate

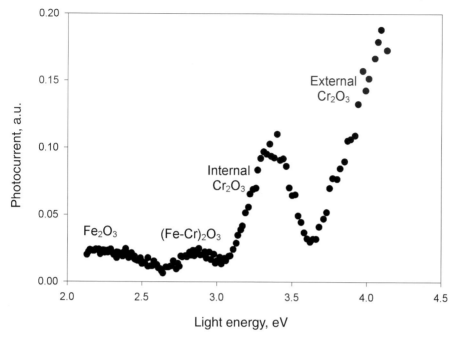

Fig. 5 *Evolution of the photocurrent vs energy curve for a ~3 mm thick oxide scale grown at 150°C under 150 mbar H_2O on the Fe–15%Cr–0.5%Si–0.2%Ti steel.*

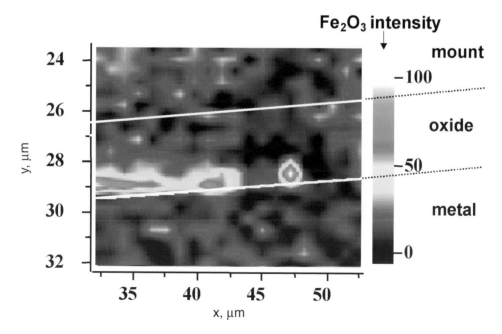

Fig. 6 *Raman spectroscopy imaging of Fe_2O_3 on the cross-section of a ~3 mm thick oxide scale grown at 900°C under 150 mbar H_2O on the Fe–15%Cr–0.5%Si–0.2%Ti steel.*

at the alloy/chromia interface. Such a nucleation needs a change in the thermodynamic state of the interface, for example the generation of a void where the oxygen partial pressure locally increases to a value in equilibrium with the gas phase. Void appearance can result from vacancy injection or mechanical decohesion.

3.2.2. Nodule growth

Contrary to nodule nucleation, which is controlled by thermodynamics and is not influenced by the nature of the oxidant, nodule growth is directly connected to the presence of H_2O in the gas phase. As previously discussed by several of us in the case of the oxidation of chromium [4], the growth of chromia scales in water vapour takes place (at least partially) by inward transport of hydroxide species arising from H_2O dissociative adsorption on the chromia surface. How these species can cause the further growth of hematite nodules, instead of forming passive chromia again, depends on their respective decomposition rates on these two oxides. It was shown recently [6] that the main parameter influencing this rate is the acidity of the oxide. Acidity of Fe_2O_3 and Cr_2O_3 can be expressed in different ways but, in any case, the figures corresponding to Fe_2O_3 always show higher acidity for this compound compared to Cr_2O_3 (Table 2).

It is therefore easy to understand the catalytic effect of Fe_2O_3 on OH decomposition leading to the fast growth of this oxide compared to Cr_2O_3. After a certain time, the Fe_2O_3 nodule can lift up and crack the passive chromia scale, and its rapid external growth can continue due to fast transport of Fe^{3+} cations but internal growth is also observed leading to the formation of chromium-containing magnetite $(Fe,Cr)_3O_4$.

3.2.3. Control of nodule growth by atmosphere changes

The mechanism described provides an understanding of why steels that are likely to suffer nodular oxidation in water vapour do not lead to any nodule formation in oxygen. In this gas, voids which may form between the alloy and the passive chromia scale are filled with O_2 which regenerates Cr_2O_3 as happens at the beginning of the exposure. Our measurements [12] show that the minimum chromium concentration

Table 2. Comparative acidity figures for Fe_2O_3 using different acidity definitions or scales

Acidity definition or scale		Fe_2O_3	Cr_2O_3
Solution definitions	pK_a of cation hydrolysis [7]	2.2	4.0
	pH of hydroxide precipitation [7]	1.9	3.6
	Cation hydration enthalpy (kJ mol^{-1}) [8]	−4061.6	−3897.2
	Point of zero charge (PZC) [8]	7.0	7.5–8.1
	Isoelectric point (IEP) [10]	5.4–6.9	7.0
Non-solution definitions	Cation ionic potential z/r^+ (nm^{-1}) [8]	4.35	3.97
	Smith's acidity [11]	−1.7	−5.2

of the dechromised zone of the Fe-15%Cr–0.5%Si–0.2%Ti steel never fell below ~13%, which corresponds to a still passivable material.

In addition, changing the atmosphere from H_2O to O_2 after the nodular corrosion had appeared, led to a rapid deceleration of the reaction rate (Fig. 7) and a blocking of matter transport at the nodule/alloy interface by a new passive chromia scale (Fig. 8). This shows again evidence for the particular effect of H_2O compared to O_2.

4. Influence of the Various Parameters Upon Lifetime

In the light of the proposed mechanism, the following important parameters controlling the more or less rapid appearance of nodular corrosion have been identified:

- Oxidation temperature,

- Water vapour pressure, both influencing the diffusion coefficient of OH species in chromia,

- Chromium content of the steel playing a role in the composition of the dechromised surface of the alloy beneath the passive chromia scale, and

- Silicon content of the steel making more or less easy the formation of an internal silica film protecting the steel substrate from OH species.

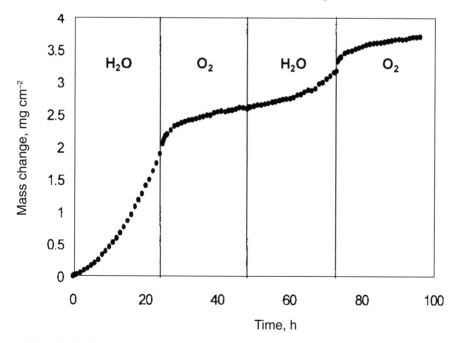

Fig. 7 *Influence of atmosphere changes on the oxidation kinetics of the Fe–15%Cr–0.5%Si–0.2%Ti steel.*

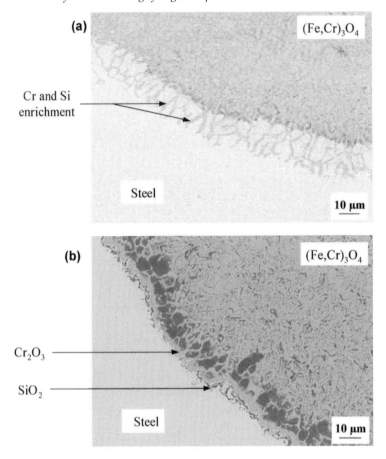

Fig. 8 *SEM observations of the cross-section of a nodules grown at 900°C: (a) during 24 h in water vapour; (b) same treatment, followed by a subsequent oxidation period of 24 h in oxygen.*

The influence of these parameters on the lifetime are discussed and a simple mathematical model is described allowing the lifetime in any condition to be predicted.

4.1. Lifetime Definition

As the oxidation in the passive regime correctly follows a parabolic rate law, the end of lifetime is taken at the point where the experimental law deviates from the parabola (see Fig. 2). At this point, nodules are only appearing on sample edges and ridges and are caused by local cracking of the passive scale. The lifetime for flat areas therefore is underestimated by such a definition.

4.2. Influence of Temperature

The graph of lifetime (LT) vs temperature (T) presented in Fig. 9 suggests a decreasing exponential shape of the form:

$$LT = \alpha \times \exp\left(\frac{\beta}{T}\right)$$

where α and β are constant factors.

Assuming the factor β can be written as:

$$\beta = \frac{E_{LT}}{R}$$

the activation energy E_{LT} lies near 213 kJ mol^{-1}. Considering the uncertainty on the lifetime determination, this value can be considered to be not too far from the measured activation energy on the parabolic rate constant (270 kJ mol^{-1}). This could possibly be connected with a constant thickness at the end of the lifetime. After calculations, it is indeed observed that this thickness is about 1.5 µm in the range 850–1000°C for the Fe–15%Cr–0.5%Si–0.2%Ti steel under 150 mbar H$_2$O.

4.3. Influence of H$_2$O Pressure

With the same idea of nodule nucleation occurring after a certain extent of reaction has been reached, water vapour pressure should have an influence through the concentration C_{OH} of OH species at the external interface of the chromia scale. The observation in Fig. 9 that lifetime is a decreasing function of H$_2$O pressure in the pressure range studied is qualitatively in good agreement but the linear shape of the evolution is more difficult to understand.

4.4. Influence of Chromium and Silicon Contents

As described in the model, the nucleation and growth of iron oxide takes place at the alloy-chromia interface and it is evident that the chemical composition of the interface plays a major role. Increasing the chromium content of the alloy leads to a higher steady-state surface concentration of the dechromised zone forcing chromia formation to occur at a higher rate and thus inhibiting nodule nucleation. Likewise, increasing the silicon content allows the development of a thicker and/or more complete silica film between the alloy and the chromia. Void formation and direct contact between hydroxide species and the alloy are more difficult and nodules nucleate and grow slower. Lengthening of lifetime with increasing both Cr and Si contents is clearly depicted in Fig. 10, but the physical meaning of the exponential shape of the curves is less evident than in the case of the temperature influence.

5. Mathematical Model for Lifetime Prediction

In order to try to predict the lifetime at any temperature, H$_2$O pressure and chromium or silicon content of the steel, a simple mathematical model was developed using four exponential functions of these parameters, deduced from the results outlined in Section 4. This choice was governed by the fact that three of the four parameters

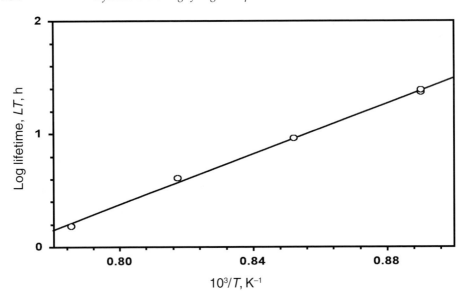

Fig. 9 *Evolution of lifetime with temperature for the Fe–15%Cr–0.5%Si–0.2%Ti steel oxidised under 150 mbar water vapour.*

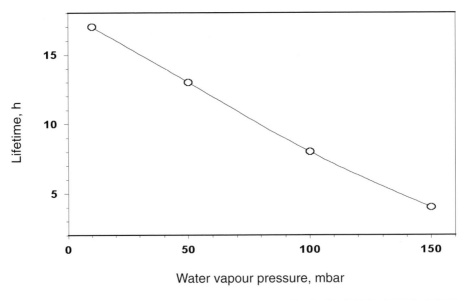

Fig. 10 *Evolution of lifetime with water vapour pressure for the Fe–15%Cr–0.5%Si–0.2%Ti steel oxidised at 950°C.*

influenced lifetime through an exponential function. Using a first order model, a good reliability between calculated and measured life time was found, as observed in Fig. 11. This could mean that no interaction exists between the four parameters under study, which is clear for temperature and pressure but less evident for Cr and Si contents, which both influence the nature of the contact between hydroxide species and the steel.

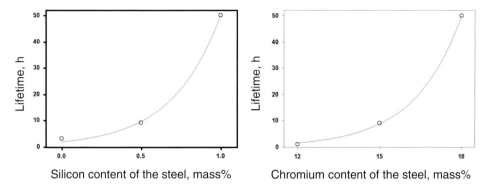

Fig. 11 *Evolution of lifetime with chromium and silicon contents of the steels. Temperature: 900°C, water vapour pressure: 150 mbar.*

6. Conclusions

Photoelectrochemical and Raman characterisation experiments have led to the conclusion that nodular corrosion of ferritic stabilised stainless steels in water vapour is initiated at the steel/chromia interface by local thermodynamic modifications followed by rapid kinetics of iron oxide(s) growth. Using the four relevant parameters — temperature, water vapour pressure, %Cr, %Si — a mathematical model was developed to predict the lifetime in any condition.

6. References

1. I. Kvernes, M. Oliveira and P. Kofstad, *Corros. Sci.*, 1977, **17**, 237–252.
2. S. Henry, A. Galerie, L. Antoni and J.-P. Petit, *Proc. 5th Int. Symp. on the Corrosion and Protection of Materials at High Temp.*, Les Embiez, May 2000, High Temperature Corrosion and Protection of Materials, *Mater. Sci. Forum*, 2001, **5**, 353–360.
3. N. Sato, *Electrochemistry of Metal and Semiconductors Electrodes*. 1998, Elsevier, Amsterdam.
4. S. Henry, J. Mougin, Y. Wouters, J. -P. Petit and A. Galerie, *Mater. High Temp.*, 2000, **17**, (2), 231–235.
5. I. Saeki, H. Konno and R. Furiuchi, *Corros. Sci.*, 1996, **38**, (1), 19–31.
6. A. Galerie, Y. Wouters and M. Caillet, *Proc. 5th Int. Symp. on the Corrosion and Protection of Materials at High Temp.*, Les Embiez, May 2000, High Temperature Corrosion and Protection of Materials, *Mater. Sci. Forum*, 2001, *Mater. Sci. Forum*, 2001, 231–238.
7. R. N. Smith and A. E. Martell, *Critical Stability Constants*. Plenum Press, New York and London, 1976.
8. J. E. Huheey, E. A. Keiter and R. L. Keiter, *Inorganic Chemistry*. HarperCollins College Publishers, New York and London, 1993.
9. P.-E. Dubois, PhD thesis, Institut National Polytechnique de Grenoble, 2000 (in French).
10. G. A. Parks, *Chem. Rev.*, 1965, **65**, 177.
11. D. W. Smith, J. *Chem. Educ.*, 1987, **64**, 480–481.
12. S. Henry, PhD thesis, Institut National Polytechnique de Grenoble, 2000 (in French).

14

The Influence of Water Vapour and Silicon on the Long Term Oxidation Behaviour of 9Cr Steels

F. DETTENWANGER, M. SCHORR, J. ELLRICH, T. WEBER and M. SCHÜTZE

Karl-Winnacker-Institut der DECHEMA e.V., Theodor-Heuss-Allee 25, D-60486 Frankfurt am Main, Germany

ABSTRACT

In order to provide oxidation resistance to Cr-steels, the Cr content in the alloy must be above a critical limit. Since recently developed 9% Cr steels are close to that limit their oxidation behaviour is a critical issue concerning service lifetime and reliability of structural components. In the present paper the influence of water vapour and the alloying element Si on the oxidation behaviour of 9% Cr steels was investigated. Seven different alloys of the ferritic–martensitic 9% Cr steel type were discontinously oxidised at 650°C for more than 7 000 h in dry air and in air containing 4% and 10% water vapour. The results show that water vapour significantly accelerates oxidation of these steels and that the Si-content of the alloys is critical for the oxidation performance, especially under wet conditions. The influence of growth stresses and Cr depletion on the long term oxidation behaviour is also discussed.

1. Introduction

In recent years the microstructure and alloy composition of the heat-resistant, ferritic–martensitic steels containing 9% Cr have been optimised for improved creep strength so that increased service temperatures are now taken into consideration [1–7]. These alloys are used in power plants, waste incineration, gasification plants and in a number of processes in the chemical and petrochemical industries. They allow heavy section components to be designed using solely ferritic materials, which have a higher thermal conductivity than the austenitic grades. However, due to the rather low amount of Cr present in the alloy, oxidation resistance may become the major life-limiting factor for these alloys. This is particularly true if lifetimes up to 100 000 h are envisaged. The oxidation behaviour of 9–12% Cr steels has been investigated extensively but several questions are still open or under discussion [8–12]. The aim of the present project is to investigate the influence of water vapour and the role of alloying additions (especially Si, W, and Mo) on the oxidation behaviour of 9% Cr steels with the intent of modelling their oxidation lifetime. The investigations focus on the two most important parameters for lifetime modelling of the steels, namely the time to breakaway and the kinetics before breakaway oxidation.

2. Experimental Procedures

The compositions of the alloys investigated are given in Table 1. The alloys are the two recently developed 9% Cr steels P91 and E911, and five laboratory versions of Nf616 with different Si-contents. Besides the Si content the alloy compositions varied with respect to the elements W and Mo, so that a possible influence of these elements on the oxidation behaviour of the steels may become evident as well. Prior to the oxidation tests the specimens were ground with 1200 SiC paper using an automated grinding machine to guarantee similar surface finishes.

In order to investigate the (quasi-)isothermal oxidation behaviour exposure tests were performed in dry air, and in air containing 4 and 10 vol.% water vapour for oxidation times of up to 7500 h. Mass change measurements were made after 300, 1000, 3500, 5500 and 7500 h. The samples were exposed at 650°C using a tube furnace with three quartz tubes, so that all three atmospheres (dry air, 4 and 10% water vapour content) were tested in the same furnace. The samples were hung in a quartz boat in a streaming atmosphere with a flow rate of about 5 Lh⁻¹.

In addition, isothermal and thermocyclic exposures in connection with acoustic emission measurements were performed for short exposure times up to 100 h. In the cyclic tests the specimens were kept for 24 h at 650°C, cooled to room temperature in about 30 min by removing the furnace from the sample and then heated up in about 7 min to 650°C by moving the hot furnace over the sample again.

3. Results and Discussion

3.1. Influence of Water Vapour and Si

The results from the discontinuous measurements of the normalised mass change as a function of oxidation time are shown in Figs 1–7. The three alloys Nf616-0, P91 and E911 showed that water vapour has a strong influence on the oxidation behaviour.

Table 1. Compositions of the alloys investigated (mass%)

Alloy	Cr	Mn	V	Ni	Mo	W	Si
E911	8.70	0.50	0.20	0.23	0.92	1.0	0.24
P91	9.00	0.55	0.22	0.36	0.93	0.00	0.40
Nf616-0	8.96	0.46	0.20	0.06	0.47	1.84	0.04
Nf616-1	8.55	0.51	0.20	0.01	0.48	1.75	0.11
Nf616-4	8.84	0.52	0.20	0.01	0.50	1.72	0.44
Nf616-6	8.98	0.53	0.21	0.01	0.51	1.72	0.57
Nf616-8	8.88	0.52	0.20	0.01	0.50	1.74	0.78

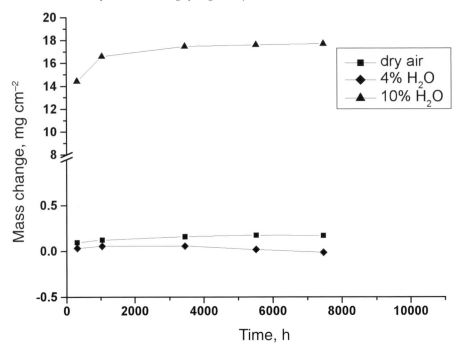

Fig. 1 *Influence of water vapour on the mass gain of E911,* T = 650°C.

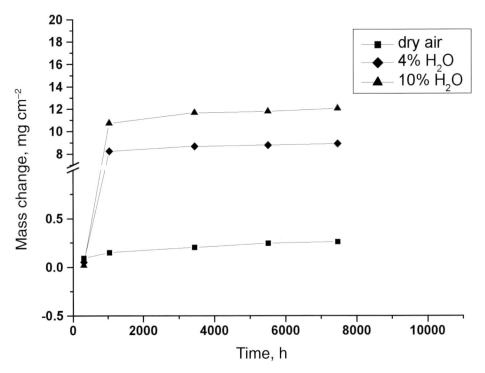

Fig. 2 *Influence of water vapour on the mass gain of P91,* T = 650°C.

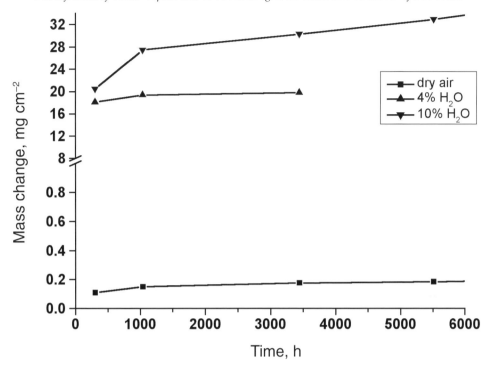

Fig. 3 *Influence of water vapour on the mass gain of Nf616-0*, T = 650°C.

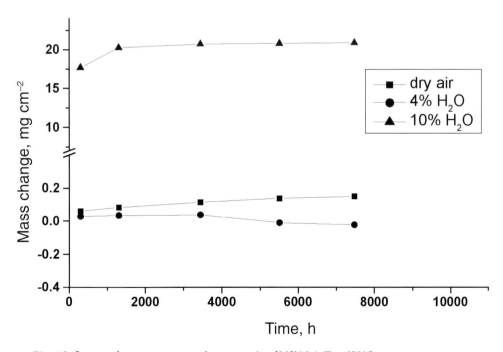

Fig. 4 *Influence of water vapour on the mass gain of Nf616-1*, T = 650°C.

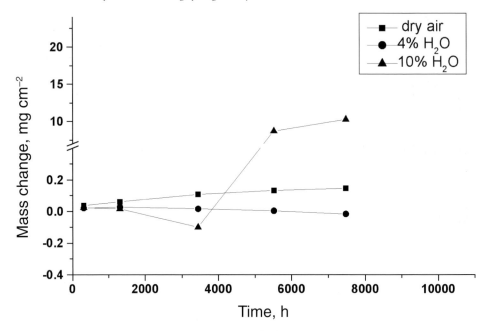

Fig. 5 *Influence of water vapour on the mass gain of Nf616-4,* T = 650°C.

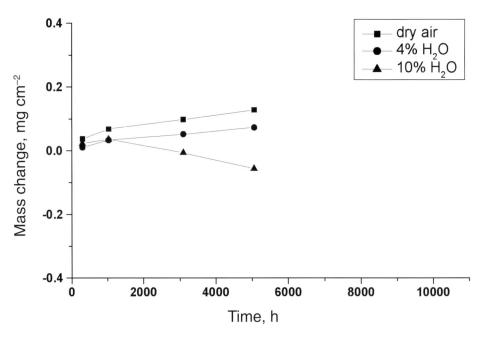

Fig. 6 *Influence of water vapour on the mass gain of Nf616-6,* T = 650°C.

All three alloys had much higher mass gains in the atmospheres containing water-vapour after a rather short oxidation time of several hundred hours. This is especially true for the alloy Nf616-0 (Fig. 3) with a Si-content of only 0.04%. This alloy exhibited

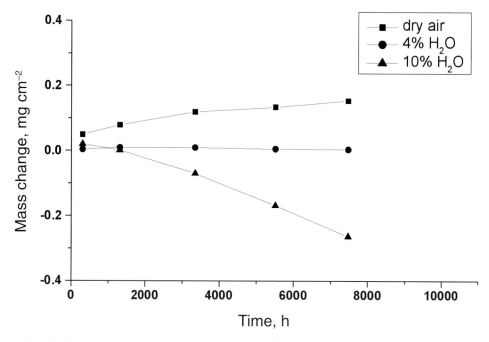

Fig. 7 *Influence of water vapor on the mass gain of Nf616-8, T = 650°C.*

a very short incubation time before breakaway oxidation and showed a mass gain of about 18 mg cm^{-2} after only 300 h in the presence of water vapour. An increase in the water vapour content led to an increase in the mass gain and to a reduction of the incubation period. The alloy E911 showed high mass gains for the atmosphere containing 10% water vapour. In the atmosphere containing 4% water vapor the alloy developed a protective chromia-based scale for oxidation times up to 5500 h. The triggering of breakaway oxidation due to the presence of water vapour is well known in the literature [e.g. 13,14].

The composition of the alloy and especially the Si content appears to be an important parameter concerning the susceptibility of the 9Cr steels to breakaway oxidation. This is demonstrated for the Nf616 series where the Si content was varied in a systematic manner. The time to breakaway increased with increasing Si content for the oxidation in moist atmospheres. The alloy containing 0.04% Si developed a non-protective oxide scale under the two moist conditions whereas increasing the Si content to as little as 0.11% resulted in protective oxide formation at a water vapour content of 4%. By further increasing the Si content to 0.44% the time to breakaway also increased for the 10% H$_2$O atmosphere to times between 4000 and 6000 h. For the high Si variants of Nf616 with 0.57 and 0.78% Si no breakaway oxidation occurred up to 7500 h either in dry or in moist atmospheres. Surprisingly, a significant mass loss was observed instead, after exposure to the water vapour containing atmospheres, especially for the atmosphere containing 10% water vapour and after longer oxidation times. Corresponding to this observation a yellow coloured powder was present at the end of the furnace tube. Analysis of this condensate in the SEM by EDX mainly revealed the presence of Cr and further chemical analysis using UV-spectroscopy showed the presence of dichromate. Thermodynamic calculations concerning the partial pressures of the possible chemical species formed under the exposure conditions were therefore performed using the CHEMSAGE-software and the corresponding data base [15]. The results are given in Table 2. It can be seen that only the dichromate has a significantly high partial pressure of about 10^{-9} bar and can therefore volatilise in significant amounts over the long time scale of the test. However, the absolute amount of volatilisation is rather small, particularly compared to the high mass gains after breakaway oxidation of Nf616-0, P91 and E911. For the Nf616 alloys with high Si contents the mass gain due to oxide formation was found to be very low so that even low mass losses from evaporation of volatile species are detectable. Whether the volatilisation of chromium or chromia by the formation of dichromate occurred only for the Nf616 types or also for the other alloys has not yet be determined, since the mass gains for the latter under water vapour-containing atmospheres were much higher than the mass losses due to evaporation. According to our calculations (Table 2), however, volatilisation of chromium-containing species generally has to be expected to occur to a lesser extent in the atmosphere containing 4% H_2O than in those containing 10% H_2O. Further calculations revealed that in pure water vapour as well as in dry air no volatilisation will occur. The evaporation of the hexavalent CrH_2O_4 is in agreement with recently published data of Asteman

Table 2. *Partial pressures of various components calculated for air containing 4% and 10% water vapour and water vapour containing 0.1% air at p = 1 bar, T = 650°C for Si, Cr and Cr_2O_3 as solid phases*

Component	Fugacity, bar, 4% H_2O	Fugacity, bar, 10% H_2O	Fugacity, bar, H_2O+0.1% air
N_2	7.7×10^{-1}	7.3×10^{-1}	9.99×10^{-1}
O_2	2.0×10^{-1}	1.8×10^{-1}	7.9×10^{-4}
H_2O	3.9×10^{-2}	9.2×10^{-2}	2.1×10^{-4}
NO	1.2×10^{-5}	1.1×10^{-5}	1.3×10^{-8}
HO	1.4×10^{-8}	2.1×10^{-8}	1.2×10^{-8}
CrH_2O_4	5.2×10^{-10}	1.1×10^{-9}	7.9×10^{-11}
HO_2	2.1×10^{-10}	3.0×10^{-10}	6.4×10^{-12}
CrO_3	1.0×10^{-11}	9.8×10^{-12}	6.2×10^{-14}
O	5.5×10^{-12}	5.2×10^{-12}	1.8×10^{-13}
H_2	6.4×10^{-13}	1.6×10^{-12}	5.0×10^{-10}
Others	$< 10^{-14}$	$< 10^{-13}$	$< 10^{-14}$

et al. [16] showing that significant chromium evaporation can occur in the presence of water vapor even at low temperatures of 600°C.

The experimental results regarding the time to breakaway as a function of Si content of the alloys are summarised in Fig. 8 for air + 10% H_2O. For the Nf616 steels at least three regions could be identified. The two alloys with the lowest Si content had a rather short incubation time whereas Si contents between about 0.2 and 0.5 mass% showed breakaway oxidation after 5000 and 6000 h respectively. Si contents above 0.5 mass% resulted in good oxidation behaviour for oxidation times well above 7500 h. Although the alloys P91 and Nf616-4 have comparable Si contents the oxidation behaviour of the alloys in atmospheres containing water vapour differs. Comparison of the oxidation behaviour of alloys E911and P91 with the Nf616 variants having a similar Si-content shows that the time to breakaway oxidation is not solely affected by the Si content of the steels. The Nf616 steels had a significantly longer incubation time compared to the P91 and E911 types. This indicates that the other alloying additions like W and possibly Mo may be also important regarding the formation of a protective chromia scale on the steels at 650°C since the microstructures of the base metals were comparable.

3.2. Influence of Growth Stress

The isothermal and thermocyclic oxidation tests of the alloys Nf616-0 and P91 at 650°C revealed another interesting aspect. Results from acoustic emission

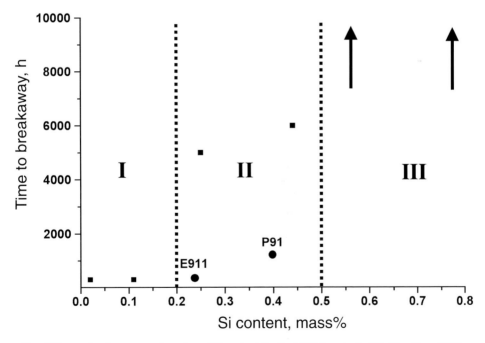

Fig. 8 *Time to breakaway as a function of Si content for the Nf616 variants (■). The alloys E911 and P91 (●) showed shorter times to breakaway compared to alloys of the Nf616 series with similar Si contents.*

measurements in isothermal tests for 100 h in dry air and in air containing 10% water vapour are shown in Figs 9 and 10. The alloy Nf616-0 showed some acoustic emission after about 65 h and stronger signals during the cooling stage, similar to the P91. However, under the moist conditions acoustic emission occurred during the whole oxidation test and was significantly higher for alloy P91. Similar results were obtained for the thermocyclic oxidation test under moist conditions in which the alloy P91 showed strong acoustic emission also during the isothermal stage (Fig.11). For Nf616-0 lower signals were obtained mainly during the cooling stage. These results indicate that the scale growth mechanism and especially the formation of growth stresses may play an important role in the oxidation behaviour of the 9Cr steels. It seems reasonable to assume that the scale growth stresses were higher on P91 than on Nf616, possibly due to the influence of alloy composition. Thus, the key for the explanation of the observed differences in oxidation behaviour of P91 and Nf616-4 (both about 0.4%Si) in air–10%H$_2$O may lie in the role of the oxide growth stresses. Therefore, *in situ* growth stress measurements in dry and moist atmospheres are presently being conducted using high temperature X-ray diffraction and monofacial oxidation deflection testing [16].

3.3. Influence of Cr Depletion

As already mentioned, the Cr content in the subsurface zone is of great importance for the oxidation behaviour of the 9Cr steels. According to Wagner's theory of alloy

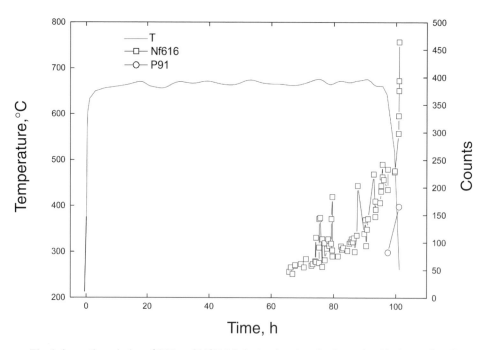

Fig. 9 *Acoustic emission of P91 and Nf616-0 during heating, isothermal oxidation and cooling in dry air.*

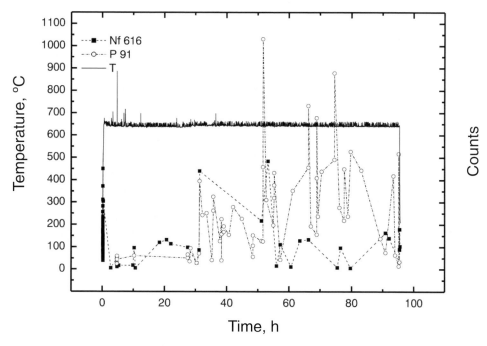

Fig. 10 *Acoustic emission of P91 and Nf616-0 during heating, isothermally oxidation and cooling in air containing 10% water vapour.*

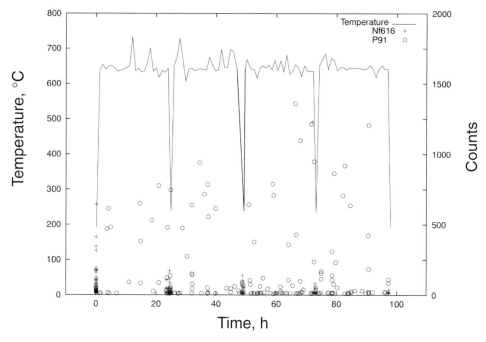

Fig. 11 *Acoustic emission of P91 and Nf616-0 during cyclic oxidation in air containing 10% water vapour.*

oxidation an alloy exhibits selective oxidation, i.e. it develops a stable, slowly growing oxide (e.g. Cr_2O_3) preventing rapid oxidation of the parent metal, if it contains a sufficiently high concentration of the corresponding solute (e.g. Cr) forming the external oxide scale. Based on this analysis two criteria or critical concentration values are of particular importance for protective scale formation. First, the solute concentration of Cr in the alloy has to be sufficiently high to produce an external, protective chromia scale and secondly, it has to be sufficiently high to maintain the growth of this external chromia scale. The latter criterion is usually not important for bulk conventional alloy components but becomes important for some intermetallic compounds or thin sections of conventional alloys as otherwise chemical failure of the protective oxide scale may occur. However, the critical Cr content for the external scale formation has to be available not only at the beginning of the oxidation process but also at later stages in the subsurface zone to prevent mechanically induced chemical failure due to potential mechanical breakdown of the external scale. Therefore, not only the starting level of the Cr content but also the time development of the Cr content beneath the oxide scale is an important parameter for the oxidation behaviour of the 9Cr steels. In the present work electron microprobe (EPMA) profiles along the metal/scale interface of metallographic cross sections were performed after the discontinuous oxidation tests to determine quantitatively the development of the Cr concentrations of the subsurface zone with oxidation time. Figures 12 and 13 show two examples of the correlation of mass gains with the Cr contents beneath the external oxide scale as a function of oxidation time with regard to protective oxide growth and breakaway oxidation. The critical Cr content in the Fe–Cr alloy was

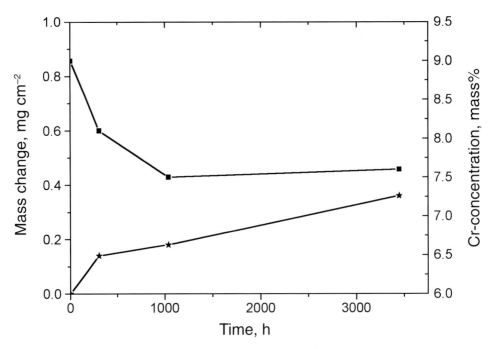

Fig. 12 *Normalised mass gain and Cr content at the metal/scale interface as a function of oxidation time, P91, 650°C, dry air.*

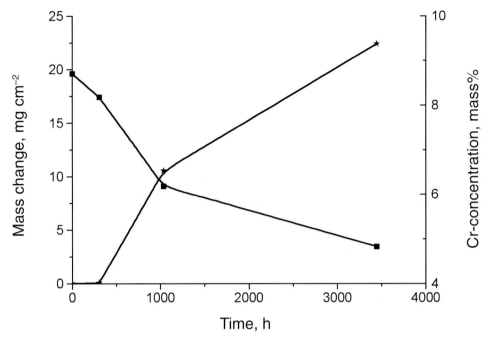

Fig. 13 *Normalised mass gain and Cr content at the metal/scale interface as a function of oxidation time, E911, 650°C, air + 10% H₂O.*

estimated according to Whittle's analysis at the prevailing temperatures to be of the order of 6.5 to 7.0 mass% [18,19]. The formation of the chromia-rich external oxide scale led to a depletion of the subsurface zone in chromium and the depletion was greater for the alloy E911 in air + 10% H_2O (showing breakaway oxidation) than for the alloy P91, which was oxidised in dry air and did not show accelerated oxidation kinetics. For E911 oxidised in moist air the Cr content dropped significantly below the estimated critical value whereas a value above 7 mass% was obtained for the P91 developing a protective oxide scale even after oxidation times longer than 3000 h. The mechanism responsible for the different depletion kinetics is not yet clear. However, knowledge of the kinetics of the Cr depletion is essential for life time prediction of the alloy. The drop of the initial Cr content N0 below the critical content NC will lead to mechanically induced chemical failure as soon as the oxide scale is damaged by external mechanical stresses, scale growth stresses, thermal stresses, etc. Further work will be focused on the time development of the depletion zone as a function of atmosphere and alloying elements. Tendencies observed in the present work are schematically summarised in Fig. 14. The development of the Cr content in the subsurface zone at a given temperature is mainly determined by the growth kinetics of the formed oxide scale which are a function of the atmosphere and alloying elements. For the high Si-containing alloys or the dry atmosphere the measured Cr content of the subsurface zone was above the estimated critical value even after long exposure times. Therefore, self-healing of scale defects can occur even after long exposures. Increasing the water vapour content of the atmosphere lead to shorter times to breakaway and faster depletion kinetics as illustrated in Fig. 14.

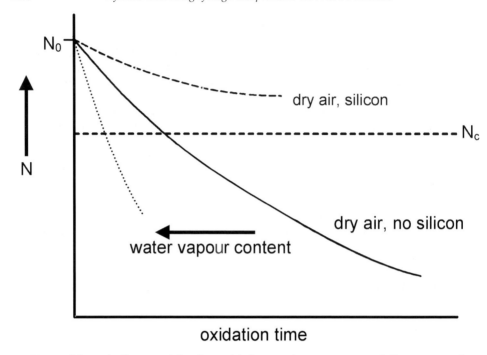

Fig. 14 *Schematic diagram of the observed influence of water vapour and Si content on the kinetics of Cr depletion for the Nf616 type alloys. N_0: initial Cr content; N_c: critical Cr content for protective chromia formation.*

4. Conclusions

The influence of water vapour content and Si on the oxidation behaviour of 9% Cr steels has been investigated at 650°C. The results show that water vapour significantly changes the oxidation behaviour and leads to breakaway oxidation, accompanied by high mass gains. However, the steels of type Nf616 with optimised composition, especially with regard to the Si content, showed much better oxidation resistance. Surprisingly, mass losses due to the formation of volatile Cr-containing species were observed for the moist conditions at 650°C. According to thermodynamic calculations evaporation of CrH_2O_4 can be expected after long oxidation times. The acoustic emission activity measured *in situ* differed for the steel types investigated, indicating a different growth mechanism and/or levels of growth stress in the oxide scales formed. The development of the Cr depletion occurring beneath the oxide scale is a critical parameter and the knowledge of its time dependence may be used for future lifetime models of 9Cr steels.

5. Acknowledgement

This work is financially supported by BMWi via Arbeitsgemeinschaft industrieller Forschungsvereinigungen e.V. (AiF) under contract no. 11581 N which is gratefully acknowledged.

References

1. R. Blum, J. Hald, W. Bendick, A. Rosselet and J. C. Vaillant, *VGB Kraftwerkstechnik* 1994, **74**, 641.
2. J. Stringer, *Heat-Resistant Materials II.*, ASM International, Materials Park Ohio, 1995, p.19.
3. M. Schirra and K. Ehrlich, *Werkstoffe für die Energietechnik.* DGM-Informationsgesellschaft, Frankfurt, 1996, p.89.
4. G. Kalwa, *Proc. Euromat '89* (**1**). DGM-Informationsgesellschaft, Oberursel 1990, p.561.
5. H. Mimura, M. Ohgami, H. Naoi and T. Fujita, *High Temperature Materials for Power Engineering 1990.* Kluwer Dordrecht 1990, p.485.
6. R. Hardt, *Werkstoffe für die Energietechnik*, DGM-Informationsgesellschaft, Frankfurt 1996, p.15.
7. K. H. Mayer, C. Berger, T. Kern and R. B. Scarlin, *Werkstoffe für die Energietechnik, DGM-Informationsgesellschaft, Frankfurt*, 1996, p.49.
8. K. Tamura, T. Sato, Y. Fukuda, K. Mitsuhata and H. Yamanouchi, *Heat-Resistant Materials II.* ASM International, Materials Park, Oh., 1995, p.33.
9. F. Schubert, M. Thiele, C. Williams and W. J. Quadakkers, *Werkstoffe für die Energietechnik, DGM-Informationsgesellschaft, Frankfurt*, 1996, p.57.
10. J. P. T. Vossen, P. Gawenda, K. Rahts, M. Röhrig, M. Schorr and M. Schütze, *Mater. High Temp.*, 1997, **14(4)**, 387.
11. M. Thiele, H. Teichmann, W. Schwarz and W. J. Quadakkers, *VGB Kraftwerkstechnik* **77**, 1997, 135.
12. H. Nickel, Y. Wouters, M. Thiele and W. J. Quadakkers, *ASME Pressure Vessels and Piping Conf.*, PVP-Vol. 359, Orlando, FL, 1997, p.269.
13. A. Rahmel and J. Tobolski, *Corros. Sci.*, 1965, **5**, 333.
14. P. Kofstad, *High Temperature Corrosion.* Elsevier Applied Science, London, 1988, p.105.
15. CHEMSAGE, GTT-Technologies, Herzogenrath, Germany.
16. H. Asteman, J.-E. Svensson, M. Norell and L.-G. Johansson, *Oxid. Met.*, 2000, **54**, 11.
17. M. Schütze, S. Ito, W. Przybilla, H. Echsler and C. Bruns, *Proc. High Temperature Corrosion and Protection 2000* (T. Narita, T. Maruyama and S. Taniguchi, eds). Hokkaido, Japan, 2000, p.19.
18. D. P. Whittle, *Oxid. Met.*, 1972, **4**, 171.
19. J. P. T. Vossen, P. Gawenda, K. Rahts, M. Röhrig, M. Schorr and M. Schütze, *Mater. High Temp.*, 1997, **14(4)**, 387.

15

Simulation of High Temperature Slurry-Erosion by an *in situ* Pulsed Laser Spallation Technique

R. OLTRA , J. C. COLSON, P. PASQUET and P. PSYLLAKI

Laboratoire de Recherche sur la Réactivité des Solides, UMR 5613 CNRS, Université de Bourgogne, B.P. 47870, 21078 Dijon Cedex, France

ABSTRACT

A pulsed laser has been used to remove oxide layers formed during oxidation of an iron-based alloy at various temperatures. This laser removal of the oxide layer was performed *in situ* on a specimen where the mass variation could be followed by a thermobalance and enables a new approach to the simultaneous oxidation and erosion of metals at high temperature to be defined.

1. Introduction

Erosion–corrosion reactions are normally followed by measuring the mass change of a metallic specimen as a function of the duration of the test [1]. But, to our knowledge, no continuous measurement of the oxidation kinetics has been performed to date since it is almost impossible to design a thermogravimetric cell within which the specimen could be eroded by slurries flowing in the gas stream, e.g. in a fluid bed or in an impinging jet.

Nevertheless, models have been proposed for describing the mass loss rate under simultaneous oxidation and erosion by solid particles [2]. Such models allow the identification of different regimes of erosion–corrosion, especially when the impacting particles are sufficiently energetic to remove the oxide film itself without ablation of the metallic substrate. However, these kinds of models are limited by the lack of knowledge about the initial rate of oxidation of the bare metal.

2. Experimental

In the Pulsed Laser Spallation Technique (PLST) a pulsed laser beam (Nd:YAG, λ = 1.06 µm, FWHM = 14.5 ns) is directed through a mask (dia. 3 mm) onto the corroding specimen, the oxidising surface of which is locally bared by a single laser pulse (Fig. 1), the laser fluence being fixed at 0.6 J cm^{-2} to avoid the ablation of the metallic substrate.

In this work, the corrosion experiments were carried on a Fe–2.25Cr–1Mo steel in a reacting chamber under an air pressure of 105 Pa and at temperatures ranging from 800 to 1000°C. Samples (8 × 4 × 0.5 mm) were cut from a rolled plate and then mechanically polished to a 1 µm diamond finish, degreased and dried in air. Energy

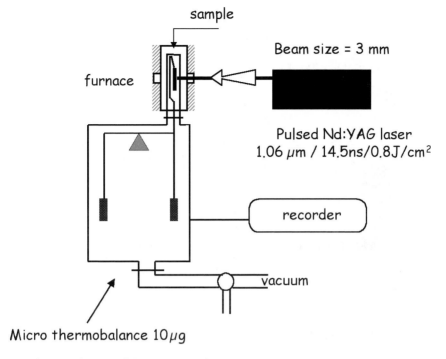

Fig. 1 *Schematic diagram of the experimental set-up.*

dispersive X-ray analysis (EDX) and scanning electron microscopy (SEM) were used to identify the corrosion products and to characterise the laser interaction.

The mass change of the tested specimen was measured by a thermogravimetric device based on a microthermobalance which was specially designed.

A conventional parabolic law [3] has been found for the corrosion regime and the chemical analysis of the oxide film suggests that it is a bilayer film which consists of an internal Fe_3O_4 layer covered by a thin layer of Fe_2O_3.

Typical thicknesses of the growing films were in the range of 5 to 30 μm depending on the temperature.

3. Laser-Material Interaction

Assuming that the oxide layer is a transparent medium to the laser wavelength (1.06 μm), then for the global description of the laser-material interaction the local thermal expansion of the metallic substrate must be taken into account. This leads to the forced bending of the oxide film (Fig. 3a), which could influence the structural integrity of the oxidised substrate in a number of ways:

(a) For laser energies lower than a critical value E_0 ($E< E_0$), the thermal expansion due to laser irradiation does not induce any damage in the system examined (regime of thermo-elastic interactions), and the film returns to its initial state without cracking (Fig. 3b).

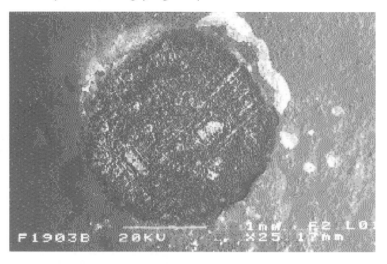

Fig. 2 *Laser impact showing the removed oxide film.*

(b) For laser energies higher than E_0 and lower than E_1 ($E_0 < E < E_1$), the laser irradiation induces a stress field in the system that leads to de-bonding at the centre of the irradiated area (regime of interfacial fracture). The size of the section of interface depends on the adhesion strength of the coated metal. Moreover, the elevated tensile stresses at the borders of the deformed interface induce crack initiation within the oxide layer, whilst the crack width depends on both the toughness and thickness of the oxide film (Fig. 3c).

(c) For laser energies higher than E_1 ($E_1 < E$), the high thermal expansion of the substrate leads to the rapid propagation of in-scale cracks at the borders of the irradiated film. The orientation of these cracks depends on both the deformation imposed by the expansion, as well as on the crystallographic characteristics of the oxide layer. The rapid propagation of interfacial cracks leads to the spallation of the irradiated oxide layer, whilst the stress relaxation taking place promotes the expulsion of the fragmented volume (Fig. 3d, regime of spallation–expulsion). This is the energy domain which was selected to simulate the erosion of the oxide layer.

4. Results

The experiments are based on a previous kinetic analysis developed to define the reaction rates by imposing constant conditions for the temperature and pressure of the reacting chamber [4,5]. These results have been summarised in Figs (4a) and (b), for a single pulsed laser removal of the oxidising surface.

A parabolic law (time-period(a), Fig. 4a), following the diffusion model described by Wagner [3], represents the oxidation regime in the absence of laser irradiation. This initial oxidation at 800°C for 40 min, is followed by a period (b) which corresponds to the oxidation regimes at various temperatures for 10 min, before the

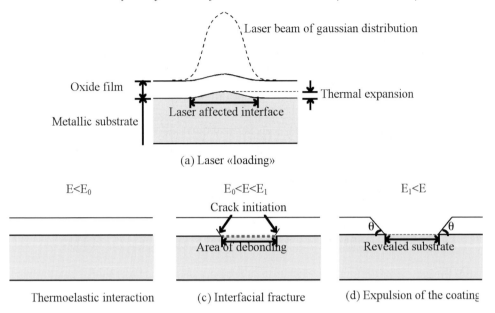

Fig. 3 *Schematic description of the different interactions between nanosecond pulsed laser irradiation and a transparent oxide layer grown on a metallic substrate: (a) Laser loading, (b) regime of elastic interactions, (c) regime of interfacial fracture and (d) regime of spallation–expulsion.*

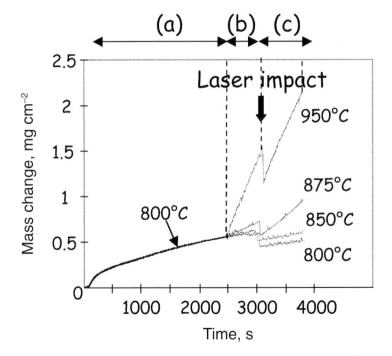

Fig. 4(a) *Plot of mass changes of a Fe–2.25Cr–1Mo steel vs time for isothermal oxidation.*

final simulation (period (c)) of the oxide film removal by a single pulsed laser irradiation. The main experimental results concern the change of the corrosion regime after the laser spallation for various oxidation rates controlled by the temperature during the period (c), i.e. the reoxidation of the bare metal surface during 15 min after laser spallation (Fig. 4b).

The mass loss of the specimen after the initial periods of corrosion correspond to the laser spallation of the oxide film grown on the surface.

From these experimental results, the values of the instantaneous rates of oxidation V_1 and V_2 defined respectively, for the unaffected surface and for the bare metal surface, have been estimated. From V_1 and V_2, it is possible to calculate the activation energy of the two oxidation regimes (Fig. 5), which gives 100 kJ mol^{-1} for the bare metal surface compared to 160 kJ mol^{-1} for the oxide film covered surface.

5. Discussion

From the experimental rate of oxidation of the bare metal surface, K_2, the total rate of corrosion of the metallic target can be calculated as a function of the temperature for the regime of erosion simulated by the laser removal. This approach could be used to predict the corrosion damage in the erosion enhanced oxidation regime in the presence of local erosion of the oxide film as shown by other authors [2].

The two corrosion regimes controlled respectively by the diffusion (cationic diffusion) and by the nucleation and growth, can be described by the following two equations:

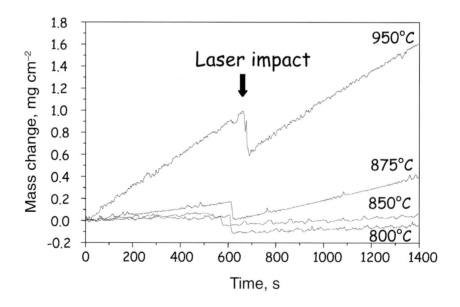

Fig. 4(b) *Plot of mass changes of a Fe–2.25Cr–1Mo steel after the laser pulse (periods b and c of Fig. 4a).*

(i) for the diffusion regime which corresponds to the non-disturbed surface of the specimen:

$$\Delta m = K_1 \, t^{1/2} + C \tag{1}$$

(ii) for the corrosion regime of the laser cleaned metallic surface:

$$\Delta m = K_2 \, t^{1/2} + C \tag{2}$$

in which C is a constant, K_2 cannot be directly defined from the experimental results since the nature of the oxide, which is growing on the cleaned surface, must be taken into account. Assuming that mainly Fe_3O_4 is formed, the mass ratio between the actual mass of oxide which is removed and the mass variation which is recorded just after the laser pulse, i.e. the mass of oxygen consumed during the re-oxidation of the metallic surface is equal to 3.62. A coefficient $\alpha = 3.62$ was then introduced in the relation (3) describing the mass-loss of a specimen of which a part is continuously bared by the laser impact:

$$\Delta m / S_0 = K_1 \, (1-(S_L / S_0))t^{1/2} - \alpha K_2 \, (S_L / S_0)t + C \tag{3}$$

where $S_0 = $ surface of the specimen and $S_L = $ continuously laser cleaned surface area.

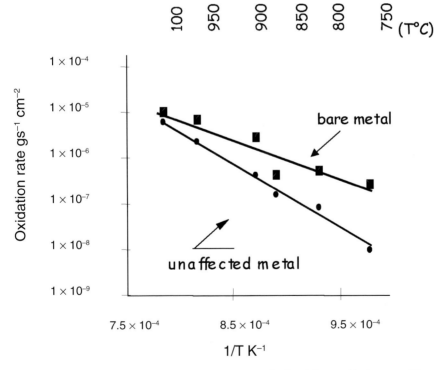

Fig. 5 *Activation energy of the oxidation reactions calculated from oxidation rates* V_1 *and* V_2 *corresponding to the experimental conditions of Fig. 4 (periods b and c).*

The relation (3) has been checked by comparing the theoretical calculation (relation (3)) with an experiment conducted in the same conditions as used previously (800°C) but with a laser impact regime fixed at 1Hz. The calculated mass variation has been compared with the experimental under-laser irradiation (Fig. 6 a and b): they are very similar thus demonstrating that the laser interaction does not have a large effect on the physical properties of the metal substrate. Consequently, as shown in Fig. 7, the model can be used to describe the erosion–corrosion regime for longer times. This variation is close to other simulated mass-loss curves presented in other studies [2].

This result confirms that the laser removal can be used to simulate corrosion-dominated regimes.

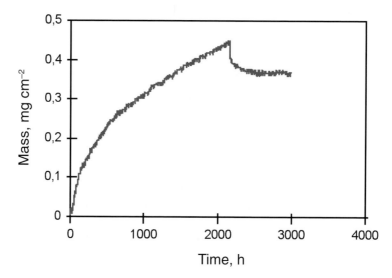

Fig. 6(a) *Mass variation during an experimental simulation of the erosion–corrosion of a Fe–2.25Cr–1Mo alloy at 800°C in air. After 2000 s laser impacts at 1 Hz.*

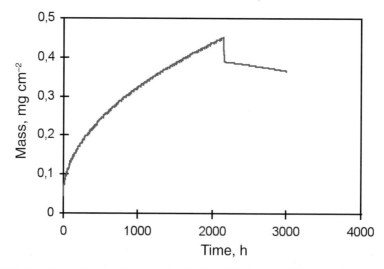

Fig. 6(b) *Calculated change of the mass for ($S_o/S_l = 0.1$). Same conditions as those in Fig. 6(a).*

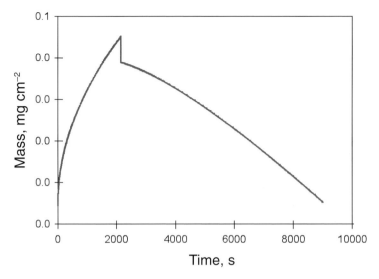

Fig. 7 *Calculated mass variation for a long period of laser spallation: laser Nd:YAG impacts at 1 Hz. Same conditions as those in Fig. 6(a).*

However, a lack of data concerning the optical properties of the oxide films which are growing precludes a complete validation of the mechanical model for the film removal. In other words, the removal of actual corrosion films grown at high temperatures could be partly due to ablation.

But the results obtained with 1Hz laser pulses demonstrate that the substrate does not seem to be grossly affected by the laser irradiation.

6. Conclusions

The application of a pulsed laser removal technique to simulate the oxidation-dominating regime during erosion–corrosion of a Fe–2.25Cr–1Mo alloy has been described.

The rate of corrosion of the bare metal surface simulating the area which is impinged by solid abrasive particles in case of slurry-erosion, was found to be higher than the corrosion rate of the non-disturbed metallic surface controlled by the diffusion regime.

Nevertheless, some work is needed to analyse the scale phase and composition of the bare metal surface after laser irradiation to validate the comparison of the pulsed laser spallation technique with the mechanical impact of solid particles.

Although improvements in our knowledge of the laser interaction with growing oxide films at high temperatures are needed, this study demonstrates that the combination of pulsed laser and conventional thermogravimetric measurements can, nevertheless, be used to simulate erosion-corrosion phenomena.

Recent experiments performed on ferritic stainless steels at room temperature, in exactly the same experimental configuration, i.e. mass-loss measurement with a microbalance [6], could be reproduced at high temperature in order to simulate the oxidation rate of stainless steels during slurry-erosion.

It can be suggested that other regimes, erosion dominating for example, can be also simulated by modifying the nature of the laser-oxide film interaction, using an ultra violet pulsed laser to perform ablation of the oxide film itself.

7. Acknowledgements

P. Pasquet and P. Psyllaki were supported by the TMR network CT88-0198 'Laser Cleaning'.

References

1. N. Birks, B. Patts and F. S. Pettit, *Trans. Mat. Res. Soc. Jpn*, 1994, **14**, 179.
2. I. G. Wright, V. K. Sethi and A. J. Markworth, *Wear*, 1995, **230**, 186–187.
3. C. Wagner, *Z. Phys. Chem.*, 1933, **B2**, 2542.
4. B. Delmon, *Bull. Soc. Chim Fr.*, 1961, 390.
5. J. C. Colson, D. Delafosse and P. Barret, *Bull. Soc. Chim. Fr.*, 1963, 687.
6. P. Psyllaki and R. Oltra, *Mater. Sci. Eng. A*, 2000, **282**, 145.

Investigation and Modelling of Specific Degradation Processes

1.3 Composites and coatings

16

Theoretical and Experimental Approach for Long Term Modelling of Oxidation and Diffusion Processes in MCrAlY Coatings

P. KRUKOVSKY, V. KOLARIK*, K. TADLYA, A. RYBNIKOV[†], I. KRYUKOV[†]
and M. JUEZ-LORENZO*

Institute of Engineering Thermophysics, 2a, Zhelyabov str., 03057 Kiev, Ukraine
*Fraunhofer-Institut für Chemische Technologie, Joseph-von-Fraunhofer Str. 7, 76327 Pfinztal, Germany
[†]Polzunov Central Boiler and Turbine Institute, Politechnicheskaya 24, 194021 St. Petersburg, Russia

ABSTRACT

A novel theoretical and experimental approach to life time modelling of MCrAlY coatings for stationary gas turbines has been undertaken using the Inverse Problem Solution (IPS) technique. With this technique feasible experimental data obtained after a defined experimental time τ_e are used as input values for the model parameters estimation. In the first stage of the approach a model is assumed, which considers the Al concentration profile across the coating. The measured Al concentration profiles are then used as input values for the estimation of the model parameters and a calculated prediction of the long term diffusion and oxidation behaviour of the coating is then performed.

Exposure experiments were carried out with an NiCoCrAlY coating containing 8% Al in air at 950°C, up to 1000 h. Additionally, *in situ* studies of the oxide formation were performed in the first 100 h by high temperature X-ray diffraction. At 950°C the coating forms a scale consisting of α-Al$_2$O$_3$ and, in the initial state, smaller amounts of θ-Al$_2$O$_3$. The oxide scale is growing continuously and no other oxides are observed. The concentration profiles of Al across the coating thickness were determined by electron microprobe analysis in the initial state and after 100, 300 and 1000 h of oxidation. The concentration profile measured after 300 h was used as input values for the estimation model parameters in order to calculate the oxide scale thickness and the Al concentration profile after 1000 h.

The model describes as a first simplified version with a high accuracy the experimentally determined aluminium depletion at the coating surface and at the coating substrate interface as well as the oxide scale thickness after 100 and 300 h. The computational prediction for 1000 h at 950°C is in good agreement with the measured data.

1. Introduction

In modern stationary gas turbines for electric power generation MCrAlY type coatings are used world-wide for protective overlay coatings against oxidation (M is Ni and/or Co) [1–6]. In future developments higher gas inlet temperatures are envisaged for a further increase of the turbine efficiency and for a more economic use of energy resources [1,2]. Higher temperatures however, lead to significantly increased oxidation and accelerated diffusion processes at the coating/substrate interface [3,4].

Although the MCrAlY coatings are thus getting closer to their limitations, a high reliability of their performance is of essential importance to fulfill the required lifetimes under the more severe conditions.

In the case of the MCrAlY coatings the protection against oxidation is achieved by the formation of a thin Al_2O_3 layer on the coating surface. The Al, which is mainly present as a β-NiAl phase in the coating, is consumed by Al_2O_3 formation at the surface and by interdiffusion at the substrate/coating interface, leading to aluminium depletion zones that increase with time and temperature. When the β-NiAl phase is completely consumed and the Al concentration reaches a critical minimum, other oxides like Cr_2O_3 and/or spinels may form besides the protective alumina, leading to internal oxidation [3].

The technological developments in stationary gas turbines for electric power generation as well as strengthened safety requirements for their operation imply the need for reliable and cost efficient lifetime prediction procedures for MCrAlY coatings. The efficiency of such a lifetime prediction procedure for industrial use however, depends not only on the reliability of the physical and mathematical model on which it is based, but also on the feasibility of the determination of the modelling parameters in practice as well as on economic aspects. The latter are at least of equal importance as a reliable model of the oxidation and diffusion processes in an MCrAlY coating.

In order to perform a reliable lifetime prediction several approaches to the modelling of oxidation and diffusion processes in MCrAlY coatings have been developed, which mostly can be divided into two types.

1. Simple power-mode or logarithmic models, which usually describe the relationship between mass change and/or thickness of the oxide scale and the time [7,8].

2. Advanced models based on the oxidation and diffusion processes (Fick's first and second law) describing the behaviour of the oxide-forming alloy elements [9–16].

The models of type 1 are fully empirical and they do not describe the physical and chemical processes, which occur in the system of oxide, coating and substrate. Therefore, their application is basically restricted to the period of time, in which the experiment is performed. An extrapolation of the relationships over time periods significantly beyond the experimental time mostly yields insufficient information.

The type 2-models describe the main physical and chemical processes during formation of the oxide scale and depletion zones. Such models are able to calculate and predict the Al and Cr concentration profiles in the coating [11,13–15] and even the β-phase volume fraction [12] and therefore deliver reliable predictions of the oxidation and diffusion processes. For a reliable quantitative application of the type 2-models however, a number of parameters, which are needed for the calculation must be known. Such parameters are, for example, the diffusion coefficients of the oxide-forming element in the alloy, the oxidation rate constant for the oxide formation and the boundary conditions. For the modelling and lifetime prediction reported in [15] for instance, special experiments and calculation techniques were used for the determination of the diffusion coefficients in Ni–Cr–Al alloys at 1100 and 1200°C.

Furthermore, experiments had to be carried out to determine the oxidation rate constants of the Al_2O_3 formation. For a cost-efficient industrial use in practice limitations may therefore arise.

In order to contribute to a practical and economic industrial use of the type 2-models a novel approach to the estimation of the model parameters was undertaken using the method of Inverse Problem Solution (IPS), which is commonly used in heat transfer science and engineering [17,18]. This method allows the model parameters to be estimated by means of data, which are determined experimentally with reasonable efforts. For obtaining calculated Al concentration profiles across the coating thickness after a time τ like those presented in [11,13–15], only the experimentally measured Al concentration profile after a defined time τ_e of exposure to temperature is needed as input value instead of diffusion coefficients and oxidation rate constant. All model parameters needed for the modelling such as the effective diffusion coefficients are then determined by the IPS technique and used for the prediction calculation.

Additionally, the oxide scale thickness is calculated as a function of time using the measured Al concentration profile as input data. Thus, the IPS technique allows an economic quantitative lifetime modelling, which is easily feasible in practice, and enables long term prediction using experimental data from some hundreds of hours of exposure time, provided that no change in the relevant mechanisms or scale spallation occurs. The purpose of the present work is to demonstrate the feasibility of the IPS technique to lifetime prediction of MCrAlY coatings.

2. Experimental

Polished samples of a 100 μm thick NiCoCrAlY coating with 8% Al forming a γ-(Ni/Co,Cr)/β-NiAl-structure were exposed at 950°C in air for 1000 h. The coating was deposited on the Ni-base alloy IN738 by Low Pressure Plasma Spraying (LPPS). Sample sets with both polished and vibro ground surfaces were used for the experiment. The samples were characterised metallographically in cross section before exposure and after 100, 300 and 1000 h. After each time interval the concentration profiles of Al across the coating thickness were determined by electron microprobe analysis every 2 to 10 μm and up to 100 μm into the base material. The window size for the element analysis was 400 μm parallel to the surface and 3 μm perpendicular to the coating surface.

In the first 100 h the oxide scale formation was followed *in situ* by high temperature X-ray diffraction. This method identifies the oxide phases *in situ* and monitors their formation as a function of time. Tha latter is achieved by measuring the diffraction peaks of the oxides as a function of time using a summing method. The resulting intensity curves $iz(t)$ show the growth of the oxide taking into account the increasing absorption of the X-rays by the growing oxide scale [19].

3. Modelling Approach

3.1. Physical Model

In the first approach to the application of the Inverse Problem Solution (IPS) to lifetime modelling of MCrAlY coatings a simplified physical model was used, assuming that

Al is initially distributed homogeneously in the coating. The fact that the Al is in reality bound in the β-NiAl phase precipitates has not so far been taken into account since previous work has shown that such a simplification leads to results that are sufficiently precise [13–15].

The oxide-forming alloying element, in this case Al, diffuses from the coating towards the oxide/coating interface. The oxygen from the gas medium diffuses in the reverse direction across the oxide scale and reacts with the Al to form Al_2O_3. During the oxide scale growth the Al concentration decreases in the near-surface region resulting in a depletion zone. At the coating/metal interface the Al diffuses from the coating into the substrate alloy driven by the concentration gradient resulting from the usually lower Al content of the substrate. An Al depletion zone similar to that at the surface is the consequence.

A typical transversal concentration profile of an oxide-forming element in the surface-near region of an alloy is shown schematically in Fig. 1. The near-surface region of the coating is divided into three main areas as a function of the distance x from the initial metal/oxide interface $x = 0$ at time $\tau = 0$, which are represented in Fig. 1:

(i)   the compact oxide scale at $x_0 < x < x_1$,

(ii) the area of a Me depletion at $x_1 < x < x_2$ and

(iii) the unaffected coating at $x > x_2$.

In the compact oxide scale at $x_0 < x < x_1$ the maximum concentration $C = C_{max}$ of the oxide forming alloying element Me is found. The physical model takes into account that during the oxidation the oxide surface x_0 moves outwards, while the boundaries x_1 and x_2 move inwards into the alloy.

3.2. Mathematical Model

In the present case the diffusion process of the alloying element (Al) and the oxygen (Ox) towards the zone of the oxide formation within the calculation domain $0 < x < x_\infty$ can be described by the following equations

$$\frac{\partial C_i}{\partial \tau} = \frac{\partial}{\partial x}\left[D_i \frac{\partial C_i}{\partial x}\right] \tag{1}$$

$$\tau > 0, \ C_i = C_i(x, \tau), \ i = \begin{cases} 1 - Ox, & x_0 \le x \le x_1 \\ 2 - Al, & x_1 < x \le x_\infty \end{cases},$$

$$D_1 = D_{Ox} = \begin{cases} D_{Ox}, & x_0 \le x \le x_1 \\ 0, & x_1 < x \le x_\infty \end{cases}, \quad D_2 = D_{Al} = \begin{cases} 0, & x_0 \le x \le x_1 \\ D_{Al}, & x_1 < x \le x_\infty \end{cases}$$

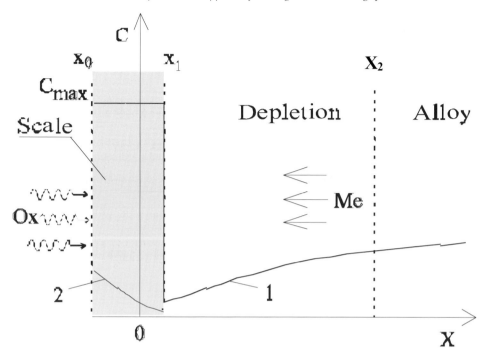

Fig. 1 *Concentration profile of the scale-forming alloying element Me (curve 1) and the oxidant Ox (curve 2).*

The initial conditions are

$$C_1(x,0) = C_{O_2}(x) = 0, \quad C_2(x,0) = C_{Al}(x) \tag{2}$$

The boundary conditions for oxidant are

$$-D_{Ox}\frac{\partial C_{Ox}(x_0,\tau)}{\partial x} = \beta_{Ox} \cdot \left[C_{Ox}(x_0,\tau) - C_{Ox}^*\right] \tag{3}$$

$$-D_{Ox}\frac{\partial C_{Ox}(x_1,\tau)}{\partial x} = J_{Ox} = k_{Ox} \cdot C_{Ox}(x_1,\tau) \cdot C_{Al}(x_1,\tau) \tag{4}$$

and for aluminium

$$-D_{Al}\frac{\partial C_{Al}(x_1,\tau)}{\partial x} = J_{Al} = k_{Al} \cdot C_{Ox}(x_1,\tau) \cdot C_{Al}(x_1,\tau) \tag{5}$$

$$\frac{\partial C_{Al}(x_\infty,\tau)}{\partial x} = 0 \tag{6}$$

where x is distance; τ is time; C is the concentration; D is the effective diffusion coefficient; β_{O_2} is the mass transfer coefficient for oxygen at the gas–oxide scale interface; $C_{O_2}^*$ is the oxygen concentration in the gas medium; j_{O_2} and j_{Al} are the mass fluxes of Al and oxygen at the boundary x_1; the coefficients k_{O_2} and k_{Al} are the kinetic rate constants for oxygen and the alloying element in (4) and (5), which are interrelated according to stoichiometry.

The outer boundary x_0 moves outwards and the boundary x_1 moves inwards into the coating with the growth of the oxide scale according to the following relationships

$$\left[C_{max} - C_{Al}(x_1, \tau)\right]\frac{dx_1}{d\tau} = J_{Al}(x_1, \tau) \tag{7}$$

$$C_{max}\frac{dx_0}{d\tau} = \frac{dx_1}{d\tau}C_{max} - \left[J_{Al}(x_1, \tau) + J_{O_2}(x_1, \tau)\right]\frac{\rho_{Coating}}{\rho_{Al_2O_3}} \tag{8}$$

where $\rho_{Coating}$ and $\rho_{Al_2O_3}$ are the densities of the coating and the oxide. The oxide scale thickness at a time τ results from the calculation of the movement of the boundaries x_0 and x_1 from their initial position at 0.

The mathematical eqns (1)–(8) are integrated by finite difference methods. This allows the most general peculiarities of the problem formulation to be taken into consideration, e.g. the multilayer, the relationships between coefficients and co-ordinate, time and concentration. During each time step a multiple solution of the equation system for the oxidant is performed and then of the equation system for the alloying element. The implicit correlation between these systems solution is carried out through the eqns (4) and (5) and is realised by means of an iteration algorithm.

4. Model Parameter Estimation by Inverse Problem Solution (IPS)

For obtaining the model parameters necessary for the application of the modelling in practice, the technique of Inverse Problem Solution (IPS) is used. The basic idea of the IPS technique is to use easily obtainable experimental data acquired after a defined experimental time τ_e as input values for the model parameter estimation. The experimental time τ_e should be as short as possible and as long as necessary for a sufficient by accurate prediction. The approach consists of four fundamental steps.

1. Design of a mathematical model or the use of an existing one.

2. Experiments with given experimental time and temperature revealing input data for the model parameter estimation.

3. Estimation of the key parameters of the mathematical model for different temperatures (calibration of mathematical model).

4.   Computational prediction of the alloying element distributions and the oxide scale formation.

The proposed mathematical model comprises four key parameters which must be estimated for the prediction of the oxidation and diffusion processes: the diffusion coefficient of aluminium D_{Al} and of oxygen D_{Ox}, the mass transfer coefficient β_{Ox} for oxygen at the surface x_0 and the kinetic rate constant for oxygen k_{Ox} (1)–(8).
Using the Inverse Problem Solution technique only the measured concentration profile C_E of Al across the coating thickness is needed as experimental input data for the calibration of the mathematical model. For the estimation of the model parameters D_{Al}, D_{Ox}, β_{Ox} and k_{Ox} the following convergence criterion is applied:

$$F = \left\{ \sum_{j=1}^{m} \left[C_{jM}(P) - C_{jE} \right]^2 / m \right\}^{0.5} \approx \delta \qquad (9)$$

where $P = P(D_{Al}, D_{Ox}, \beta_{Ox}, k_{Ox})$ is the unknown model parameter vector, $C_{j,iM}$ and $C_{j,iE}$ are the calculated and measured Al concentrations at the jth measuring point of the concentration profile at a given position x, time τ and temperature T, m is the number of the measuring points, δ is the average square root error. The fit procedure of the model parameters is iterated until F reaches the given limit of δ, which is 0.8% in the present case. The scheme of the approach is represented in Fig. 2.

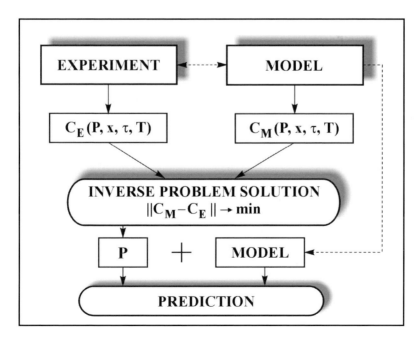

Fig. 2 *Scheme of computational and experimental approach for oxidation and diffusion processes prediction using the Inverse Problem Solution (IPS) method.*

For performing the lifetime modelling approach by the Inverse Problem Solution with the selected MCrAlY coating the following assumptions were made:

- One alloying element only, in the present case the aluminium, forms the oxide scale.

- Oxidation and diffusion mechanism and model parameters do not change with time.

- The oxide scale formation occurs at the boundary x_1 [13,20].

- The diffusion coefficients of the alloying element D_{Me} and the oxidant D_{Ox} are constant and are effective empirical parameters obtained by short-time experiments. The empirical character of these coefficients implies that they consider the real complexity of the processes, although the model (1)–(8) is sufficiently simple.

- In the first stage of the model a homogeneous distribution of Al was assumed in the coating. The fact that the aluminium is bound in the β-NiAl phases was not taken into account.

5. Results

The coating forms a γ/β-structure or, more precisely, a matrix of γ-(Ni/Co,Cr) containing a homogeneous distribution of β-NiAl precipitations. After 100 h of oxidation at 950°C a β-phase depletion zone of 10 μm is observed in the sub-surface region and in the interdiffusion zone (Fig. 3). The oxide scale after 100 h at 950°C shows an average thickness of about 1.8 μm, a good adherence and no internal oxidation is observed (Fig. 4).

It follows from *in situ* high temperature X-ray diffraction studies that the coating forms α-Al_2O_3 and θ-Al_2O_3 in the studied time period of 100 h. No other oxides were detected. The growth of α-Al_2O_3 and θ-Al_2O_3 at 950°C in the first 100 h was determined as a function of time and plotted by means of intensity curves $iz(t)$, which are explained in [19]. The coating with a polished surface shows a continuous formation of α-Al_2O_3 and in the first 50 h an enhanced θ-Al_2O_3-formation with comparable kinetics. On further oxidation the formation of α-Al_2O_3 keeps increasing and the θ-Al_2O_3-formation slows down. The amount of θ-Al_2O_3 seems to reach a maximum at 100 h (Fig. 5).

On the coating with a vibro ground surface the process of θ-Al_2O_3 formation and transformation into α-Al_2O_3 is accelerated. Already by 50 h a maximum of θ-Al_2O_3 is reached, and on further oxidation θ-Al_2O_3 begins to disappear by transformation into α-Al_2O_3. At the same time an enhanced α-Al_2O_3-formation is observed, which is due to the transformation of the faster growing θ-Al_2O_3 into α-Al_2O_3 (Fig. 6).

On comparing the oxide formation on both surfaces it is noticed that on the vibro ground surface the amount of θ-Al_2O_3 after 50 h corresponds to that after 100 h on the polished surface. Also the amounts of α-Al_2O_3 are similarly comparable. At 50 h

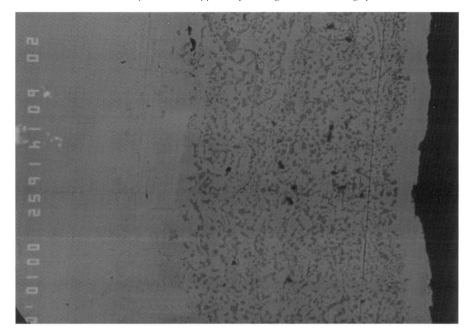

Fig. 3 *NiCoCrAl Y coating with 8% Al in cross section after 100 h at 950 °C.*

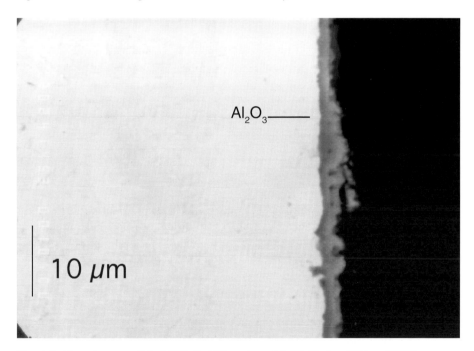

Fig. 4 *Oxide scale on a polished NiCoCrAlY coating with 8%Al after 100 h at 950 °C.*

however, the α-Al_2O_3-formation increases significantly on the vibro ground surface (Figs 5 and 6).

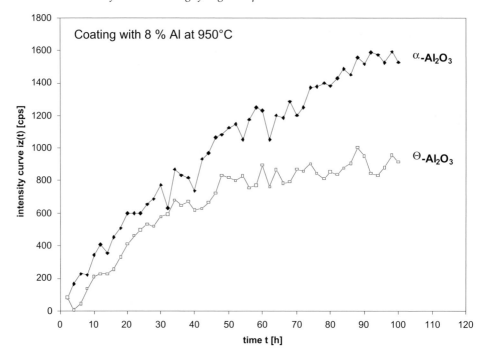

Fig. 5 *Diffraction peak intensities iz(t) of α-Al₂O₃ and θ-Al₂O₃ on a polished NiCoCrAlY coating with 8%Al at 950°C as a function of time.*

The Al concentration profile across the coating was determined by electron microprobe analysis. The measured concentration values $C_{jE}(P, x, \tau, T)$ were used to estimate the coefficients D_{Al}, D_{Ox}, β_{Ox} and k_{Ox} of the model described by eqns (1)–(8) using the Inverse Problem Solution method [18,21]. The concentration profiles determined after exposure times of 100 and 300 h were used for the estimation of the model parameter. The resulting estimated parameter values are:

$$D_{Al} = 6 \cdot 10^{-17} \text{ m}^2\text{s}^{-1}, \quad D_{Ox} = 4 \cdot 10^{-18} \text{ m}^2\text{s}^{-1}, \quad \beta_{Ox} = 1.0 \cdot 10^{-9} \text{ ms}^{-1}, \quad k_{Ox} = 3 \cdot 10^{-5}\text{s}^{-1}.$$

With these parameter values and using the model described by eqns (1)–(8) the Al concentration profiles across the coating as well as the oxide scale thicknesses were calculated for oxidation times of 100 and 300 h at 950°C. A prediction for 1000 h was calculated with the model parameters estimated from the data after 300 h and compared with experimentally obtained Al concentration profile and scale thickness.

The curves calculated for 100, 300 and 1000 h were plotted together with the measured Al concentration profiles in one diagram (Fig. 7). A sufficient agreement of the measured and calculated Al concentration profiles is observed. The Al concentrations in the zone of unaffected γ/β-structure are accurately calculated and predicted and the increasing Al depletion in the subsurface and in the interdiffusion zone is shown. In the subsurface β-phase depletion zone an Al concentration of 3.5% was measured. In the measured concentration profile a sharp edge is observed at the

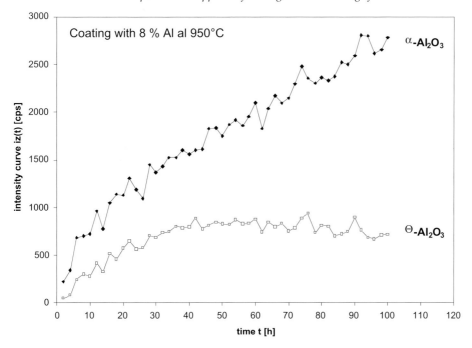

Fig. 6 *Diffraction peak intensities iz(t) of α-Al₂O₃ and θ-Al₂O₃ on a vibro ground NiCoCrAlY coating with 8%Al at 950°C as a function of time.*

boundary between the unaffected γ/β-structure and the β-phase depletion zone, especially after 1000 h. The assumption of Al being initially homogeneously distributed leads to a continuously decreasing Al concentration towards the surface according to the physical model in Fig. 1. The Al concentration decreases in an analogous way at the coating/subtrate interface reaching a concentration of 3.4% in the substrate IN738 substrate which corresponds to the Al content of this alloy. The increase of the measured Al concentration at the coating/substrate interface at $x = 120\ \mu m$ after 1000 h is due to a local oxidation (Fig. 7).

The calculation and prediction of the oxide scale thickness is shown in Fig. 8, which is a magnification of the surface-near region in Fig. 7. With the measured Al concentration profile as input values and the calculation of the movement of the boundaries x_0 and x_1 in the model from Fig. 1 using eqns (7) and (8) the oxide scale thickness is calculated as a function of time. The Figure shows that the oxide scale grows both outward as well as inward (moving boundaries x_0 and x_1 in Fig. 1), which results from the quotient of the alloy and oxide densities in eqn (8) of the model.

The oxide scale thicknesses calculated by the modelling are in agreement with those measured from micrographs (Table 1). With the calculated thicknesses a parabolic rate constant of

$$k_p = 2.95 \cdot 10^{-13}\ \mathrm{g^2\ cm^{-4}\ s^{-1}}$$

was determined for the Al₂O₃-formation yielding a predicted oxide scale thickness of 8.5 μm after 5000 h.

Table 1. *Measured and calculated oxide scale thicknesses after different times at 950°C*

950°C	Oxidation time			
	100 h	**300 h**	**1000 h**	**5000 h**
Measured oxide scale thickness x_{mes} in µm	1.8	2.0	3.5	–
Calculated oxide scale thickness x_{cal} in µm	1.0	2.1	4.0	8.5

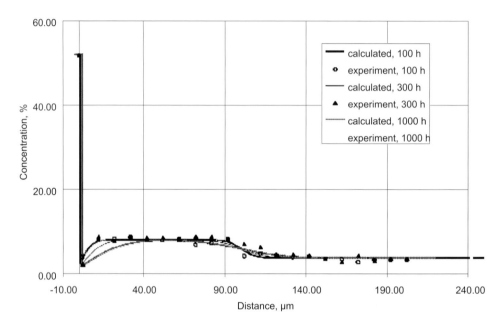

Fig. 7 *Measured and calculated Al concentration profile across the coating and across the first 140 µm of the base material for 950°C.*

6. Discussion

The present study shows that the approach by Inverse Problem Solution (IPS) to the lifetime prediction of MCrAlY coatings is a useful additional tool for the application of existing as well as newly developed models. The IPS provides a technically easily feasible way to obtain input values for the model parameter estimation. The Al concentration profiles needed for the IPS-method were obtained by electron microprobe analysis and no complex procedures had to be applied to determine diffusion coefficients and no oxidation rate constants had to be determined.

The diffusion coefficients estimated by the IPS and used in the model are effective diffusion coefficients representing the real processes in the coating under the given conditions. This means that processes altering the diffusion profiles like chromium diffusion from the bulk into the β-phase depletion zone and the influence of Co on the diffusion processes are included in these effective diffusion coefficients, which are based on real measured Al concentration profiles. In this way also the influence

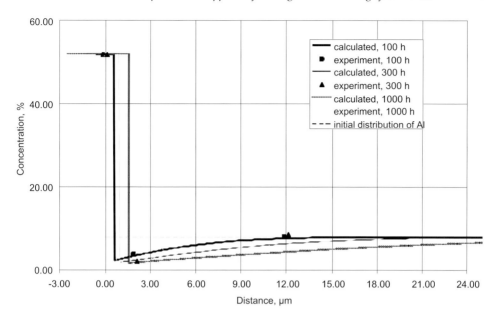

Fig. 8 *Measured and calculated Al concentration profile across the coating at 950°C in the subsurface region; magnified detail from Fig. 7.*

of the coating surface performance on the oxidation behaviour like that shown in Figs 5 and 6 and reported in [19] is taken into account. Such influences are however not considered when calculated diffusion coefficients are used as input data for modelling.

Using IPS with only experimentally measured Al concentration profiles as input values for model parameters estimation, the model calculates and predicts the growth of the alumina scale with sufficient accuracy. It stands out however, that after 100 h the measured oxide scale thickness is higher than the calculated value (Table 1). This is obviously due to the θ-Al$_2$O$_3$-formation in the initial state of the oxidation that was observed *in situ* by high temperature X-ray diffraction (Figs 5 and 6). The faster forming θ-Al$_2$O$_3$ is transformed into α-Al$_2$O$_3$ thus leading to a higher oxide scale thickness than is calculated by the model which considers no alumina phase transitions. How far this initial deviation in the approximately first 300 h will affect the reliability of the long term prediction for more than 15000 h has to be worked out in further research.

In order to test the feasibility of the Inverse Problem Solution (IPS) and the mathematical model used for lifetime modelling and prediction of MCrAlY coatings the Al concentration profile was used instead of the β-phase concentration profile in this first stage of the research work. Although this provides a simplification the agreement between the experimentally determined and the calculated as well as predicted results is high. The alumina scale thicknesses are in good agreement (Table 1), the measured width of the unaffected γ/β-structure zone coincides with the calculated one considering the points of inflection of the curves as zone boundaries, and the total Al concentration in the latter zone is well calculated and predicted (Fig. 7).

The main weak point of the use of Al concentration profiles is that the model is unable to describe the sharp boundaries between the depletion zones and the unaffected γ/β-structure zone (Fig. 7). Once the feasibility of the IPS approach has been demonstrated however, further research work will consider the β-phase concentration profiles and an assessment of the prediction reliability will be performed taking into account possible factors altering the diffusion processes.

7. Conclusions

In this first step of the research work the application of Inverse Problem Solution (IPS) to the life time modelling of MCrAlY coatings has been showed to be a suitable and cost efficient method for life time prediction in industrial practice. The necessary input data are easily obtained by economic routine analysis and no complex estimation procedures of the model parameters are necessary. The IPS method can be applied to every existing physical and mathematical model as well as to newly developed models, thus providing a wide application field for the method. However further work is needed using models, which take into account the binding of Al in the β-NiAl phase and other diffusion profile altering processes for a more precise description of the complex oxidation and diffusion processes in MCrAlY coatings.

8. Acknowledgements

The authors would like to thank NATO for the support of the research work, Werner Stamm and Norbert Czech from the Siemens Power Generation for supporting the work with sample material and discussion and Serge Alpérine for his contributions in numerous discussions.

References

1. N. Czech, F. Schmitz and W. Stamm, *Surf. Coat. Technol.*, 1994, **68–69**, 17.
2. N. Czech, F. Schmitz and W. Stamm, *Surf. Coat. Technol.*, 1995, **76–77**, 28.
3. R. Streif in *High Temperature Corrosion 3* (R. Streiff, J. Stringer, R.C. Krutenat and M. Caillet, eds). J. Phys. (France) IV (1993), Colloque C9, vol. 3, pp. 17–41.
4. A. Strawbridge and P. Y. Hou, *Mater. High Temp.*, 1994, **12**, 177.
5. N. Czech, F. Schmitz and W. Stamm, *VGB Kraftwerkstechnik* 1997, **77**, 221.
6. W. Beele, N. Czech, W. J. Quadakkers and W. Stamm, *Surf. Coat. Technol.*, 1997, **94–95**, 41.
7. C. T. Sims and W. C. Hagel, *The Superalloys*, New York (1972).
8. S. M. Meier, D. M. Nissleu, K. D. Sheffler and T. A. Cruse, *J. Eng. Gas Turbines Power.*, 1992, **114**, 250–257.
9. C. Wagner, *Corros. Sci.*, 1968, **8**, 889–893.
10. G. L. Wulf, M. B. McGirr and G. R. Wallwork, *Corros. Sci.*, 1969, **9**, 739–754.
11. J. A. Nesbitt, R.W. Heckel, *Thin Solid Films*, 1984, **119**, 281–290.
12. E. Y. Lee, D. M. Chartier, R. R. Biedermann and R. D. Sisson, Jr., *Surf. Coat. Technol.*, 1987, **32**, 19–39.

13. J. A. Nesbitt, Numerical Modeling of high-temperature corrosion processes, *Oxid. Met.*, 1996, **44**, 309–338.

14. J. A. Nesbitt and R.W. Heckel, Interdiffusion in Ni-rich, Ni–Cr–Al alloys at 1100 and 1200°C: Part I. Diffusion paths and microstructures, *Met. Trans.*, 1987, **18A**, 2061–2073.

15. J. A. Nesbitt and R. W. Heckel, Interdiffusion in Ni-rich, Ni–Cr–Al alloys at 1100 and 1200°C: Part II. Diffusion coefficients and predicted poncentration profiles, *Met. Trans.* 1987, **18A**, 2075–2086.

16. J. A. Nesbitt and R. W. Heckel, Predicting diffusion paths and interface motion in $\gamma/\gamma+\beta^+$, Ni–Cr–Al diffusion couples, *Met. Trans.*, 1987, **18A**, 2087–2094.

17. J. Beck and B. Blackwell, *Inverse Heat Conduction*. Wiley, New York, 1985.

18. P. G. Krukovsky and E. S. Kartavova, *Inverse Problems in Engineering. Theory and Practice.* ASME 1999, 403–408.

19. N. Czech, M. Juez-Lorenzo, V. Kolarik and W. Stamm, *Surf. Coat. Technol.*, 1998, **108–109**, 36–42.

20. E. W. A. Young and J. H. W. de Wit, *Oxid. Met.*, 1986, **26**, 351–361.

21. P. G. Krukovsky, *Proc. 30th National Heat Transfer Conf.*, ASME (United Eng. Center), (W. J. Bryan and J. V. Beck, eds). PV 312–10, New York, 1995, pp.107–112.

17

Development of Type II Hot Corrosion in Solid Fuel Fired Gas Turbines

N. J. SIMMS, P. J. SMITH, A. ENCINAS-OROPESA, S. RYDER, J. R. NICHOLLS
and J. E. OAKEY

Power Generation Technology Centre, Cranfield University, Cranfield, Bedfordshire, MK43 0AL, UK

ABSTRACT

The solid fuel fired combined cycle power systems that are currently being developed all utilise industrial gas turbines. The successful development of these systems up to economically viable power stations relies on their gas turbines giving high efficiencies together with acceptable performance and availability. In particular, the vane and blade materials for the hot combustion gas paths must give acceptable and predictable in-service life times.

This paper reports the results of a series of detailed laboratory studies of the effect of deposit composition, deposition flux and gas chemistry on the development and rate of growth of type II hot corrosion pits. The test conditions were targeted to span the range of conditions that were anticipated to occur in gas turbines incorporated into solid fuel combined cycles: covering realistic ranges of SO_x and HCl partial pressures, as well as deposits containing Na, K, Pb and Zn. The well established 'deposit' recoat test method was used for this work, as it allowed accurate control of the 'deposit' composition and 'deposition flux'. All the tests were carried out in controlled atmosphere furnaces in the temperature range 650–700°C.

The corrosion damage observed was characterised in terms of metal loss at a minimum of 24 locations around each sample. Thus, a distribution of metal damage was obtained for each material at specific exposure times, 'deposit' compositions, 'deposition fluxes', gas compositions and temperature. The characteristics of these metal loss distributions changed with exposure time and conditions, as the type II hot corrosion pits nucleated and grew. These distributions were analysed using a probabilistic approach that allowed characteristic parameters to be evaluated for each combination of material and exposure condition. The change in the characteristic parameters with exposure time and conditions allows the development of models of type II hot corrosion.

1. Introduction

Many of the solid fuel fired advanced power generation systems currently being developed in Europe, Japan and the USA are combined cycle systems based on gasification and/or combustion and utilising both gas and steam turbines [1–3]. Such systems were originally developed to enable coal to be used to generate electricity with greatly increased efficiency and much lower environmental emissions (specifically CO_2, SO_x, NO_x and particulates) than with conventional pulverised coal systems. However it has subsequently been found that many such systems can also be co-fired with biomass and/or waste products offering further advantages in that the systems can partially use renewable CO_2 neutral fuels and also be used as a method for waste disposal in addition to power generation. In all such combined cycle systems the performance of the gas turbine is vital to the overall plant efficiency and economic viability.

Industrial gas turbines have been developed to fire on a wide variety of fuels, ranging from natural gas to sour gases and heavy fuel oils. The degradation of materials in such systems has been the subject of many investigations during the past 40 years, as operating conditions have developed and/or fuels have changed, and the potential problems which may be encountered in gas and oil fired gas turbines have been well characterised [4,5]. Many similar types of materials degradation can be expected in gas turbines using solid fuel derived fuels, as some of the contaminant species are the same as for oil and/or gas fired systems. However, the levels of contamination are different, there are additional as well as absent species and the sources/forms of the contaminant species also differ. As a result, fuel and air quality standards produced for gas turbines fired on more traditional fuels [6] need to be thoroughly reviewed and revised [7], to take into account the significant differences with these new fuel compositions, as well as the damage rates that will be acceptable for these new power systems.

The environments found within the hot gas paths of gas turbines depend on the contaminants present in the fuel and air entering the turbine, as well as the turbine operating conditions. Fuel gases derived from solid fuels such as coal, biomass and waste and their mixes have the potential to cause both erosion and corrosion damage to components in the hot gas path of the gas turbine depending on the balance of contaminants within such gases. Fuel derived particles can cause either erosion damage or deposition depending on their size, and composition, as well as aerofoil design and operating conditions. Corrosion can result from the combined effects of gaseous species (e.g. SO_X and HCl) and deposits formed by condensation from the vapour phase (e.g. alkalis and other trace metal species) and/or particle impaction and sticking. The corrosion damage mode is highly dependent on the local component environment. Conventionally, the metal vapour species of most concern were alkalis in gas turbines fired on clean fuels (either as fuel contaminants or via the combustion air) or vanadium from heavy fuel oils. In coal fired systems, as well as significant levels of both SO_X and HCl, combustion derived gases tend to have significant levels of alkali metals, whereas gasifier derived fuel gases (when used with hot gas cleaning processes) have higher levels of heavy metals, e.g. lead and zinc [8]. Co-firing with biomass or waste fuels changes the levels of all the contaminants, e.g. for trace metals, wheat straw has higher potassium levels and sewage sludges have higher zinc levels [8]. The effects of different fuels on the levels of contaminant vapour phase species, deposition fluxes and deposit compositions (e.g. melting points) requires careful consideration for each process and fuel.

These degradation modes of materials can be life limiting for such components as first stage vanes and blades, in addition to the creep and fatigue processes which limit blade lives in the longer term. For new fuels to be used in gas turbines, it is necessary that critical components are manufactured from appropriate materials and that these materials give predictable in-service performance. Thus, one aspect of supporting the development and introduction of new solid fuel fired combined cycle power generation systems, is to develop statistically based predictive corrosion models for gas turbine degradation, under appropriate ranges of deposition/corrosion conditions, to allow the realistic prediction of potential blade and vane lives.

The gas cleaning systems in solid fuel fired combined cycle power generation systems are predominantly aimed at reducing the emissions of potentially damaging gaseous species and dust to the environment [9]. However, such systems could also be used to remove or control the gaseous species and dust that are potentially damaging to the gas turbine components. The gas cleaning requirements to meet these two potential objectives are not necessarily the same, in fact some proposed gasifier fuel gas cleaning systems would both remove and add contaminant species to the fuel gases and actually reduce gas turbine lives. A better knowledge of the dependence of materials degradation on contaminant species and their various interactions, would allow the requirements of such gas cleaning systems to be more thoroughly defined and thus optimised to minimise environmental emissions and maximise gas turbine component lives.

This paper reports some of the results obtained in a series of laboratory hot corrosion tests that form part of an on-going systematic investigation of the effect of realistic levels of contaminants and other exposure parameters on the corrosion of candidate gas turbine materials for use in solid fuel fired systems. Initially these studies were targeted at conditions anticipated in gas turbines in coal combustion systems [10], but more recent and on-going tests have targeted gas turbines fired on fuel gases derived from gasified coal, waste and biomass. The tests reported in this paper are only those carried out at 650 and 700°C (typical type II hot corrosion temperatures) and the results are restricted to those obtained on a standard base alloy, IN738LC.

One of the aims of this test work is the development of empirically based predictive models of the corrosive degradation of gas turbine materials in 'dirty' gas conditions [10–14]. In order to develop and test models of corrosion degradation, an important activity is dimensional metrology of the test samples before and after their exposure to quantify the materials degradation in terms of accurate and statistically useful metal losses [10,11,16,17]. This approach to dimensional metrology is now becoming more widely adopted and has been accepted as the approach to use in the current European COST522 activities [18]. It also forms part of the basis for the EU sponsored TESTCORR activity aimed at developing guidelines for high temperature corrosion testing [19].

2. Experimental

2.1. Hot Corrosion Testing

A series of laboratory tests have been carried out using the well established 'deposit replenishment' technique [11–13] to investigate systematically the effects of different 'deposit' compositions and gas partial pressures on the corrosion behaviour of gas turbine materials.

During the 12 year period (to date) over which these tests have been carried out, the test apparatus has gradually evolved, but all the tests have been carried out in controlled atmosphere furnaces in experimental systems that allowed close control of the inlet gases from pre-mixed gas bottles and the analysis of the outlet gases. An example of the test apparatus configuration is given in Fig. 1, for a horizontal tube furnace. For a vertical tube furnace, the variation in the apparatus in terms of sample exposure location is shown in Fig. 2 (gas inlet supply systems and outlet analysis systems are the same).

The exposure conditions used in these tests are given in Tables 1 and 2. In all the tests the samples were given a thermal cycle every 20, or 100 h, depending on the test series. The surface deposits were renewed on each cycle by spraying solution(s) containing the required deposit on to the sample surfaces; different fluxes were obtained by spraying different weights of deposit onto the surfaces. The levels of SO_X and HCl selected for these atmospheric pressure corrosion tests were chosen to cover the ranges of SO_X and HCl partial pressures anticipated in gas turbines operating in solid fuel fired combined cycle gas turbines (Fig. 3); these ranges were refined over the course of the test programme.

In all these tests solid cylindrical samples have been used and a common metrology method adopted (section 2.2). A range of base alloys and coatings were (and continue to be) used in these tests; IN738LC (16 Cr–8.5 Co–3.4 Al–3.4 Ti–2.6 W–1.7 Ta–1.7 Mo–0.9 Nb–0.11 C–0.05 Zr–0.01 mass% B-balance Ni) has been used as the reference base alloy material throughout the test programme and is the only material for which results are reported in this paper.

2.2. Sample Metrology

The aim of the sample metrology and analysis process was to produce a set of metal loss data for each sample exposed in the laboratory tests. The development and

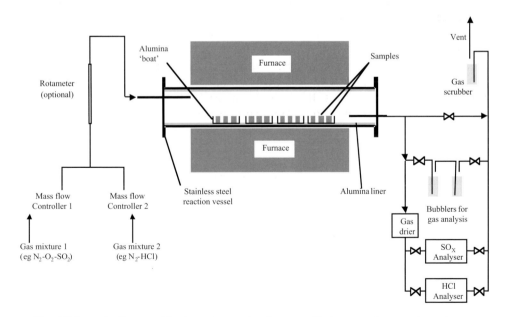

Fig. 1 *Schematic diagram of horizontal corrosion furnace with dry corrosion gas atmosphere.*

Table 1. Exposure conditions for laboratory tests

Temperature	650, 750°C
Exposure times	100, 200/300, 500, 1000, 2000
Gas composition (see also Fig. 3)	
SO_X	40–3500 vol. ppm
HCl	60–4200 vol. ppm
Air	Balance
Initial deposit compositions	A–K (see Table 2)
Deposition fluxes	0.15–500 μm cm^{-2} h^{-1}
Deposit re-coat intervals	20 or 100 h

Table 2. Initial deposit compositions

Deposit Identification	Initial deposit composition, mole%				Initial deposit composition, mass%			
	Na_2SO_4	K_2SO_4	$ZnCl_2$	$PbCl_2$	Na_2SO_4	K_2SO_4	$ZnCl_2$	$PbCl_2$
A	80	20	–	–	76.5	23.5	–	–
B	90	10	–	–	88.0	12.0	–	–
C	50	50	–	–	44.9	55.1	–	–
D	10	90	–	–	8.3	91.7	–	–
E	40	10	20	30	30.7	9.4	14.7	45.1
F	40	10	30	20	33.3	10.2	23.9	32.6
G	40	10	50	–	39.9	12.2	47.9	–
H	40	10	–	50	26.6	8.2	–	65.2
I	12	3	55	30	9.4	2.9	41.5	46.2
J	60	15	–	25	47.1	14.5	–	38.4
K	30	7	–	63	18.5	5.3	–	76.2

application of this metrology method has been reported previously [e.g. 10,11]. The methodology is based on the accurate measurement of sample dimensions before and after exposure.

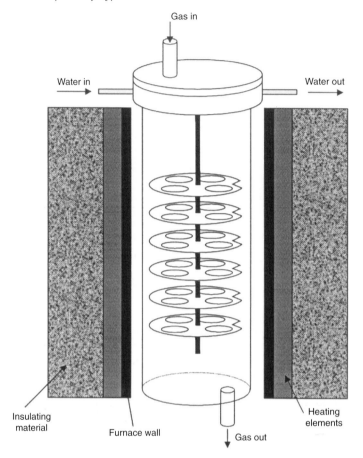

Fig. 2 *Schematic diagram of vertical controlled atmosphere corrosion furnace.*

For the cylindrical samples used in these laboratory tests, the sample dimensions were measured at particular orientations around a central plane using an accurately calibrated micrometer (± 2 µm) prior to exposure. After exposure, polished cross-sections were prepared close to and parallel to the pre-exposure measurement plane. An essential part of this process was the use of a jig to hold the samples whilst they were being mounted (Fig. 4); this ensured that the samples could be held firmly whilst retaining their surface deposit/corrosion products and be cut perpendicularly to their cylinder axes at a controlled distance from the pre-exposure measurement plane.

The polished cross-sections were measured using either a projection microscope or an optical microscope/image analysis system both with accurately calibrated *x–y* stages (±2 µm). A generic set of post-exposure measurements was made that included the position of the metal surface and other features of the corrosive degradation (Fig. 5). For these laboratory samples, a minimum of 24 data points per set were obtained for statistical analysis and use in the model development processes. During the post exposure metrology process, checks were made to determine which corrosion mechanism was operating to ensure that consistent corrosion models were developed.

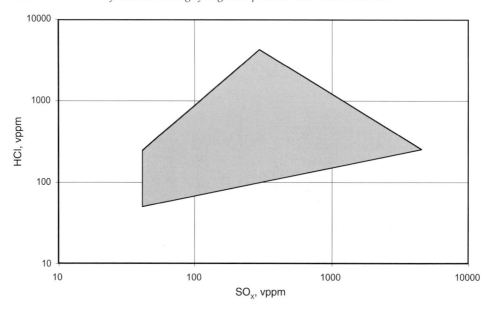

Fig. 3 Ranges of SO$_X$ and HCl anticipated in a various solid fuel fired combined cycle power systems. (a) side view; (b) plan view.

3. Results, Modelling and Discussions

The results obtained from these tests on IN738LC show a complex interaction of the different exposure variables. It has been found that the deposition fluxes, deposit compositions and SO$_X$ partial pressures dominate the corrosion behaviour of the materials, with less effect of HCl partial pressures. The effect of the number, and frequency, of the thermal cycles that are an intrinsic part of this type of corrosion testing was limited to the number of initiation sites of the corrosion damage after short exposure times, and had no significant effect on the performances of the materials in the longer term. For simplicity, the effects of the main variables and their interactions will be described in turn.

3.1. Time Dependency of Hot Corrosion Damage

The corrosion damage observed on IN738LC in these hot corrosion tests at 650 and 700°C varied, as anticipated, between localised pitting, developing at low values of 'deposition flux' and more broad front corrosion damage at high values of 'deposition flux'.

These effects can be illustrated by plotting the measured corrosion damage data in the form of probability plots. In this form of plot, the measured damage data are ordered and then plotted as a function of their probability of not being exceeded [16]. Figures 6 and 7 illustrate the results for a series of samples exposed to a 'deposit' mix of 80/20 mole% (Na/K)$_2$SO$_4$, at 700°C with deposition fluxes of 1.5 and 5 µg cm^{-2}h^{-1} respectively. The data in Fig. 6 show relatively low uniform damage for the first 200 h and then the development of a number of deeper pits. This is a clear indication of an incubation time for the development of the pitting damage at this

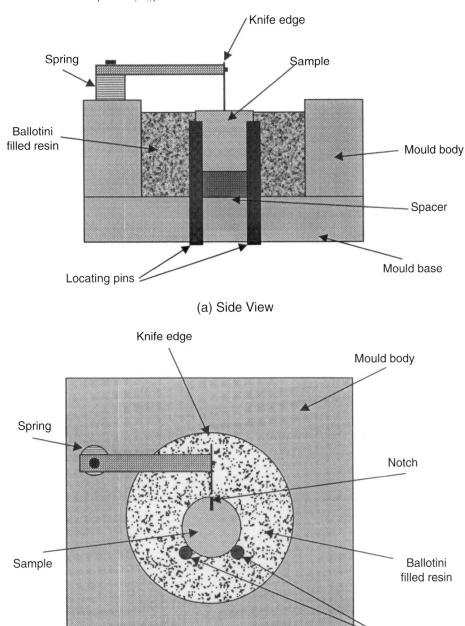

(a) Side View

(b) Plan View

Fig. 4 *Post-exposure laboratory sample mounting jig.*

low 'deposition flux'. It also illustrates the erroneous results that would be gained from using a mass gain method, which gives an average value of corrosion damage, to assess the corrosion damage that occurs at low 'deposition fluxes'. The data in

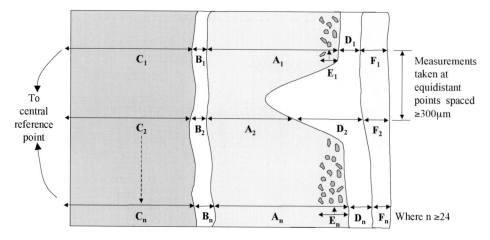

A = Coating thickness, B = Interdiffusion zone, C = Substrate thickness
D = Surface oxide thickness, E = Internal corrosion depth, F = Deposit thickness

Fig. 5 *Schematic diagram of features measured during post-exposure metrology.*

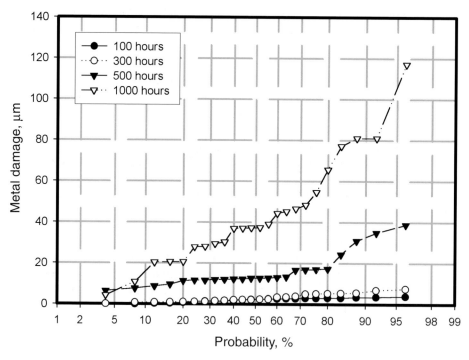

Fig. 6 *Evolution of the distribution of metal loss measurements from IN738LC exposed at 700°C, $SO_x = 260$ vol. ppm, HCl = 300 vol. ppm, flux of 80/20 mole% $(Na/K)_2SO_4 = 1.5$ µg cm^{-2} h^{-1}.*

Fig. 7 show a uniform distribution of corrosion damage developing around the samples from the start of the exposures. In this case there is no evidence of a incubation period. The difference between the maximum measured damage and the median

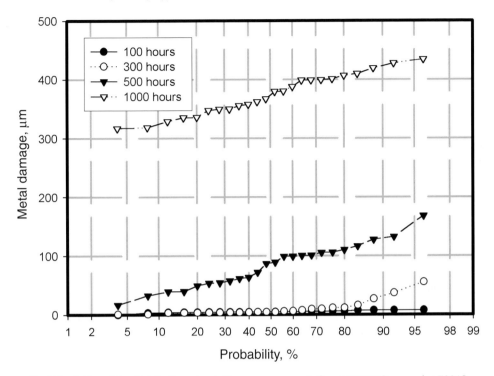

Fig. 7 *Evolution of the distribution of metal loss measurements from IN738LC exposed at 700°C, $SO_X = 260$ vol. ppm, $HCl = 300$ vol. ppm, flux of 80/20 mole% $(Na/K)_2SO_4 = 1.5$ μg cm^{-2} h^{-1}.*

damage, whilst significant (at ~ 16% after 1000 h exposure) is not as large as for the samples exposed at low 'deposition fluxes' (at ~216% after 1000 h exposure).

This effect of deposition flux on the development of hot corrosion damage has been observed for a larger number of 'deposit compositions' (eight = A–H in Table 2). It is illustrated in Figures 8 and 9 (for 'deposition fluxes' of 5 and 15 μg cm^{-2} h^{-1}), in which 'maximum' (= damage with a 4% probability of being exceeded) corrosion damage values are plotted as a function of exposure time. Again, the samples with lower 'deposition fluxes' have a clear incubation period for all the different deposit compositions, but this is significantly shorter for the lead- and zinc-containing deposits. At the higher 'deposition fluxes', there are clearly differences between the deposit compositions (see below) but the development of the 'maximum' hot corrosion damage is approximately linear with time.

These results show that it is important to carry out hot corrosion tests for at least 500–1000 h in order to be able to evaluate the different stages in the development of the hot corrosion damage, especially for the more realistic lower deposition flux exposure conditions.

3.2. Effect of Deposition Fluxes

The 'maximum corrosion damage' results obtained for IN738LC at 700°C with a range of 'deposition fluxes' of a 'deposit' mix of 80/20 mole% $(Na/K)_2SO_4$ are

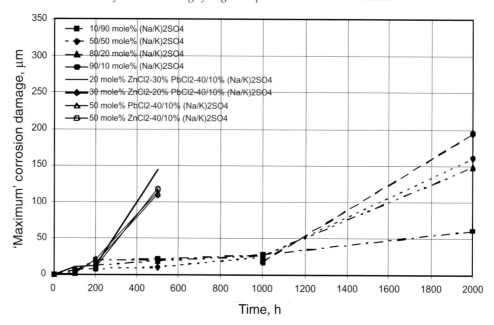

Fig. 8 *Effect of different deposit compositions 'maximum' corrosion damage to IN738LC at 700°C (deposition flux = 5 µg cm^{-2} h^{-1}, SO$_X$ = 2000 vol. ppm, HCl = 350 vol. ppm).*

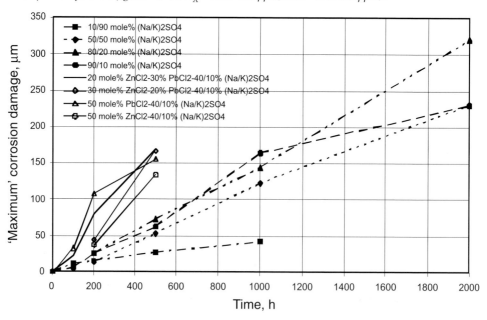

Fig. 9 *Effect of different deposit compositions on 'maximum' corrosion damage to IN738LC at 700°C (deposition flux = 15 µg cm^{-2} h^{-1}, SO$_X$ = 2000 vol. ppm, HCl = 350 vol. ppm).*

illustrated in Fig. 10. These results are typical of the response of gas turbine materials, with the dependence of corrosion rate on deposition flux being approximately sigmoidal with three distinct behaviour regimes:

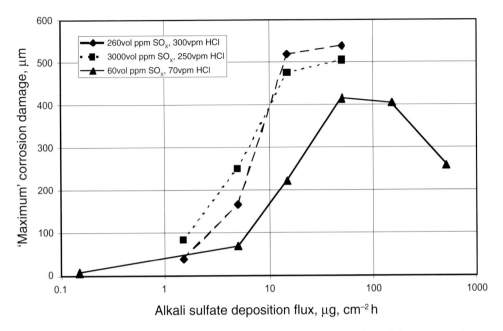

Fig. 10 *Effect of alkali sulfate flux and gaseous reactants on 'maximum' metal damage recession of IN738LC at 700°C [21].*

- for low deposition fluxes there were low corrosion rates;

- at intermediate deposition fluxes, much faster corrosion rates were found with a dependence on deposition flux close to linear;

- at high deposition fluxes, a thick scale/deposit layer was found which effectively creates 'a buried in ash' scenario and slightly reduces the corrosion rate with further increases in flux.

3.3. Effect SO_X and HCl Concentrations

The effect of varying SO_X and HCl concentrations on the hot corrosion damage of IN738LC at 650 and 700°C were evaluated from a series of tests in which each of these contaminants was varied for the same 'deposit' composition (80/20 mole% $(Na/K)_2SO_4$) and a series of 'deposition fluxes'. It was found that the effect of SO_X levels was much more significant than HCl. The corrosion damage (whether expressed as 'maximum' or median values) increased with increasing SO_X levels, but the relationship was not linear, it was closer to logarithmic, with changes between lower levels (e.g. 60–300 vol. ppm) being more significant than changes between higher levels (e.g. 300–3000 vol. ppm). In addition, the effect was more pronounced at lower values of 'deposition flux', as illustrated in Fig. 10.

3.4. Effect of Deposit Composition

The 'deposit' composition used for the majority of the test work was 80/20 mole% $(Na/K)_2SO_4$ (deposit A in Table 2). This deposit was chosen as it corresponded to the lowest melting point composition in the Na–K sulfate mixtures and was close to deposit compositions measured on cooled probes exposed in a coal fired PFBC plant downstream of a hot gas cleaning system. The effect of other mixtures of Na–K sulfates (deposits B–D in Table 2) were investigated to check if they caused significant differences to the corrosion rates. The results for 'maximum corrosion damage' are illustrated for IN738LC in Figs 8 and 9. In addition, the effects of varying mixes of lead and zinc (in addition to 80/20 mole% $(Na/K)_2SO_4$ — i.e. deposits E–K in Table 2) were investigated. The results for maximum corrosion damage to IN738LC given in Figs 8 and 9 show that Pb and Zn additions have significantly more effect than variations in alkali mixes, although these are still significant [e.g. 20]. It should be noted that the lead and zinc, whilst sprayed on to the samples as chlorides, convert to sulfates during the course of the exposures. This is in line with thermodynamic predictions that the volatile chloride present in the gas stream would condense as sulfates on components operating with surface temperatures below their dewpoints.

3.5. Modelling Corrosion Damage

From the growing set of materials performance results, models of corrosion behaviour have been produced from statistical analysis of the metal loss data sets to indicate the sensitivity of corrosion damage to the various exposure parameters [11–14]. These models can be based on several measures of the corrosion behaviour observed, for example 'median metal loss', or 'maximum metal loss' (e.g. damage with a 4% probability of being exceeded) or for type II hot corrosion on the most likely pit depths. In the latter case, a Gumbel type I model of maxima is used to analyse the extreme metal loss corrosion data identified through an initial statistical analysis of each data set [16,17]. The corrosion models for the chosen measure are then generated by curve fitting with respect to the well-characterised exposure parameters. With the currently available data, each model is material and temperature specific.

An early part of this study [11–14] used an extreme value analysis (ie for most likely pit depths) of type II alkali sulfate hot corrosion behaviour evaluated for the upper part of regime (i) and regime (ii) (i.e. alkali sulfate deposition fluxes of 3–30 $\mu g/cm^2$ h^{-1}), at a set temperature and HCl partial pressure. This work demonstrated that, within such limits, the corrosion rate, expressed as the rate of growth of the most likely pit depth, had a parabolic dependence on alkali sulfate deposition flux and a logarithmic dependence on SO_2 partial pressure, for the limited range of conditions studied [e.g. 11,13].

With the much more extensive set of corrosion performance data now available, covering wider ranges of deposition fluxes, gas compositions (SO_x and HCl) and deposit compositions (as illustrated in Figs 6–10), more general corrosion models can be generated. It has been found that a sigmoidal dependence on deposition flux provides a good framework for isothermal corrosion models for a range of materials [21] (Figs 6–10 illustrate the main features for IN738LC), the model therefore takes the form:

$$\text{Corrosion rate} = \frac{a-d}{\left(1+\left(x/c\right)^{b}\right)^{e}} + d \tag{1}$$

where a represents the maximum corrosion rate; d represents the minimum corrosion rate; b governs the slope of the curve in the deposition flux limited regime; c represents the curve inflection point within the deposition flux limited regime; e governs the asymmetry, or skewedness, of the curve with a value of 2.6 (there is a more gentle transition from regime (i) to (ii) than from regime (ii) to (iii), e.g. Fig. 10); x is the effective deposition flux (i.e. the sum of the individual deposition fluxes weighed by their effect relative to alkali sulfate (with a 4:1 ratio of Na:K) — as illustrated in Figs 8 and 9).

Parameters a to d show limited logarithmic dependencies on SO_X and HCl, with the dependence on SO_X being more significant, as illustrated in Fig. 10 — for simplicity, these dependencies can be modelled independently of the flux dependent regime.

These models continue to be developed as more data become available and the assumptions that have had to be made to simplify the analysis of the data become more refined.

4. Conclusions

A series of laboratory corrosion tests has been carried out using the well established 'deposit replenishment' technique in controlled atmosphere furnaces. Test have been carried out to investigate the effects of 'deposition flux', 'deposit' composition (Na, K, Pb, Zn), gas contaminant partial pressures (SO_X and HCl) and exposure times. The performance of the samples has been determined using dimensional metrology on samples before and after exposure to obtain statistical measures of the extent of corrosion damage. The results obtained from these tests, as illustrated using IN738LC, show complex interaction of the different exposure variables. It has been found that the deposition fluxes, deposit compositions and SO_X partial pressures dominate the corrosion behaviour of the materials, with less effect from HCl partial pressures. The rates of corrosion can be modelled by applying statistical analysis of the metal damage measurements, with the resulting parameters fitted as a function of environmental variables, using a sigmoidal flux dependence.

6. Acknowledgements

Funding has been provided by British Coal Corporation, UK Department of Trade and Industry (Clean(er) Coal Programme contracts C/05/00135 and C/05/00266), European Commission JOULE programme (contract JOUF-0022), European Coal and Steel Community programme (contracts 7220-ED/069 and 7220-PR/053).

References

1. W. Schlachter and G. H. Gessinger, in *High Temperature Materials for Power Engineering 1990*, (E. Bachelet *et al.* eds). Kluwer, 1990, pp.1–24.

2. *Proc. Corrosion in Advanced Power Plants*, Special Issue of *Mater. High Temp.*, 1997, p.14.

3. Proc. 1st Int. Workshop on Materials for Coal Gasification Power Plant, in *Materials for Coal Gasification Power Plant*, Special Issue of *Mater. High Temp.*, 1993, p.11.

4. C.T. Sims., N. S. Stoloff and W.C. Hagel, *Superalloys II*. Wiley, New York, 1987.

5. Hot Corrosion Standards, Test Procedures and Performance, *High Temp. Tech.*, 1989, 7 (4).

6. ASTM D2880, Standard Specification for Gas Turbine Fuel Oils (1990).

7. M. Decorso, D. Anson, R. Newby, R. Wenglarz and I. G Wright, *Int. Gas Turbine and Aeroengine Congr.*, Birmingham, UK, ASME Paper 96-GT-76 (ASME, 1996).

8. "Co-gasification of Coal/Biomass and Coal/Waste Mixtures", Final Report EC APAS Contract COAL-CT92-0001, University of Stuttgart, Germany (1995).

9. J. E. Oakey and N. J. Simms, in *Materials for Advanced Power Engineering 1998* (J. Lecomte-Beckers *et al.*, eds). Forschungszentrum Julich GmbH, 1998, pp.651–662.

10. N. J. Simms and J. E. Oakey, in *Microscopy of Oxidation — 3*. The Institute of Materials, 1997, pp.647–658.

11. N. J. Simms, J. E. Oakey and J. R. Nicholls, *Mater. High Temp.*, 2000, **17** (2) 355–362.

12. N. J. Simms, J. E. Oakey, D. J. Stephenson, P. J. Smith and J. R. Nicholls, *Wear*, 1995, 186–187, 247–255.

13. J. R. Nicholls, P. J. Smith and J. E. Oakey, in *Materials for Advanced Power Engineering 1994*, (D. Coutsouradis *et al.*, eds). Kluwer, 1994, pp.1273–1289.

14. N. J. Simms, S. G. Ryder, and J. E. Oakey, "Effects of Contaminants on Materials Performance in Industrial Gas Turbines for Combined cycle Power Plants", Final Report on UK DTI Clean Coal Contract C/00135 (1997).

15. D. H. Allen, J. E. Oakey and B. Scarlin, in Materials for Advanced Power Engineering 1998, (J. Lecomte-Beckers *et al.*, eds). Forschungszentrum Julich GmbH, 1998, pp.1825–1839.

16. J. R. Nicholls and P. Hancock, in *High Temperature Corrosion* (R. A. Rapp, ed.). NACE, 1983, pp.198–210.

17. J. R. Nicholls, N. J. Simms and J. E. Oakey, in *Quantitative Microscopy of High Temperature Materials* (A. Strang and J. Cawley, eds). Published IOM Communications, 2001.

18. D. H. Allen, J. E. Oakey and B. Scalin, in *Materials for Advanced Power Engineering 1998* (J. Lecomte-Beckers *et al.*, eds). Forschungszentrum Julich GmbH, 1998, pp.1825–1839.

19. 'Draft Code of Practice for Discontinuous Corrosion Testing in High Temperature Gaseous Atmospheres', European Commission Project SMT4-CT95-2001 'TESTCORR', ERA Technology, UK (2000).

20. C. Leyens, I. G. Wright, B. A. Pint and P. F. Tortorelli, in *Cyclic Oxidation of High Temperature Materials* (M. Schütze and W. J. Quadakkers, eds). Publication No. 27 in European Federation of Corrosion Series, Published by The Institute of Materials, London, 1999, pp.169–186.

21. N. J. Simms, J. R. Nicholls and J. E. Oakey, *Mater. Sci. Forum*, to be published.

18

Influence of the Salt Composition on the Hot Corrosion Behaviour of Gas Turbine Materials

B. WASCHBÜSCH and H. P. BOSSMANN

ALSTOM Power, Dept. GTEM.E, Haselstrasse 16, CH-5401 Baden, Switzerland

ABSTRACT

The hot corrosion behaviour of three base materials, IN738®, CM247® and CMSX-4®*, and also SV20, a NiCrAlY-coating material, was studied in air and 300 ppm SO_2 at 800°C. For the simulation of gas-fired turbines with back-up fuel, a salt-spraying test was chosen and the salt was applied once before exposure to temperature. For each material, specimens with Na_2SO_4 and Na_2SO_4/K_2SO_4 were exposed for 500 and 1000 h. NaCl-containing salts were only applied to CMSX-4. The present investigation has established that the addition of K_2SO_4 and/or NaCl to Na_2SO_4 causes shorter incubation periods and higher corrosion rates. IN738 has shown a good resistance against hot corrosion with incubation times up to 700 h. The corrosion resistance of CM247 and CMSX-4 was very poor and the incubation times were less than 45 h. In a corrosive environment, both alloys have to be protected by an oxidation- and corrosion-resistant overlay coating. SV20 has exhibited an excellent corrosion resistance with incubation times >1000 h.

1. Introduction

Hot corrosion in gas turbines is caused by deposits, which are formed by impurities of the combustion air. The composition and the dew point of such deposits depend on the pressure and the amount of impurities. Some thermodynamic calculations of the dew point of different salts have been reported [1]. With a given composition of fuel and air, the dew point and the composition of the deposits can be calculated and the hot corrosion risk can be evaluated at each place in the gas turbine.

Stationary gas turbines have in addition to the standard fuel a back-up fuel to guarantee the production of electricity. For gas-fired turbines, diesel No. 2 is often used as back-up fuel. The turbine is usually fired with back-up fuel for several hours per month. As a general rule, the amount of impurities in diesel is higher than in gas. This leads to an increased risk of corrosion. To reproduce the hot corrosion in the laboratory, a salt-spraying test [2–4] was chosen because the amount of salt in the atmosphere during the test can be well defined. For the simulation of gas-fired turbines with a diesel back-up fuel, the salt was applied once before exposure to temperature. The results of such corrosion tests can be implemented in the lifetime prediction modelling for components in gas turbines.

---

*IN738® is a registered trademark of Special Metals Corporation; CM247® and CMSX-4® are registered trademarks of Cannon-Muskegon Corporation.

In order to discuss the hot corrosion of superalloys it is convenient to define two stages — the incubation stage and the propagation stage [5]. The incubation stage is characterised by a behaviour close to the behaviour without deposit. During the propagation stage, the initially formed oxide scale is no longer protective and the corrosion rate is increased. The deposit affects the time during which the alumina or chromia scales are protective towards the superalloys.

2. Experimental

Three commercial base material alloys, IN738, CM247 and CMSX-4, and also SV20 were tested in this study. Their chemical compositions are given in Table 1. IN738 was investigated as conventionally cast material, CM247 as directional solidified material, CMSX-4 as single crystal material. The specimens of CM247 and CMSX-4 were cut perpendicular to <001>. Powder of the coating material SV20 was hot isostatically pressed to bulk material and then cut into specimens. All the specimens were manufactured by spark erosion to the dimensions $10 \times 10 \times 1$ mm.

Na_2SO_4-coated and Na_2SO_4/K_2SO_4-coated specimens of these materials were exposed at 800°C for 500 and 1000 h. The salt mixture of Na_2SO_4/K_2SO_4 was composed of 80mol% Na_2SO_4 and 20mol% K_2SO_4, which is a composition near the eutectic point. The melting point of this salt mixture was determined by differential thermoanalysis to be 831°C. The composition and the melting points [6] of the applied salts are listed in Table 2. Salts containing NaCl were only applied to CMSX-4. These specimens were exposed for 500 h at 800°C. The reference state was a specimen without salt deposit oxidised in the same atmosphere.

The specimens were ground to 800grit and ultrasonically cleaned in ethanol for 5 min. The salt was applied to the specimen, heated on a plate at ~80°C, the water-soluble salts being sprayed on the specimen as a 0.02 mol-solution. The water evaporates rapidly and the salt forms a thin salt layer on the specimen. Only one side was coated. A salt deposit of 1 mg cm^{-2} (±0.03 mg) was used as the standard. The specimens were isothermally exposed to synthetic air and 300 ppm SO_2. The flow rate was 60 Lh^{-1}. The specimens were pulled out of the furnace every 2 days and cooled down to room temperature to measure the mass change.

After the exposure in the corrosive environment, the surface of the corrosion product was investigated by X-ray diffraction and electron microprobe (EMPA) and

Table 1. Chemical composition of the investigated materials in mass%

mass%	Ni	Co	Cr	Al	Y	Si	Ta	Ti	Nb	Hf	Re	W	Mo
IN738	62.6	8	16	3.4	–	–	1.7	3.3	1	–	–	2.5	1.5
CM247	62.3	9	8	5.6	–	–	3.2	0.7	–	1.4	–	9.5	0.5
CMSX-4	60.4	10	6,6	5.5	–	–	6.5	1	–	0.1	3	6.4	0.6
SV20 (NiCrAlY)	Bal.	–	23–27	3–7	0.2–2	1–3	0.2–2						

Table 2. *Composition and melting points of the applied salts. The NaCl-containing salts were only applied on CMSX-4 specimens*

Salt mixture	Na$_2$SO$_4$	K$_2$SO$_4$	NaCl	Melting point
1	100%			884°C [6]
2	80mol%	20mol%		831°C [1]
3	48mol%		52mol%	628°C [6]
4	40.5mol%	15.6mol%	43.9mol%	~635°C [6]

then cross sections of the corroded specimens were prepared. The metallographic preparation was performed without any water, to preserve the water soluble corrosion product on the specimens. The corrosion product was characterised by EMPA and the depth of the corrosion was measured.

3. Results

3.1. Reference Samples Without Salt (Oxidation in Air + 300 ppm SO$_2$)

Oxidation treatment was performed in the same atmosphere (air + 300 ppm SO$_2$) without salt deposit. The four investigated materials can be divided into two groups: IN738 is a chromia-former, CM247, CMSX-4 and the NiCrAlY-coating SV20 are alumina-formers. The specific mass gain of IN738, CMSX-4 and CM247 is plotted versus time in Figs 1, 2 and 3, respectively. The parabolic rate constant (k_p-value) of

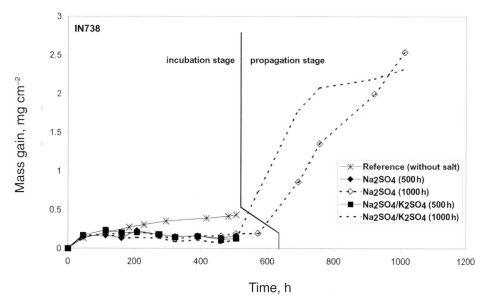

Fig. 1 *Specific mass gain of IN738 in dry air and 300 ppm SO$_2$ at 800°C.*

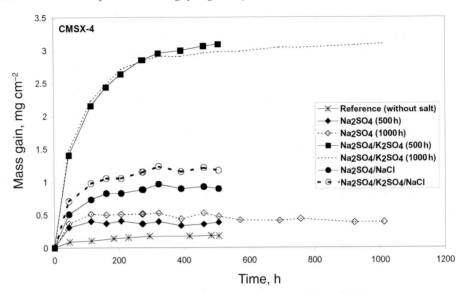

Fig. 2 *Specific mass gain of CMSX-4 in dry air and 300 ppm SO$_2$ at 800°C.*

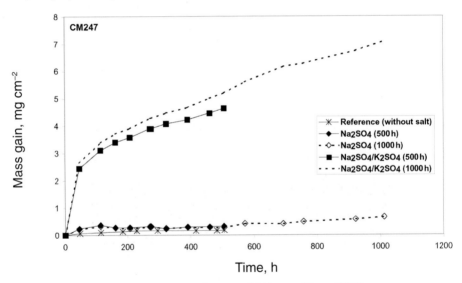

Fig. 3 *Specific mass gain of CM247 in dry air and 300 ppm SO$_2$ at 800°C.*

oxidation in air + 300 ppm SO$_2$ one similar to those reported in [1]. These reference samples show no evidence of internal corrosion in the form of sulfidation.

3.2. Samples with Salt Deposit

3.2.1. IN738

For both salt coatings, IN738 shows the typical behaviour for type I corrosion with incubation and propagation stages. The specific mass gain of the salt coated surface is plotted vs time in Fig. 1. Up to 200 h, the mass gain of the salt-coated specimens is comparable to that of the reference sample. Until the end of the incubation stage, the salt-coated specimens show a small mass loss, which could be explained by evaporation of sulfates. The incubation period takes 600–700 h for the Na_2SO_4-coating and 500–600 h for Na_2SO_4/K_2SO_4-coated sample (Table 3). Within this period the material shows protective behaviour. During the propagation stage the corrosion rate is nearly linear. After ~800 h, the Na_2SO_4/K_2SO_4-coated sample no longer follows linear kinetics. Partial spallation of corrosion products could be a reason for this behaviour.

Both salts induce the same corrosion morphology. During the incubation stage, this consists only of a small internal sulfidation of chromium (Fig. 4). Oxides of Cr and Ti make up the outer oxide scale. Underneath the oxide scale, aluminium is oxidised internally. After 1000 h (propagation stage), corrosion occurs only directly below the salt and starts preferentially on the precipitates of carbides of heavy elements such as W, Ta, Nb. The outer scale still consists of oxides of Cr and Ti

Table 3. Duration of the incubation time as function of the applied salt composition

Salt mixture / Material	1 Na_2SO_4	2 Na_2SO_4/K_2SO_4	3 $Na_2SO_4/NaCl$	4 $Na_2SO_4/K_2SO_4/NaCl$
IN738	600–700 h	500–600 h		
CMSX-4	<45 h	<45 h	<45 h	<45 h
CM247	<45 h	<45 h		
SV20	>1000 h	>1000 h		

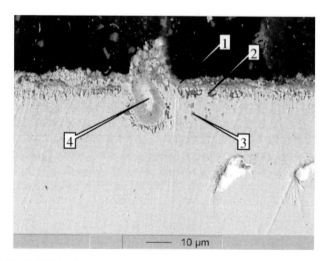

Fig. 4 IN738 800 °C/ 507 h, air + 300 ppm SO_2, Na_2SO_4-deposit; 1: Cr- and Ti-oxides; 2: internal oxidation of Al; 3: Cr-sulfides; 4: precipitate of W, Ta, Nb.

(Fig. 5) and covers a mixture of oxides of Cr, Ti, Ni and heavy metals like W, Ta, Nb. A small scale of alumina separates these mixed oxides from the metal. Under these oxides, a zone of internal sulfidation of Cr and Ni appears.

3.2.2. CMSX-4

After 500 h, the corrosion of the Na_2SO_4-coated sample is localised directly below the applied salt (Fig. 6). It consists mainly of internal sulfidation of Cr and first corrosion pits of about 8 µm. After 1000 h, the corrosion is more pronounced (Fig. 7) and the maximum depth reaches 40 µm. The corrosion has a pitting-type morphology as reported in [1]. The external scale is mainly composed of NiO. Oxides of Cr and Al form the darker scale below. The pit consists of mixed oxides. The brighter areas

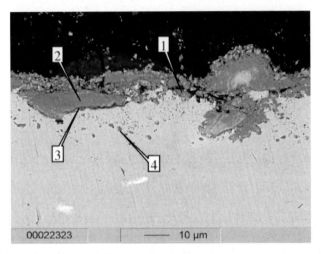

Fig. 5 *IN738 800°C 1000 h, air + 300 ppm SO_2, Na_2SO_4/K_2SO_4-deposit, 1: Cr- and Ti-oxides; 2: mixed oxides of Cr, T, Ni, W, Ta, Nb; 3: alumina; 4: Cr-, Ni-sulfides.*

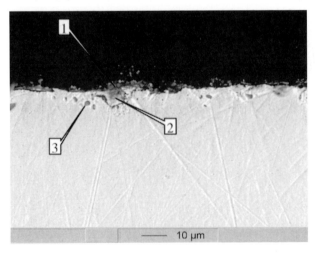

Fig. 6 *CMSX-4, 800°C/500 h, air + 300 ppm SO_2, Na_2SO_4-deposit; 1: Na_2SO_4; 2: corrosion pit; 3: Cr-sulfides.*

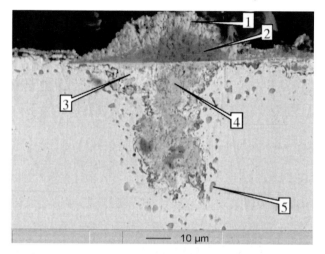

Fig. 7 *CMSX-4, 800°C/1000 h, air + 300 ppm SO$_2$, Na$_2$SO$_4$- deposit; 1: Ni-oxides; 2: Cr-, Al-oxides; 3: Ni-W-oxides; 4: Al-Cr-oxides; 5: Cr-sulfides.*

are Ni–W–oxides, the dark ones Al–Cr–oxides. Around the pits a zone of internal sulfidation of chromium is located. The growing of the pits cannot be detected in the mass gain curve (Fig. 2). If one refers to the definition of propagation stage given in [5], the duration of the incubation stage is shorter than 45 h for all CMSX-4 samples irrespective of the applied salt.

The corrosion is more severe, if other salts, especially K$_2$SO$_4$, are present. Although the morphology and the composition of the corrosion are the same for the Na$_2$SO$_4$-coated alloy after 1000 h as for the Na$_2$SO$_4$/K$_2$SO$_4$-coated sample (Fig. 8) after 500 h. The Na$_2$SO$_4$/K$_2$SO$_4$-salt on CMSX-4 leads to the highest mass gain and the deepest corrosion compared to the other salts. The eutectic mixture of Na$_2$SO$_4$ and NaCl

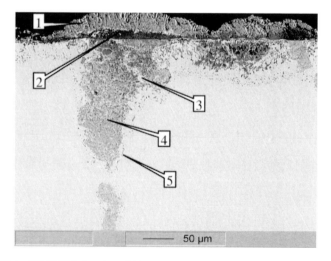

Fig. 8 *CMSX-4, 800°C/500 h, air + 300 ppm SO$_2$, Na$_2$SO$_4$/K$_2$SO$_4$-deposit; 1: Ni-oxides; 2: Cr-, Al-oxides; 3: Ni-W-oxides; 4: Al-Cr-oxides; 5: Cr-sulfides.*

increases the corrosion rate in a severe manner compared to pure Na_2SO_4 (Fig. 9). The composition of the corrosion product is comparable to the samples without NaCl. No chlorine could be identified in an element mapping of the corrosion product. The mixture of Na_2SO_4, K_2SO_4 and NaCl causes a higher specific mass gain (Fig. 2) and a similar pit depth (Fig. 10) compared with the eutectic mixture of Na_2SO_4 and NaCl.

3.2.3. CM247

CM247 and CMSX-4 show similar hot corrosion behaviour. The corrosion caused by Na_2SO_4 (Fig. 11) is less pronounced compared to CMSX-4 (Fig. 7). As before, the addition of K_2SO_4 increases the corrosion (Fig. 12). The composition of the pits is comparable to the Na_2SO_4-coated sample of CMSX-4, described above. The specific

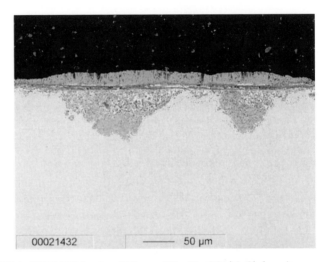

Fig. 9 CMSX-4, 800°C/500 h, air + 300 ppm SO_2, Na_2SO_4/NaCl-deposit.

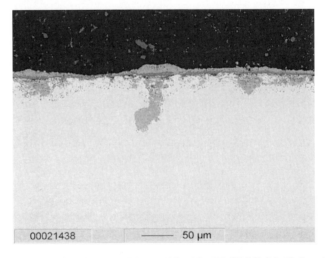

Fig. 10 CMSX-4, 800°C/500 h, air + 300 ppm SO_2, Na_2SO_4/K_2SO_4/NaCl-deposit.

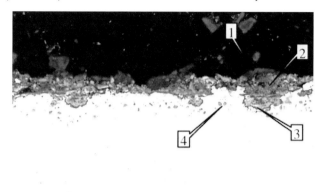

Fig. 11 *CM247, 800°C/1000 h, air + 300 ppm SO$_2$, Na$_2$SO$_4$-deposit; 1: Na$_2$SO$_4$; 2: Al–Cr–oxides; 3: Ni-W-oxides; 4: Cr-sulfides.*

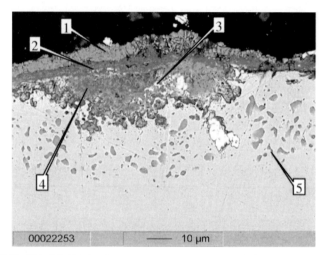

Fig. 12 *CM247, 800°C/1000 h, air + 300 ppm SO$_2$, Na$_2$SO$_4$/K$_2$SO$_4$- deposit; 1: Ni-oxide; 2: Al–Cr–oxides; 3: Ni–W-oxides; 4: Al–Cr-oxides; 5: Cr-sulfides.*

mass gain curve (Fig. 3) shows a comparable behaviour to that of CMSX-4. The specific mass gain of the samples with salt is significantly higher than for the reference sample without salt. Thus, the duration of the incubation stage can be set to <45 h (Table 3).

3.2.4. SV20
Up to 1000 h, SV20 is still in the incubation stage. The only sign of corrosion is the localised formation of Y-sulfides below the salt (Fig. 13). The Y-sulfides are growing with exposure time (Fig. 14). Yttrium can act as a sulfur-getter because the Y-sulfides are thermodynamically more stable than Cr-sulfides [7]. Between the Y-sulfides and the applied salt, a small oxide scale consisting of Cr- and Al-oxides is present (1–2 µm).

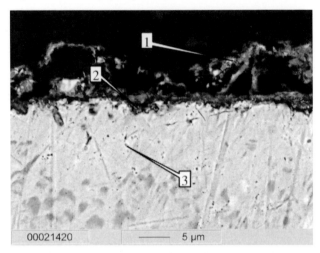

Fig. 13 *SV20, 800°C/500 h, air + 300 ppm SO$_2$, Na$_2$SO$_4$-deposit; 1: Na$_2$SO$_4$; 2: Al–Cr–oxides; 3: Y-sulfides.*

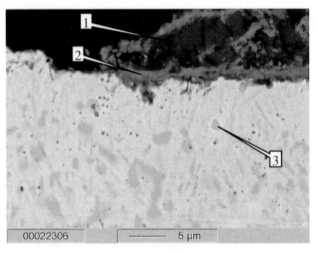

Fig. 14 *SV20, 800°C/1000 h, air + 300 ppm SO$_2$, Na$_2$SO$_4$-deposit; 1: Na$_2$SO$_4$; 2: Al–Cr–oxides; 3: Y-sulfides.*

4. Discussion

The reference samples show no internal corrosion such as, sulfidation. All the salt-coated samples suffer from type I corrosion. The incubation period and the corrosion rate depend on the applied salt (Table 3) and the material. The four tested materials can be ranked with decreasing corrosion resistance in the order: SV20 > IN738 > CM 247, CMSX-4. As a first approach, the corrosion resistance depends on the Cr-content. SV20 was developed to achieve a good oxidation and corrosion resistance [8], which is caused by the high chromium-content and the absence of heavy elements. The high chromium-content and the aluminium content result in the formation of a protective oxide scale. Yttrium improves the oxidation behaviour and acts as sulfur-getter by the formation of Y-sulfides. The chemical composition of SV20 leads to a long incubation time of greater than 1000 h under the present conditions. To shorten the incubation time, the supply of salt can be increased, e.g. by re-applying the salt every 1–20 h.

IN738 with 16mass% Cr, is a chromia-former and shows quite good corrosion resistance. The mass gain curve (Fig. 1) indicates well the transition between the incubation and propagation stages. Fryburg *et al.* [9] have investigated the hot corrosion behaviour of Na_2SO_4-coated IN738 in pure O_2 at 975°C. The corrosion mechanism proposed in [9] can be summarised as follows. During the incubation stage, Na_2SO_4 dissolves Cr_2O_3, the protective oxide scale, via basic fluxing to form Na_2CrO_4. Na_2CrO_4 reacts with WO_3 and MoO_3 to form Na_2MoO_4 and Na_2WO_4. This results in a molten phase of MoO_3–WO_3/ Na_2MoO_4–Na_2WO_4. This molten phase leads to a change of mechanism to alloy-induced acidic fluxing, which is characterised by a linear corrosion rate. Whereas the basic fluxing needs a continuous supply of sulfate, the alloy-induced acidic fluxing is self-sustaining [10]. Afterwards the corrosion rate decreases due to conversion of the molten MoO_3–WO_3/Na_2MoO_4–Na_2WO_4 to $NiMoO_4$ and $NiWO_4$. In the present investigation, $NiWO_4$ is found in the corrosion product. This indicates that even at 800°C, the corrosion could proceed by the mechanism proposed by Fryburg *et al.* [9]. The validation of this assumption needs further investigations.

CMSX-4 and CM247 are two base materials with a chemical composition optimised for an adequate mechanical behaviour. Both alloys have a low chromium content (Table 1) and are alumina-formers at higher temperatures. As the reference samples show, aluminium is oxidised internally and cannot form a continuous scale. Different oxides and chromia compose the outer scale. The incubation period of the samples with salt deposit ends before the first measure point at 45 h. To determine the exact incubation time, short-term exposures need to be done. The short incubation period of the salt-coated specimens can be explained as follows. The sulfur diffuses into the metal and forms chromium sulfides, which induce chromium depletion below the surface. At this moment the chromium- and aluminium- contents below the surface are not high enough to form a protective oxide scale. This leads to a transition to the propagation stage with corrosion of the base material. In the corrosion products of CMSX-4 and CM247, a W-rich component was identified as $NiWO_4$ by X-ray diffraction. The high tungsten content of 6.4–9.5mass% seems to be detrimental for the hot corrosion resistance. A change from basic fluxing to alloy-induced acidic fluxing could be responsible, as proposed for IN738 [9] or for the alumina-former such as B-1900 and NASA-TRW VIA [11].

The addition of K_2SO_4 to Na_2SO_4 results in shorter incubation periods (Table 3) and higher corrosion rates (Figs 1–3). The lower melting point compared to pure Na_2SO_4 could be a reason for this behaviour. The addition of K_2SO_4 to Na_2SO_4 causes the highest corrosion rate even if the melting point is only lowered by ~50°C. An influence of K_2SO_4 on the corrosion mechanism could not be identified in this study. The composition and morphology of the corrosion product is comparable to that caused by pure Na_2SO_4.

Even the addition of NaCl to Na_2SO_4 or to Na_2SO_4 /K_2SO_4 does not lead to a different corrosion product. NaCl needs a certain partial pressure of Cl-compounds to be thermodynamically stable. Thus, a part of NaCl could evaporate during the experiment or react with SO_2 to Na_2SO_4. This could explain the absence of Cl-compounds in the corrosion product. Other authors [12] propose, that NaCl affects the mechanism and is absent in the product due to volatile metal chlorides. Detailed investigations on the incubation period are needed to describe exactly the influence of NaCl.

5. Conclusions

The present investigation has established that the addition of K_2SO_4 and/or NaCl to Na_2SO_4 causes shorter incubation periods and higher corrosion rates. Even if Na and S are the prime relevant corrosion species in a stationary gas turbine, K and Cl are often present and cannot be neglected for hot corrosion. Especially, power plants close to the sea can experience high amounts of NaCl in the air. To get a corrosion test close to the conditions in the turbine, sulfur, sodium, potassium and chlorine should be considered. For a corrosion-free operation, the maximum impurity amount and not the mean value in the hot gas at the turbine inlet is crucial.

IN738 and SV20 are appropriate materials to be used in turbines with diesel as back-up fuel. Due to the incubation time of 600–700 h for IN738, the salt could be removed, e.g. by washing procedures, and corrosion could be avoided. The corrosion resistance of CM247 and CMSX-4 is very poor and the incubation times are less than 45 h. This leads to different consequences for the use in gas turbines. Deposition of salts must be completely avoided, if these materials are to be used without a corrosion-resistant coating. In all other cases, CM247 and CMSX-4 must be protected by an oxidation- and corrosion-resistant overlay coating, like SV20.

The salt-spraying test is an appropriate test method for hot corrosion testing in gas turbines. Corrosion rates and mechanisms can be studied for the elaboration of lifetime prediction models. For the simulation of oil-fired gas turbines or for the testing of highly corrosion-resistant materials such as SV20, the test should be modified with a continuous supply of salt, e.g. by reapplying the salt every 1–20 h.

References

1. B. Waschbüsch, H. P. Bossmann and L. Singheiser, Proc. EUROCORR 2000, London, UK, Paper 014-80 (2000). Published on CD ROM, The Institute of Materials, London, 2000.

2. K. L. Luthra and D. A. Shores, Mechanism of Na_2SO_4 induced corrosion at 600–900°C, *J. Electrochem. Soc.*, 1980, **127**, (10), 2202–2210.

3. K. L. Luthra, Kinetics of the low temperature hot corrosion of Co–Cr–Al-alloys, *J. Electrochem. Soc.*, 1985, **132**, (6), 1293–1298.

4. K. L. Luthra and J. H. Wood, High chromium cobalt-base coatings for low temperature hot corrosion, *Thin Solid Films*, 1984, **119**, 271–280.

5. F. S. Pettit and G. H. Meier, Oxidation and hot corrosion of superalloys, in *Superalloys 1984*, (M. Gell *et al.*, eds). The Metallurgical Society of AIME, Warrendale, PA, 1984, pp.651–687.

6. E. K. Akopov and A. G. Bergman, *Zhur. Neorg. Khim.*, 1959, **4**, (7), 1655.

7. L. Singheiser, Untersuchung zur Reduktion der Hochtemperaturkorrosion metallischer Werkstoffe durch legierungstechnische Maßnahmen und Beschichtungen, *Habilitationsschrift*, Erlangen 1991.

8. L. Singheiser, Europäische Patentschrift, Veröffentlichungsnr.: 0 318 803 B1 (1993).

9. G. C. Fryburg, F. J. Kohl and C. A. Stearns, Chemical reactions involved in the initiation of hot corrosion of IN738, *J. Electrochem. Soc.*, 1984, **131**, (12), 2985–2997.

10. J. A. Goebel, F. S. Pettit and G. W. Goward, Mechanisms for the hot corrosion of nickel-base alloys, *Met. Trans.*, 1973, **4**, 261–278.

11. G. C. Fryburg, F. J. Kohl, C. A. Stearns and W. L. Fielder, Chemical reactions involved in the initiation of hot corrosion of B-1900 and NASA-TRW VIA, *J. Electrochem. Soc.*, 1982, **129**, (3), 571–585.

12. D. A. Shores, D. W. McKee and H. S. Spacil, The effect of NaCl on high temperature hot corrosion, General Electric Report No. 76CRD253 (1976) 1-7.

19

High-Temperature Cyclic Oxidation Behaviour of a Hot-Dip Aluminium-Coated 12%Cr Stabilised Ferritic Stainless Steel

L. ANTONI and B. BAROUX

Usinor R&D, Ugine Research Center, Ugine, France

ABSTRACT

The long term high-temperature cyclic oxidation behaviour of a hot-dip aluminium-coated 12%Cr titanium stabilised ferritic stainless steel has been investigated in air between 650 and 850°C. The degradation depths remain small (≤ 40 µm) while keeping a nice visual aspect even after 1200 cycles (equivalent to 400 h at the test temperature). The most important part of the degradation occurs during the first cycles and develops slowly thereafter when inter-diffusion between aluminium coating and stainless steel leads to the formation of an Fe–Cr–Si–Al coating, with a gradient of Al and Si concentration from the surface to the base metal.

Two types of degradation are observed: thin cracks within the coating due to thermal cycling and localised oxidation of the underlying steel due to the applied initial bending deformation of the samples. Owing to Al diffusion into the metal, the former defects are rapidly stopped by the formation of a protective alumina scale. The high cyclic oxidation resistance of the underlying Fe12CrTi grade ensures a slow growth of the latter.

1. Introduction

The introduction of catalytic exhaust lines to comply with state regulations in the automotive industry, and the even stricter environment standards on car gas emissions expected for the year 2002, imply the need for complex and new designs for exhaust systems. Moreover, greater time warranties are offered on such parts so that they can no longer be considered as consumable items of a vehicle. This generates ever-growing demands from car manufacturers for durable, heat-resistant and corrosion-resistant materials, such as stainless steels.

The search for performance at a reduced price puts the focus more specifically on high quality ferritic stainless steels. In that respect, and to meet market needs, aluminium-coated 12%Cr steel has been developed [1,2]. Hot-dip aluminium coatings on stainless steels combine the advantages of a suitable corrosion and oxidation resistance of aluminium with the advantageous mechanical behaviour of stainless steels at high temperature. Moreover, even if the coating fails, the underlying standard titanium stabilised ferritic stainless steel Fe12CrTi ensures a sufficient corrosion resistance that will limit the degradation of the assemblage while maintaining an aesthetically flawless aspect. The metallurgical structure and mechanical properties of the base material ensure good formability of hot-dip aluminium coated Fe12CrTi.

Standard welding techniques can also easily be applied. Tubes are already available where the welding seam is protected by spraying molten aluminium on the welding zone. These features give hot-dip aluminium coated Fe12CrTi excellent suitability for transformations commonly found in manufacturing of exhaust lines: tube forming, deep-drawing, etc. They are thus proposed today for the intermediate and cold parts of the automotive exhaust systems.

This paper addresses more specifically the service conditions of hot-dip aluminium-coated Fe12CrTi. This topic has indeed hardly ever been scientifically studied. Some literature is available [3–7], but this mostly focuses on hot-dip aluminium coated carbon steel corroded or oxidised under continuous conditions. The aim of this contribution is to investigate the long term high temperature cyclic oxidation behaviour of these aluminium coated stainless steels. When designing materials for exhaust systems, the resistance to thermal cycling more than the isothermal resistance has to be taken into consideration since they are subjected to frequent temperature variations [8], and this can strongly affect their lifetime in service conditions [9].

The degradation depths measured after cyclic oxidation in air up to 1200 cycles (equivalent to 400 h at the test temperature) at 650, 750 and 850°C are presented. The comparative performance is discussed with regard to the types of degradation observed, the evolution of the coating and the interdiffusion between the aluminium–silicon coating and the underlying stainless steel.

2. Materials and Experimental Procedure

2.1. Materials

A standard titanium stabilised 12%Cr ferritic stainless steel Fe12CrTi (1.4512, AISI 409) was employed as base material (chemical composition (wt-%): 11.5Cr–0.5Si–0.2Mn–0.05C–0.16Ti wt%). The adherent and regular coating (*ca.* 20 µm thick on both sides) was obtained by immersion in a metallic molten bath of Al–10%Si [10]. The micrographic section (Fig. 1) showed that the coating was in fact composed of two layers, with an Fe–Cr–Si–Al intermetallic compound between the stainless steel and the coating. This intermetallic compound was very brittle and cracks appeared rapidly when the material was subjected to deformation.

Rectangular samples 20×30 mm^2 were punched out of a 1.1 mm thick industrial plate. Samples were cleaned in alcohol-acetone with an ultrasonic cleaner. A 2.5% bending deformation was imposed to the hot-dip aluminium coated samples to reproduce the metal deformation during tube manufacturing. This allowed the oxidation resistance to be investigated taking into account the possible breaking or spalling of the coating. Three samples of each selected grade were then put in a high-temperature resistant metallic sample holder and placed in the furnace.

2.2. Cyclic Oxidation

The objective of the cyclic test was to reproduce service conditions of a normal exhaust system, especially during starting and stopping of the vehicle, with short alternating heating and cooling of the engine. Standard vertical furnaces were used. Temperature

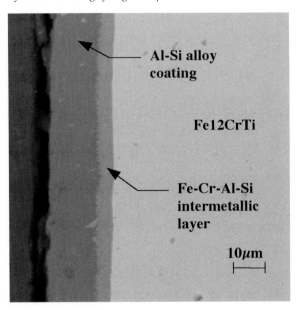

Fig. 1 *Cross-section micrography (SEM) of the aluminised Fe12CrTi grade.*

was regulated directly with a thermocouple placed in the middle of the sample holder. The temperature homogeneity was carefully controlled in the central zone to make sure that all samples went through the same heating cycle. A special automated apparatus was added to introduce the samples successively into the furnace and take them out for intermediate cooling cycles [8].

The typical heating/cooling cycle comprised a 20 min heating stage (time count started only when the samples reached the testing temperature), and a 5 min rapid cooling stage under air jets. Heating time from room to test temperature was between 15 and 17 min, while cooling to room temperature (303 K) lasted about 2 min. Total holding times up to 400 h were selected, i.e. up to 1200 heating cycles. New samples were used after each measurement between 1 and 1200 cycles. Tests were performed in laboratory air at 650, 750 and 850°C, which are realistic temperatures measured on the outer surfaces of parts behind the automotive manifold.

2.3. Data Processing

Standard techniques such as mass changes or metal losses could not be implemented on aluminum coated materials since they would only provide information about the formation of the alumina protective layer which is not representative of the degradation of such coated materials [1]. Therefore, micrographic observations of cross-sections of the samples were used. The aluminum coating is considered as sacrificial. The degradation of the material corresponds to the occurrence of defects or oxidation affecting the underlying stainless steel. The difference between the non-affected metal thickness after cyclic oxidation and the initial metal thickness gave the thickness of the affected metal by cyclic oxidation. These values were measured by image analysis on Scanning Electron Microscope (SEM) pictures.

SEM/EDX and Electron MicroProbe Analysis (EMPA) on cross sections have also been carried out to investigate the type of oxides formed, the nature of the coating and the inter-diffusion depth between the coating and the stainless steel.

3. Results

3.1. Degradation Kinetics

The influence of thermal cycling on maximum loss of non-affected metal thickness (i.e. maximum metal damage from both sides of the samples) is presented in Fig. 2. At 750 and 850°C, the degradation escalated in the first hundred cycles before reaching a quasi steady-state after 150 cycles. This increase was even sharper and the steady-state degradation value even lower as the temperature was raised. Some experiments at 750°C were continued up to 2400 cycles (i.e. 800 h at test temperature) without any deviation from the steady-state value.

At 650°C, the material behaved slightly differently. The initial degradation in the first cycles remained high (50% of the total degradation was reached after 150 cycles) but rose slower than at the previous temperatures. Moreover, no plateau arose and the decrease of non-affected metal thickness appeared to follow a nearly linear

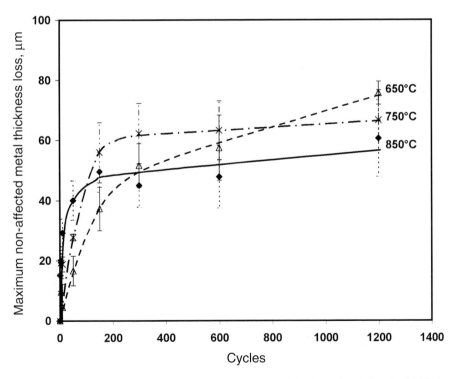

Fig. 2 *Evolution of the maximum loss of non-affected metal thickness (total from both sides) vs cycles under cyclic oxidation in air at 650, 750 and 850°C for the hot-dip aluminium coated Fe12CrTi.*

evolution between 300 and 1200 cycles. After around 800 cycles, this led to a higher non-affected metal thickness loss than for the higher investigated temperatures.

Nevertheless, this degradation remained low with regard to the initial material thickness (1.1 mm). After 1200 cycles, the maximum losses of non-affected metal thickness were respectively 76, 67 and 60 µm at 650, 750 and 850°C, i.e. less than 40 µm for each sample side.

Thus, even after a mechanical deformation, the hot-dip aluminium coated Fe12CrTi steel appeared resistant to long term severe cyclic oxidation conditions up to 850°C.

These results indicating that the cyclic oxidation resistance of an aluminium coated stainless steel is improved when increasing the temperature from 650 to 850°C might appear as quite surprising. The observation of cross-sections described below may clarify this phenomenon.

4. Surface Morphologies and Analysis

4.1. Surface Visual Aspect

The visual observation of the different samples revealed that the surface remained homogeneous whatever the testing conditions. The aluminium coating endowed the surface samples with a greyish colouring which turned continuously darker when raising the number of thermal cycles or the temperature but remained quite similar to the initial appearance. This continuous and homogenous evolution of the visual perception of the material meets perfectly with today's requirements of the automotive industry.

4.2. Modes of Degradation

SEM and optical cross-section micrographies of the samples were carried out to follow the evolution of the coating and the intermetallic layer. Two types of defects, described below, were observed whatever the test temperature: transversal thin cracks within the coating (Fig. 3a) and localised oxidation of the underlying steel (Fig. 3b).

The inter-diffusion between aluminium coating and stainless steel occurred rapidly at such temperatures. It was of course even quicker as the temperature increased. It led to the formation of a continuous Fe–Cr–Al–Si intermetallic layer, as illustrated in Fig. 3(a) after only 5 thermal cycles at 750°C, with a gradient of Fe concentration from the steel to the surface and a gradient of Al and Si concentration from the surface to the base metal.

The coating can be described from the surface to the underlying Fe12CrTi steel by the following: an outer Fe–Al–Si layer which corresponds to the Fe diffusion into the initial Al–Si coating. As this disappeared with time, the underlying layer (composed of a regular Fe–Al layer in which Cr–Si nodules precipitated) increased. The Fe concentration rose with the high temperature holding time. The Fe–Cr–Al–Si intermetallic layer below can be divided into 2 parts. The outer one is rich in Si (~4–6%) and Al (gradient between 47 and 8%). The inner one corresponded to a sharp decrease of the Si content up to the nominal concentration of the stainless steel (0.5%), the continuous decrease of Al and the base metal Cr content (10–11%). This difference in the intermetallic layer tended to disappear with increases in temperature and thermal cycle numbers (Fig. 3b), indicating an homogenisation by the diffusion

A

Fe12CrTi

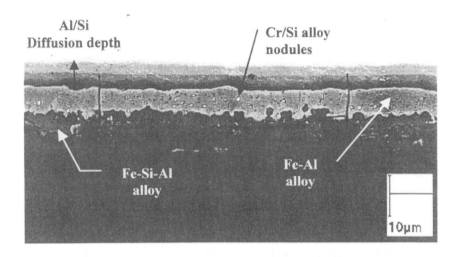

B

Fe12CrTi

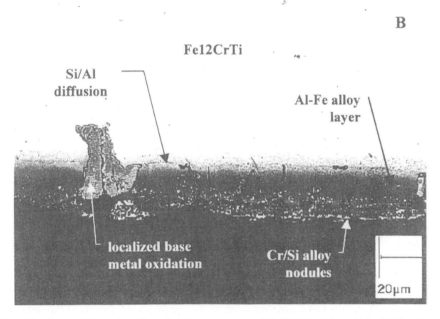

Fig. 3 *Surface morphologies of hot-dip aluminium coated Fe12CrTi after cyclic oxidation in air at 750°C (a) cracks in the coating after 5 thermal cycles; (b) cracks and localised base metal oxidation after 50 cycles.*

of Al and Si into the stainless steel. The latter was, of course, observed even deeper as the temperature was increased (Table 1). Al diffusion remained low (15 μm) at 650°C even after 1200 cycles while it occurred throughout nearly all of the sample after 1200 cycles at 850°C.

4.3. Transverse Surface Cracks

Small transverse cracks located close to the material surface have been observed (Fig. 3a). These cracks generally stopped near the border between the 2 parts of the inner intermetallic layer described above, i.e. where Al and Si contents returned to low values. This corresponded to the decrease in the brittleness of the intermetallic layer. This was confirmed by precise microanalysis of the samples (Fig. 4). It also showed that a protective aluminium oxide layer was formed inside the cracks, therefore avoiding a catastrophic spreading of oxidation but also keeping a suitable wet corrosion resistance. As a matter of fact, only low internal oxidation of aluminium was observed, mainly along grain boundaries.

Further study of the formation of cracks showed that they were formed in the first stages of the test. This is illustrated by the development of the density of these cracks with regard to the test temperature and the number of thermal cycles. Figure 5 shows this devlopment for the inner side of the bent sample. Similar shapes and densities were measured on the outer side. The densities were even more important on this external side and reached a stable value even quicker as the test temperature was raised. The latter occurred instantaneously at 850°C, between 50 and 100 cycles at 750°C and only after 300 cycles at 650°C.

Table 1. *Depth of aluminium diffusion after 150 and 1200 cycles (corresponding to 50 h and 400 h at test temperature) measured by EMPA*

	650°C	750°C	850°C
150 cycles	<10 μm	50 μm	180 μm
1200 cycles	15 μm	200 μm	500 μm

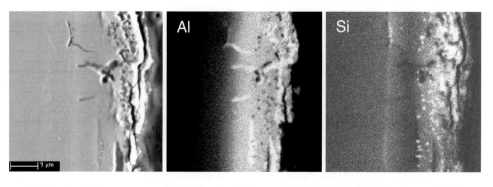

Fig. 4 *EMPA X-ray maps on the hot-dip aluminium coated Fe12CrTi after 50 cycles at 750°C.*

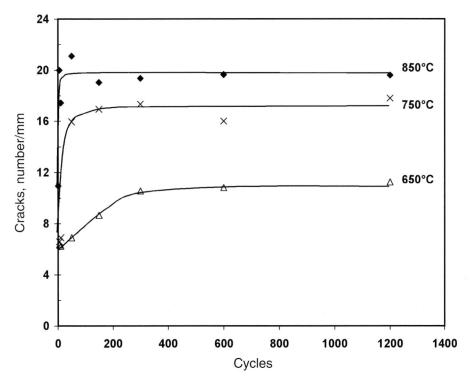

Fig. 5 *Evolution of the density of cracks (number/mm) in the inner side of the coating versus cycles under cyclic oxidation in air at 650, 750 and 850°C for the hot-dip aluminium coated Fe12CrTi.*

4.4. Localised Fe12CrTi Oxidation

The second type of defect observed was the presence of localised oxidation of the underlying metal (Fig. 3b). This defect developed much deeper than the previous cracks. They appeared as the most detrimental and corresponded to the evolution of the maximal loss of non-affected metal thickness (Fig.2).

X-ray maps have been produced by EMPA in order to investigate the nature of the oxide formed (Fig. 6). The result was consistent with the Fe–Cr oxides observed at these temperatures on a Fe12CrTi stainless steel [11]. The outer part was composed of iron oxide, whereas the inner part of the oxide was enriched in Cr. Silica could also be observed at the oxide/metal interface. These oxides ensured a low oxidation rate for the stainless steel which resulted in the appearance of a quasi steady-state degradation value, even at 750 and 850°C (Fig. 2). Alumina formed within the diffusion depth of aluminium bordering the oxide formation and therefore prevented any lateral propagation. It also allowed the formation of a protective alumina scale at the bottom of the localised oxidation after longer holding times at 750 and 850°C owing to deeper Al diffusion (Table 1).

The density of this localised oxidation remained much lower than that of the cracks described previously. Unlike the fine cracks, this density differed between the outer

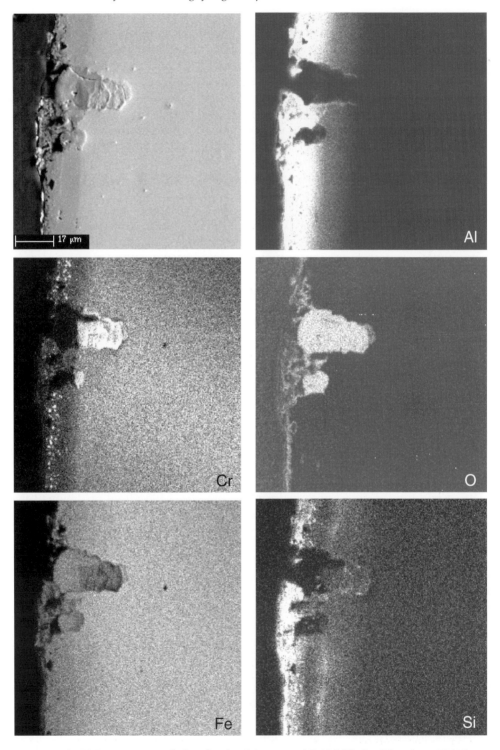

Fig. 6 *EMPA X-ray maps on the hot-dip aluminium coated Fe12CrTi after 50 cycles at 750°C.*

and the inner side of the sample. At the outer side, the density remained constant whatever the temperature from the first cycles. It measured respectively 5, 2.2 and 1.7 localised oxidation occurrences per mm at 650, 750 and 850°C.

The development at the inner side is shown in Fig. 7. Whereas it remained quite constant at 750 and 850°C and much lower than at the outer side with 0.5 per mm, the density was still rising at 650°C. The resulting shape appeared quite similar to the evolution of the maximum of the non-affected metal thickness loss (Fig. 1). This result is explained by the continuous appearance of localised oxidation at the inner side, giving damage much deeper than the crack depths previously mentioned,which affects the overall measure of the non-affected metal thickness across the sample, especially when 2 cases of localised oxidation are on opposite sides.

5. Discussion

The lifetime of high temperature materials in corrosive environments depends strongly on their resistance to oxidation. Therefore, the technical application of these materials requires the formation of a protective slow growing and adherent oxide scale. The resistance to thermal cycling is a particularly important consideration [9], especially when considering materials for automotive applications [8] in which there

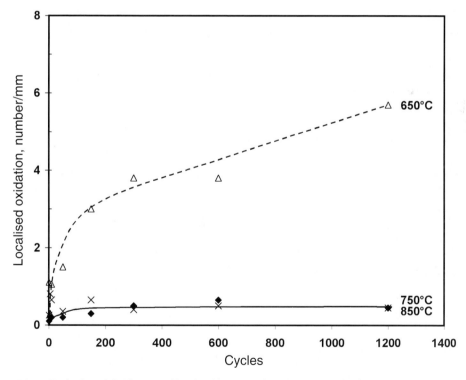

Fig. 7 *Evolution of the density of localised base metal oxidation (number/mm) in the inner side of the coating versus cycles under cyclic oxidation in air at 650, 750 and 850 °C for the hot-dip aluminium coated Fe12CrTi.*

are hot-dip aluminium coated Fe12CrTi steels. The results described in this paper indicate that the investigated hot-dip aluminium coated Fe12CrTi steel behaves well even on deformed samples under severe thermal cycling and is thus suitable for use in an automotive exhaust line behind the manifold. Several questions still remain open: the origin of the two types of degradation, the effect of the initial deformation and of the thermal cycling on the degradation and finally the different kinetics of degradation with regard to the temperature.

5.1. Origin of the Two Types of Degradation

As mentioned before, at high temperatures, diffusion of aluminium quickly formed an homogeneous layer of Fe–Cr–Al–Si intermetallic. The coated material actually behaved as a multi-layer composite with different high temperature physical properties of the base material, the interdiffusion zone layer, and the protective oxide scale. Cyclic oxidation enhanced the thermo-mechanical stressing of the surface, leading to the formation of cracks. An earlier study [5] reported the coefficients of thermal expansion (CTE) of Al–Fe alloys. They were all at least 1.5 times higher than that of Fe12CrTi. During thermal cycles, both the temperature gradient through the sample and the difference in CTE between the coating and the base material induced strain/compression cycles of the outer layer, promoting the formation of cracks at the surface and their inward progression. This explanation is consistent with the fact that these cracks were observed on both sides and with a similar density of cracks. It is also consistent with the lower crack density when decreasing the temperature from 850 to 650°C. The lower the test temperature, the lower the difference in the CTE and therefore the lower the thermal stresses. Moreover, these stresses were relaxed by the formation of the first cracks, thus explaining the stability of the crack density from the first cycles. For longer holding times, as inter-diffusion of iron and aluminium proceeded, the CTE gradient was reduced, and the progression of cracks decreased.

Using the same thought process with the results obtained from the localised oxidation of the underlying metal, it can be deduced that the thermal cycling cannot be at the reason for these defects. They are more likely to be linked to the initial bending deformation. As mentioned previously, the initial intermetallic layer was very brittle. The bending deformation applied tensile stresses to the outer side of the coating which can be relaxed by transversal cracks across this intermetallic layer as observed on cross-section examination. Compressive stresses were applied at the inner side, which led to significantly fewer defects. Some cracks have been observed, but they were more diagonal. This result explains the higher density of these defects at the outer side of the samples.

To investigate further the origin of this localised oxidation, two complementary sets of oxidation tests were carried out. The first, continuous oxidation up to 400 h, was performed on bent samples at the same temperatures. Cross-section observations also revealed the presence of such localised base metal oxidation with a quite similar density (Table 2). The second, cyclic oxidation up to 1200 cycles, was carried out at 750 and 850°C on non-deformed samples. No base metal oxidation was observed on the two samples but transversal cracks were observed with a density of respectively 18 and 20.2 cracks/mm which correspond very well with the values obtained on the deformed samples (Fig. 5). These results confirmed that the initial bending was the

Table 2. *Density of localised base metal oxidation (number/mm) after 400 h of continuous oxidation in air for the hot-dip aluminium coated Fe12CrTi*

	650°C	750°C	850°C
Internal side	1.0	0.3	0.3
External side	5.1	2.0	1.3

cause of the observed base metal oxidation by cracking the thin Fe–Cr–Al–Si intermetallic layer. Fortunately, the high temperature resistance of the Fe12CrTi grade strongly limited the progression of this degradation.

5.2. Effect of Thermal Cycling

The complementary continuous oxidation experiments also allowed us to quantify the effect of thermal cycling on the high temperature behaviour of these hot-dip aluminium coated materials. The maximum losses of non-affected metal thickness after 400 hours were respectively 33, 25 and 27 µm at 650, 750 and 850°C. These are almost half of those measured under cyclic conditions. It highlighted the well-known detrimental effect of thermal cycling on the oxidation resistance of materials [9] and the importance when looking for a lifetime prediction of using meaningful testing conditions with regard to the application under consideration. This higher degradation was caused by the thermal expansion difference between metal and oxide and the consecutive failure of the localised oxides. During cooling, the base metal contracts more than the oxide due to its higher CTE [8]. The oxide is subjected to tensile stresses which can be relieved by cracking as illustrated in Fig. 3(b). During the following isothermal hold, damaged areas are protected by a newly oxidised layer on the exposed metal leading to a deeper oxide penetration.

5.3. Influence of Temperature on Degradation Kinetics

The measured kinetics (Fig. 1) surprisingly indicate that between 650 and 850°C the long term degradation was even lower and remained stable even quicker as the temperature was increased. In fact, this appears to be the result of a competition between the inter-diffusion kinetics and the oxidation kinetics of the metal.

As shown above, the maximum degradation was caused by base metal oxidation due to initial cracking of the Fe–Cr–Al–Si intermetallic layer. The resulting difference with regard to the applied temperature therefore found its origin in the base metal oxidation kinetics. EMPA analysis (Table 1) showed that alloying between coating and stainless steel, and especially diffusion of Al into the stainless steel remained very low at 650°C (<20 µm after 400 h). The majority of the initial cracks allowing the contact of the underlying metal with air then led to the direct oxidation of the stainless steel. Therefore, it resulted in an initially higher and stationary density of base metal oxidation at the outer sample side which increased thereafter from cycle to cycle.

On the contrary, at 750 and 850°C, the rapid formation of a thicker intermetallic layer as a result of Al and Si diffusion led to a quicker establishment of the protective alumina layer and thus to less localised base metal oxidation. However, the brittleness of this intermetallic combined with the higher stresses induced by the thermal cycling and the difference of CTE between the intermetallic and the metal enhanced the prolongation of some of these defects up to the underlying metal during the first cycles. It resulted in a more pronounced and deeper oxidation during the first cycles at 850 than at 750°C. Nevertheless, as with the above mentioned fine cracks, the CTE gradient reduced more quickly at 850°C as inter-diffusion of iron and aluminium was faster, and so a steady state was reached earlier at 850°C. Moreover, high temperature favoured the rapid formation of a protective chromia layer on the Fe12CrTi stainless steel [11]. This oxide protection and the continuous Al and Si diffusion into the steel consequently led to a lower degradation value at 850°C.

As mentioned above, the cracks at the inner side induced by the bending deformation were less numerous and subjected to compressive stresses. The base metal oxidation was thus less favourable. However, the cyclic conditions and the difference of CTE between coating and metal favoured the opening of the cracks. Combined with the slow alloying at 650°C, it resulted in the oxidation of the underlying metal and explained the resulting density development shown in Fig. 7. The quicker inter-diffusion at 750 and 850°C enabled the reduction of the thermal dilation gradient and the formation of alumina protective scale leading to the smaller and more stationary localised oxidation density.

Even with Fe12CrTi oxidation kinetics being the lowest at 650°C, the absence of alumina protection led to the continuation of the base metal oxidation, especially at the inner sample side. Finally, the maximum non-affected metal thickness loss at 650°C crossed the quasi-stationary values measured at 750 and 850°C after about 300 and 800 cycles respectively. An initial high temperature treatment therefore appears favourable to limit the degradation kinetics while also forming an outer protective alumina scale ensuring a good corrosion resistance and finally to prolong further the lifetime of these kinds of materials.

6. Conclusions

Using a special test designed to simulate carefully the service conditions of exhaust lines, long term high temperature cyclic oxidation in air between 650 and 850°C of hot-dip aluminium coated Fe12CrTi has been investigated and led to following conclusions :

1. Two different types of degradation were observed:

    •   thin cracks within the coating due to thermal cycling. Owing to Al diffusion into the metal, these defects were rapidly stopped by the formation of a protective alumina scale, and

    •   localised oxidation of the underlying steel, due to initial bending deformation of the samples (reproducing the metal deformation during tube manufacturing), was the origin of the deepest degradation.

2. The most important part of the degradation occurred during the first cycles and developed slowly thereafter when interdiffusion between the aluminium coating and stainless steel led to the formation of an Fe–Cr–Si–Al coating, with a gradient of Al and Si concentration from the surface to the base metal.

3. As a result of the good cyclic oxidation behaviour of the underlying Fe12CrTi grade and to Al and Si diffusion, the degradation depths remained small (≤ 40 μm for each side) while maintaining a nice visual aspect even after 1200 cycles (i.e. 400 h at the test temperature).

4. Increasing the temperature favoured this inter-diffusion phenomenon and the resulted in less deep degradation and favoured the formation of an outer alumina scale ensuring good corrosion resistance. It therefore appeared favourable to prolong further the lifetime of these kinds of materials.

All in all, with good formability, good weldability, and service properties, hot-dip aluminium coated Fe12CrTi appears to be a high-quality stainless steel, tailored to meet today's stringent requirements of the intermediate and down parts of automotive exhaust lines.

References

1. R. Craen, L. Antoni, J. Ragot, J. Y. Cogne and V. Dusseque, in *Stainless Steel '99, Science and Market 3rd European Congress Proceedings, Vol. 2: Innovation in Processes and Products*, pp. 305–314, Sardinia, Italy, 6–9 June, 1999, Publ. Associazione Italiana di Metallurgia, Piazzale Rodolfo Morandi, 2, Milano, I–20121, Italy, 1999.
2. J. A. Douthett, *Automot. Eng.*, 1995, November, 45–49.
3. L. Bednar and R. A. Edwards, *Proc. SAE*, Detroit, 1998, p.195–199.
4. A. Ando, Y. Hattori, N. Hatanaka and T. Kittaka, *Corrosion '91*. Paper No. 384, NACE, Houston, TX, 1991.
5. M. Ryabov, *Aluminizing of Steels*. Oxonian Press, New Delhi, 1985.
6. U. Etzold, U. Heidtmann, G. Neba and W. Warnecke, *Stahl Eisen*, 1991, **111**, 12, 111–116.
7. T. Yamada and H. Kawase, *Tetsu-to-Hagané*, 1986, **72**, 1021–1028.
8. L. Antoni and J. M. Herbelin, in EFC Working Party Report on Cyclic Oxidation of High Temperature Materials: Mechanisms, Testing Methods, Characterisation and Life Time Estimation (M. Schütz and W. J. Quadakkers, eds). Publication No. 27 in European Federation of Corrosion series. Published by The Institute of Materials, London, 1999, p.187–197.
9. EFC Working Party Report on Cyclic Oxidation of High Temperature Materials: Mechanisms, Testing Methods, Characterisation and Life Time Estimation (M. Schütze and W. J. Quadakkers, eds). Publication No. 27 in European Federation of Corrosion series. Published by The Institute of Materials, London, 1999.
10. Ugine and Sollac, European patent 0467749, 1990.
11. F. Armanet and J. H. Davidson, *Stainless Steels* (P. Lacombe, B. Baroux and G. Beranger, eds). Les Editions de Physique, 1993, pp.435–473.

20

Void Nucleation and Growth at the (Pd,Ni)Al Coating/Alumina Scale Interface During High Temperature Oxidation and Relation to Oxide Scale Spallation

D. OQUAB and D. MONCEAU

CIRIMAT UMR 5085 CNRS/UPS/INPT, ENSIACET, 31077 Toulouse Cedex 4, France

ABSTRACT

β-NiAl (B2-structure) is used as a bond coat in TBC systems because it offers a good resistance to high temperature oxidation with the formation of a compact alumina scale. Pt has been added to NiAl to increase the resistance to hot corrosion and to improve the scale adherence. Pd-modified β-NiAl coatings were also developed with a cost reduction objective and to reduce the precipitation of brittle intermetallic phases. All these coatings may form prismatic voids at the metal/oxide interface during oxidation. This work reports SEM observations of such interfacial voids in palladium-modified nickel aluminide oxidation. The effect of the coating grain crystallographic orientation on underscale void shape and on scale spalling is presented. The substrate surface of the void under the scale together with the oxide inner surface has been detailed.

1. Introduction

Aluminides may form prismatic voids at the alumina scale/alloy interface, after short-term high temperature oxidation. This phenomenon has been observed in massive materials as well as coatings. The appearance of such interfacial voids has been reported on MCrAlY [1,2], Ni_3Al [3,4], FeAl [5], NiAl [6–9], and on Pd modified NiAl [10–13]. The prismatic shape of these cavities has always been supposed to be related to the alloy (or coating) grain crystallographic orientation, and this was proven by diffraction in the TEM on oxidised NiAl+Zr single crystals [8]. *Ex situ* SEM observations presently show how the void shape varies from grain to grain, but also how the substrate orientation influences the contact area between the oxide scale and the alloy and, in turn, how it influences the extent of spalling during cooling. *In situ* SEM observations permit the separation of the effect of the substrate grain orientation on void nucleation from that on growth kinetics [13]. Void nucleation and growth are also related to initial surface finish and defects. Alumina scale spallation then depends on the initial surface preparation and on NiAl crystallographic orientation, in addition to other possible factors such as interface chemistry and alumina microstructure. Obviously, alumina scale spallation depends also on the mechanical stresses developed during cooling and/or during oxide growth. Corrugated alumina surfaces have been observed in the adherent areas.

These SEM observations are consistent with an elastic deformation of the oxide scale and this deformation could be used as an *in situ* stress probe.

2. Materials and Experimental Procedures

The aluminide bond coats investigated were provided by SNECMA SERVICE (Chatellerault, France). Pd-modified Ni aluminides were deposited on IN100 substrates according to the following three step process:

1. Electrolytic deposition of a Pd-Ni alloy;

2. Interdiffusion treatment at 850°C; and

3. Vapour phase aluminisation (VPA).

The microstructure of the external surface of as-received bond coats is illustrated in Fig. 1(a). The bond coats deposited on disc specimens were characterised by a regular equiaxed microstructure with an average grain size of about 50 μm, which is approximately equal to the bond coat thickness. Figure 1(b) reveals that the bond coat grain boundaries were marked by a 5 to 10 μm wide bulged zone.

The composition profiles across the bond coats and the IN100 substrate after heat treatments and oxidation showed that the aluminium content close to the surface is over 50at.%. Palladium concentration is about 9at.% at the surface and reaches about 12at.% before decreasing in the IN100 substrate direction [11].

The isothermal oxidation tests of specimens were performed on a SETARAM™ TAG 24S thermobalance or in a simple furnace, under air atmosphere at 900°C, for a period of 6 h.

Ex situ observations were performed on a LEO 435VP scanning electron microscope system in conventional mode along with a PGT (imix-PC) system for the EDS analysis.

In situ isothermal oxidation was carried out in an environmental scanning electron microscope XL30 ESEM-FEG from Philips. The accelerating voltage used for the

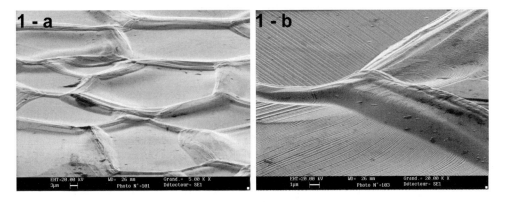

Fig. 1 *SEM views showing the morphology of as-received (Ni, Pd) Al bond coat (tilted view).*

in situ study was 30 kV. The oxidation temperature of 950°C was obtained with a Philips hot stage 1000, rated with a maximum temperature of 1000°C. The temperature was measured close to the heating region and the sample temperature may have been slightly lower than the measured temperature. The air pressure used in this experiment was 2.5 mbar. The air atmosphere was renewed several times during the oxidation of 6 h to avoid oxygen deficiency.

3. Results and Discussion

3.1. Growth Mechanism of the Oxide Layer

Oxidation of Pd modified NiAl at 900°C for 6 h under air atmosphere leads to a compact alumina layer about 250 nm thick (Fig. 2a and 2b). At the external surface of the oxide layer (Fig. 2a) the small crystallites are possibly the transient alumina phases already characterised by XRD in previous work [10,11]. According to the morphology

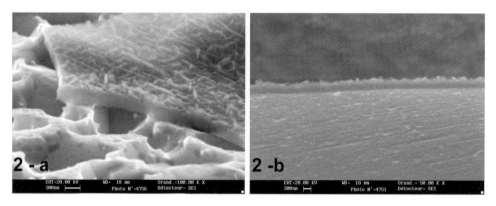

Fig. 2(a) *Oxide morphology in a partially spalled region (tilted view); (b): spalled oxide showing an uniform thickness (inner surface tilted view) (6 h in air at 900°C).*

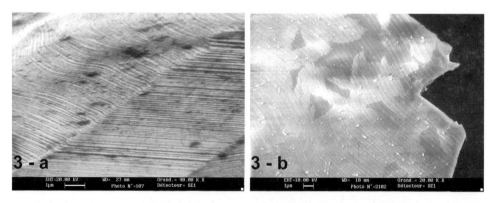

Fig. 3 *Comparison of the as-received (Ni, Pd) Al surface and the oxide layer inner surface. (a) as-received (Ni, Pd) Al surface; (b) oxide's inner surface (6 h in air at 900°C).*

of the inner surface of the oxide, the alumina growth proceeds by outward diffusion of aluminium through the oxide layer. Indeed, the inner oxide scale morphology (Fig. 3a) reproduces clearly the morphology of the initial substrate surface (Fig. 3b).

3.2. Interfacial Voids and Aluminum Transport

The uniformity of oxide scale thickness (i.e. independently of underlying voids) is quite surprising, as has already been observed [6]. Moreover, the density of the transient oxide crystallites seems also to be uniform (Fig. 2). As the oxide growth occurs at the oxide/gas interface (see section 3.1), this observation agrees with the model of aluminium evaporation and transport through the voids [6, 9] with kinetics high enough not to be a rate limiting step in the oxide scale growth overall kinetics. This observation is consistent with previously measured oxidation kinetics that do not slow down when pores are formed [11]. Etching-like steps (Fig. 6a) seem to agree with such an aluminium evaporation mechanism.

3.3. Nucleation and Growth of Interfacial Voids

3.3.1. Void growth mechanism

In situ oxidation in the ESEM (Fig. 5 and 6) leads to cavities, similar to those observed after standard oxidation and cooling (Fig. 2a). SIMS analysis (intensity profiles and imaging mode with a spatial resolution better than 0.1 μm) shows that a very thin pure alumina layer is formed, uniformly covering the alloy surface over the voids, in agreement with SEM observations [12,13]. Several mechanisms of void formation were or could be proposed:

(i)   a Kirkendall effect due to the different diffusivities between Ni and Al in β-NiAl, as proposed in Ni$_3$Al [3] or NiAl [6,9];

(ii) vacancy condensation during cooling, similar to Anthony's experiment in aluminium alloys [14]; and

(iii) vacancy condensation due to the large change of the equilibrium vacancy concentration [15] when the aluminium content in the intermetallic NiAl decreases during oxidation.

Our *in situ* SEM imaging definitively proves that cavity nucleation arises at the very early stage of high temperature oxidation [13]. Voids even nucleate during the heating period (about 25°C/min) and are already visible at 650°C. Void shape, size and density do not change during cooling.

3.3.2. Effect of the NiAl substrate crystallographic orientation

Void formation is observed beneath the oxide layer as shown in Fig. 2(a). The void morphology and size depend on the of the alloy grain orientation. Figure 4 shows the surface morphology in spalled regions, in different (Ni,Pd)Al grains. It appears that the metal/oxide anchorage area is dependent on the crystallographic orientation of the metal grain (Fig. 4(a), 4(b) and 4(c)). This observation could explain why spalling is observed on some grains and not on others. Before complete spallation occurs, the metal under the oxide layer in the anchorage regions is subjected to plastic deformation. When the anchorage surface is small enough, metal rupture is reached. Figure 4(d) illustrates such a rupture. Void density also depends strongly on NiAl grain

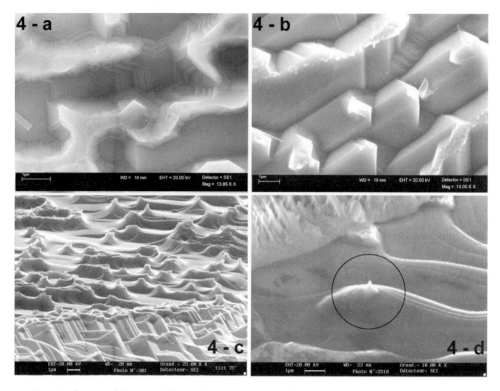

Fig. 4 *Influence of the crystallographic orientation on the morphology of voids and anchorage surface: the orientation allows (a) large or (b) small metal/oxide anchorage surfaces; (c) shows both side of a grain boundary (tilted view); (d) illustrates a plastic deformation of the sub-layer metal due to the effect of local stresses (tilted view) (6 h in air at 900°C).*

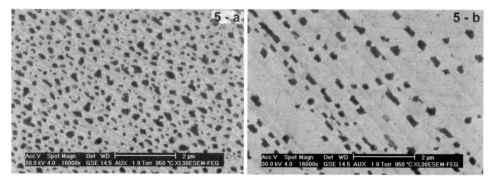

Fig. 5 *ESEM micrographs showing the influence of grain orientation on void formation and void surface density after (a) 78 min, and (b) 90 min (approx. 950°C in air vacuum).*

crystallographic orientation, as can be seen on ESEM micrographs in Fig. 5. During *in situ* oxidation, observation of void growth was possible due to the difference of secondary electron emissivity of the area corresponding to the cavity surface and out of the cavities.

3.3.3. *Void growth kinetics*

Void growth kinetics were followed from the image analysis of the *in situ* SEM images [13]. General observations by Smialek on NiAl [6] or Liu and Gao on Ni_3Al [4] at room temperature after interrupted oxidation experiments were confirmed, but the *in situ* ESEM observations allows a more complete analysis because of the possibility of following the growth kinetics of individually selected voids. It was shown [13] that the growth kinetics are neither linear nor parabolic and that the surface fraction "s" of the cavities is not proportional to the oxide scale thickness. It can be seen on the videotape of the experiment that some cavities are continuously growing whereas some others of the same initial size and geometry are not growing (see labels "NG" on Fig. 6a and 6b). It should be noted that some coalescence occurs as can be seen in Fig. 6. Then, it is difficult to build a quantitative model (such as the one found in ref. [4]) because the ratio of the total surface area of voids to their total volume is not constant. Then, 3D topography of the surface after spallation is necessary in order to establish a quantitative correlation between the total void volume and the oxide scale thickness, which is directly related to the aluminium consumed in the intermetallic.

3.4. Elastic Deformation of the Alumina Scale

The SEM observations of oxidised samples presented in Figs 7 and 8 show the partial decohesion of the oxide layer (Fig. 7b). Observed at higher magnification, it appears that the oxide layer is smoothly convoluted (Fig. 7c) in the adherent areas while it presents a totally planar surface when spalled (Figs 7d and 8b). The measurement of the amplitude of the convolution was not easy to carry out by means of SEM or interferometric analysis. However, from the SEM micrographs, we can estimate the flambé height to be about 100 to 200 nm. Then, the flat surface of the spalled oxide layer seems to be evidence of the elastic deformation of the adherent oxide scale.

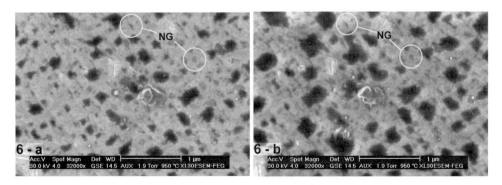

Fig. 6 In situ *ESEM micrographs showing the interfacial voids during oxidation after (a) 112 min and (b) 304 min. The object at the centre of micrographs was used to locate the region under study during the 6th experiment. The voids were not growing in areas marked NG (see text) (approx. 950°C in air vacuum).*

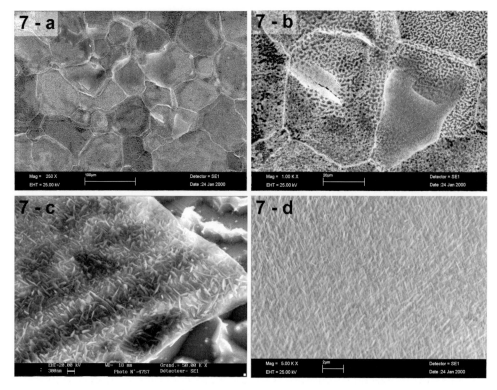

Fig. 7 *(a) Oxidised sample's surface; (b) partially spalled and relaxed oxide layer on a grain; (c) adherent oxide layer showing a convoluted surface; and (d) spalled oxide presents a totally planar surface (6 h in air at 900°C).*

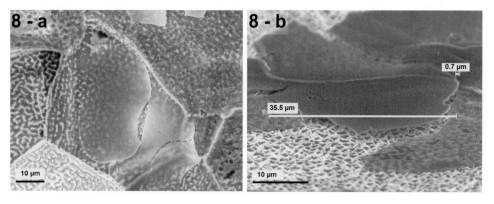

Fig. 8 *Measurement of the length variation of a spalled oxide flake. (a) Partially spalled oxide layer showing a larger surface than the original place of the oxide; (b) the measured scale length of 36.5 μm is 0.70 μm larger (i.e. 2% larger) than the length of the original (tilted view) (6 h in air at 900°C).*

Also, the spalled oxide has a larger surface area than the adherent layer as shown in Fig. 8(b). The linear length variation between adherent and non-adherent oxide measured in a partially spalled area reveals a difference of about 2%.

Because spalled oxide appears to be more planar, the hypothesis of an elastic deformation of the scale during cooling was tested, using a finite element analysis [12]. It was found that the calculated tensile stress compatible with the observed deformation was very large (11 GPa) and that the amplitude of the corrugation was incompatible with the observed radial deformation values (17% instead of 2%). These deformations are also much higher than the deformation which can be predicted from the differences in thermal expansion of the alumina scale and its nickel aluminide substrate ($\approx$0.46% instead of 2%). Further work is underway to understand these discrepancies.

4. Conclusions

Oxide scale spallation results from a loss of adherence combined with high stresses developed during cooling and/or during scale growth:

- Concerning the adherence, this work reports a cause of weakening of the oxide scale/coating interface when interfacial voids formed after short term oxidation at relatively low temperature. The use of the *in situ* ESEM technique has definitively proven that these interfacial voids nucleate in the very early stage of (Ni,Pd)Al oxidation. This is important for the initial oxidation (for example, for the NiAl coating pre-oxidation in the EB-PVD apparatus before zirconia deposition, for TBC systems) but also for the re-oxidation of coatings after spallation during cycling exposures, for which similar observations with NiAl and Pt–NiAl are underway. Some coalescence of voids is observed during the first hour at high temperature. The surface area of voids at the metal/scale interface depends on NiAl substrate crystallographic orientation. This qualitative relationship found between oxide scale adherence and crystallographic orientation of the substrate need to be quantified on NiAl single crystals.

- Concerning the stresses in the oxide scale, an *in situ* stress gauge was identified.

SEM observations of Pd-modified nickel aluminum coating oxidised in air has produced insights into oxide growth, interfacial void formation and spalling mechanisms:

1. Cationic growth of the scale.

2. Rapid transport of aluminum through the sub-scale cavities in these oxidation conditions.

3. Substrate grain orientation effect through sub-scale cavity density.

4. The metallic anchorage points experiencing plastic deformation before complete oxide separation from the substrate.

5. Elastic deformation of the oxide scale after cooling.

References

1. J. K. Tien and F. S. Pettit, *Metall. Trans.*, 1972, **3**, 1587.
2. F. A. Golightly, F. H. Stott and G. C. Wood, *Oxid. Met.*, 1976, **10**, 163.
3. J. D. Kuenzy and D. L. Douglass, *Oxid. Met.*, 1974, **8**, 139.
4. Z. Liu and W. Gao, *Intermetallics*, 2000, **8**, 1385–1391.
5. H. J. Grabke, *Intermetallics*, 1999, **7**, 1153–1158.
6. J. L. Smialek, *Metall. Trans. A*, 1978, **9A**, 309.
7. J. K. Doychak, in *Int. Congr. on Metallic Corrosion*, Vol. 1, p. 35, Toronto, 1984.
8. J. Doychak and M. Rühle, *Oxid. Met.*, 1989, **31**, 431.
9. M. W. Brumm and H. J. Grabke, *Corros. Sci.*, 1993, **34**, 547.
10. D. Monceau, A. Boudot-Miquet, K. Bouhanek, R. Peraldi, A. Malie, F. Crabos and B. Pieraggi, *J. Phys. (France) IV*, 2000, **10**, 167.
11. D. Monceau, K. Bouhanek, R. Peraldi, A. Malie and B. Pieraggi, *J. Mater. Res.*, 2000, **15**, (3) 665–675.
12. D. Oquab and D. Monceau, *Mater. Sci. Forum*, 2001, in press.
13. D. Oquab and D. Monceau, *Scripta Mat.*, 2001, **44**, 2741–2746.
14. Cited in J. Philibert, "Atom movements diffusion and mass transport in solids", Les Editions de Physique, Les Ullis (France), 1991, p.318.
15. M. P. Gururajan and T. A. Abinandanan, *Intermetallics*, 2000, **8**, 759.

21

Investigations on the Lifetime of Alumina-forming Ti–Al–Ag Coatings

L. NIEWOLAK, V. SHEMET, A. GIL, L. SINGHEISER and W. J. QUADAKKERS

Research Centre Jülich, Institute for Materials and Processes in Energy Systems IWV-2,
D-52425 Jülich, Germany

ABSTRACT

One important hindrance to the use of γ-TiAl alloys at high temperatures is their relatively poor oxidation resistance and their sensitivity towards environmentally induced embrittlement. The latter effect is also of major concern during elevated temperature application of Ti-based alloys. These problems can principally be avoided by the application of oxidation resistant, alumina-forming coatings. In the present paper the possibility of using a Ag-containing TiAl alloy as a coating material for Ti and high-strength γ-TiAl-alloys is being studied. First, oxidation tests in different environments were carried out, to compare the isothermal and cyclic oxidation resistance of a selected Ag-containing coating alloy with that of two, new generation, high strength γ-titanium aluminides. Subsequently, magnetron sputtering was used to apply the TiAl–Ag coating to a γ-TiAl alloy (Ti–45Al–8Nb–0.2C) and to pure titanium. The coated materials were tested at 800°C in case of the γ-TiAl substrate, and at 600°C in case of the Ti substrate. The results illustrate, that the coatings in all the cases studied formed a protective alumina surface scale even up to the longest studied test times of 1000 h. The interdiffusion processes between coating and base material, which eventually determine the coating life, were studied by SEM/EDX investigation after different exposure times.

1. Introduction

A large number of investigations have been carried out world-wide aiming at the development of new lightweight materials with high strength and suitable oxidation resistance at elevated temperatures. One new class of materials, which was extensively studied in recent years, is based on the intermetallic phase γ-TiAl. Materials of this type are being considered as possible structural materials e.g. in advanced gas turbine engines, because they combine low density with high strength even at elevated temperatures [1,2]. One important hindrance in the use of γ-TiAl alloys as high temperature material is their relatively poor oxidation resistance and their sensitivity towards environmentally induced embrittlement caused by oxygen dissolution. The latter effect also occurs in elevated temperature application of Ti-based alloys [3,4].

Several studies have shown that the oxidation resistance of γ-TiAl based alloys can substantially be improved by suitable alloying additions [4,5]. Promising results have been obtained with alloying elements such as Nb, Ta, Re, and W [5]. By using these alloying additions reduced scale growth rates could be obtained, although still substantially higher than those commonly found for alumina-forming alloys. Brady *et al.* [6,7] reported that Ti–Al–Cr alloys with Cr contents of 10–20 at.%. form protective alumina scales during exposure at temperatures up to 900°C. However, due to poor mechanical properties (e.g. poor creep resistance and/or brittleness at room temperature) [6,7] these alloys seem to be unsuitable as construction materials. Recently it was shown that alumina formation on γ-TiAl can also be obtained by small additions of Ag. Shemet *et al.* [8] showed that alloy Ti–50Al–2Ag (concentration in at.%) forms surface scales with excellent protective properties during long term cyclic oxidation at 800°C in air up to 6000 h [9]. An advantage of the Ag-containing materials is that they are less brittle than the high Cr-containing TiAl alloys. However, it is believed that the Ag-containing alloys possess too poor creep resistance for their use as high temperature construction materials. In applications which require optimised mechanical properties in combination with high oxidation resistance it will thus be necessary to combine a high strength α_2/γ-TiAl (e.g. on the basis of TiAl-(Cr,Mn)-(Ta,Nb)-(C,B))[10] or Ti-based alloys with an oxidation resistant coating.

In the present paper, the suitability of Ti–Al–Ag alloys as coating material for high-strength γ-TiAl intermetallics and Ti-based alloys will be discussed. In the first part of the study the fundamental differences between the oxidation behaviour of Ti–Al–Ag alloys and common binary and ternary TiAl alloys will be described. Subsequently, the suitability of a selected Ti–Al–Ag alloy as coating material on a high strength TiAl–Nb–C alloy and on pure Ti will be evaluated.

2. Experimental Procedure

2.1. Oxidation Testing of Cast and Wrought Alloys

The first series of oxidation experiments was carried out on two commercial γ-TiAl based alloys, and a model TiAl alloy with silver addition. The commercial alloys were two representatives of new, high strength γ-TiAl-based materials. The first (supplied by Plansee GmbH), which possessed a full lamellar microstructure, had a composition given by Ti-46.5 at.% Al and contained Cr-, Nb-, Ta-, B-additions. The alloy was supplied in form of 1-mm sheet. The second (supplied by GKSS), is a representative of the high Nb-containing TiAl alloys with a composition given by Ti–45Al–8Nb–0.2C (additions in at.%). The model alloy Ti–50Al–2Ag (additions in at.%) alloy was produced by levitation induction melting from high purity elements Ti-99.98, Al-99.999 and Ag-99.99 wt%). After melting, the ingots were homogenised for 250 h at 950°C in a quartz glass capsule, which was evacuated down to 10^{-5} mbar. To avoid reaction between alloy and quartz capsule the specimens were wrapped in a molybdenum foil. Samples with nominal dimensions $10 \times 10 \times 1$ mm were machined from the ingots and the rolled sheets and subsequently polished up to 800-grit surface finish. The oxidation experiments were carried out in synthetic air and in Ar–4% H_2O at 800°C using a microbalance (Setaram TG 50). The specimens were heated to test temperature with a rate of 90K/min, kept at temperature for 100 h, and subsequently furnace cooled. The water vapour was added to the gas by flowing argon through distilled water, which was kept at constant temperature (30°C). In addition to the isothermal exposures, cyclic oxidation tests were carried out in laboratory air at 800°C. In these experiments mass changes were measured in time intervals of 100 h after removing the samples from the furnace and subsequently cooling to room temperature.

2.2. Coating Manufacturing and Testing

For the coating tests, alloy Ti–45Al–8Nb–0.2C (composition in at.%) and pure Ti were selected as base materials. Samples were prepared in form of square coupons, 10 × 10 × 2 mm in size, with rounded corners. After cutting, the coupons were ground with SiC abrasive paper up to 4000 grit surface finish and ultrasonically cleaned in ethanol. The coatings were manufactured using Radio Frequency (RF) Magnetron Sputtering in a commercial device (Z 400 produced by Leybold-Hereaus). The MS-sputtering facility was a planar magnetron with a 75-mm diameter target. The target was made of alloy Ti–48Al–2Ag, supplied by GfE Metalle und Materialien GmbH. The cleaned base material samples were placed in the magnetron chamber which was subsequently evacuated to 3×10^{-5} mbar. To increase adhesion between coating and substrate, the sample surfaces were pre-sputtered before coating. The substrates were then covered with two layers of the Ti–Al–Ag coating. The coating had a total thickness of approximately 30 μm. The coated samples were oxidised in air or in Ar-20%O_2 at temperatures in the range 500–800°C depending on the substrate material. Microstructure and composition of the coatings before and after oxidation were investigated using XRD (X-ray diffraction), OM (optical metallography), SEM/EDX (scanning electron microscopy with energy dispersive X-ray analysis) and SEM/BSE (Scanning electron microscopy with back scattered electrons). Elemental profiles in the interdiffusion zones between coating and substrate material were measured by EDX at 12 kV accelerating voltage taking 100 s analysis time in each point. The 12 kV accelerating voltage appeared to be an optimum compromise between achievable lateral resolution and detection limits for each element. The collected signal came from a near-sphere volume of 2 μm in width and 1.3 μm in depth.

3. Results and Discussion

3.1. Oxidation Behaviour of Cast and Wrought γ-TiAl based Alloys

In general, the Ti–50Al–2Ag alloy exhibited far better oxidation resistance both in air and in Ar-4%H_2O than the commercial alloys (Figs 1, 2). In both atmospheres the oxidation rates for Ti–50Al–2Ag were similar to those observed for γ-Al_2O_3-forming materials. The scale microstructure on the Ti–50Al–2Ag alloy after 1000 h exposure in air at 800°C is shown in Fig. 3. XRD and SEM/EDX investigations verified that the surface scale exclusively consisted of γ-Al_2O_3. Beneath this alumina scale the sub-surface layer consisted of Z-phase [11]. The SEM/EDX examination of this phase revealed a composition of Ti–29.9Al–22.6O–2.5Ag (in at.%) which is in good agreement with previously reported data [11–14]. The composition and microstructure of the scale formed in Ar-4%H_2O were similar to those formed in air. Indications were found that water vapour promoted the growth of metastable Al_2O_3 phases in the early stages of oxidation. These phases formed whisker and platelike oxide morphologies on the sample surfaces as has frequently been observed in other studies

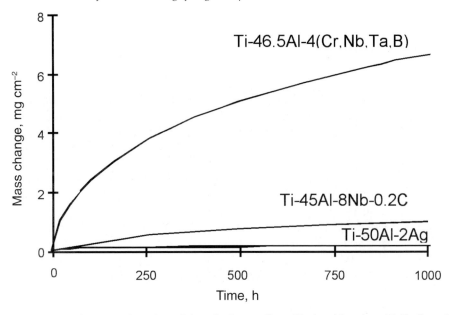

Fig. 1 *Mass change as function of time during cyclic oxidation of various TiAl alloys in air at 800°C.*

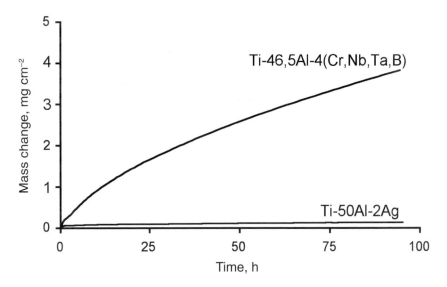

Fig. 2 *Mass change as function of time during isothermal oxidation of α_2/γ-TiAl alloy and Ag-containing alloy in Ar-4%H_2O at 800°C.*

[15,16]. The oxidation tests in the atmosphere containing the water vapour showed that the formation of the metastable aluminium oxides did not substantially deteriorate the long-term protective properties of the alumina based scale on the Ti–50Al–2Ag alloy.

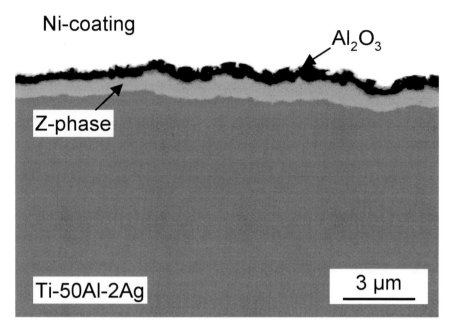

Fig. 3 SEM/BSE image showing microstructure of the oxide scale on Ti–50Al–2Ag alloy after 1000 h exposure in air at 800°C.

The commercial Ti–46.5Al–4(Cr,Nb,Ta,B) alloy formed mixed TiO_2/Al_2O_3 scales in both atmospheres (Fig. 4). The presence of water vapour appeared to increase the oxidation rate of this alloy, in agreement with results in reference [17] showing that the growth rates of mixed scales strongly depends on the presence of water vapour in the gas. It is presently not completely understood which mechanisms are responsible for this enhanced oxidation. A general observation was that the Ag-containing alloy which formed an alumina based scale, is less susceptible to the presence of the water vapour in the reaction gas than alloys forming mixed scales. Due to the fact that water vapour is always present in air and in exhaust or combustion gases, the Ti–50Al–2Ag thus seems to be an excellent material for protection of γ-TiAl and Ti-based alloys.

3.2. Oxidation and Interdiffusion Properties of Alumina-forming Ti–Al–Ag Coatings

The chemical composition of the coating which was obtained by RF magnetron-sputtering, was Ti–(48–50)Al–2Ag. The difference in chemical composition between target alloy and coating is probably related to differences in transport rate of the various elements in the gas phase during sputtering. Applying a pre-sputtering process appeared to increase significantly the adhesion of the coating to the substrate. The Ti–50Al–2Ag coatings possessed excellent adherence to both substrate materials, i.e. Ti–45Al–8Nb–0.2C and Ti, and they were nearly free from pores or cracks. An additional beneficial property was that propagation of the few cracks, which

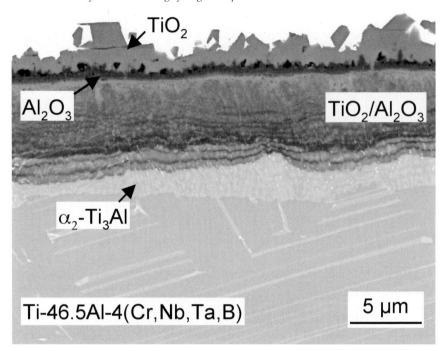

Fig. 4 *SEM/BSE image showing microstructure of the oxide scale on Ti–46.5Al–4(Cr,Nb,Ta,B) alloy after 100 h oxidation in Ar-4%H₂O at 800°C.*

sometimes occurred during annealing, could be stopped by applying a two layer coating system. Cracks, which were initiated at the coating surface, penetrated only in the outer layer of the coating material and seemed to be stopped at the interface between the two layers (Fig. 5). The as-deposited coatings were highly textured as indicated by XRD (Fig. 6). During annealing at high temperatures the textured coating materials re-crystallised (Fig. 6) resulting in increased ductility and a decreased hardness indicated by hardness measurements (Figs 7, 8). The oxidation tests carried out in Ar-20%O_2 showed that the Ti–Al–Ag coating on the Ti–45Al–8Nb–0.2C alloy is an excellent alumina former at 800°C up to the maximum achieved exposure times of 1000 h (Fig. 5, 9). The coating formed the same type of alumina-based scale as observed in case of the cast Ti–50Al–2Ag alloy (Fig. 3). The small bright lines visible within the coating zone are due to small cracks which, however, re-healed during further exposure. During the re-healing process the space in the cracks became filled with Z-phase [11–14].

The coated Ti-samples oxidised at 500–600°C (Fig. 10) formed on the surface a very protective oxide layer, which was so thin that it could hardly be characterised by XRD. Generally it can be said that the Ti–50Al–2Ag coating not only substantially improved the oxidation resistance of Ti and the studied γ-TiAl based alloy, but it also exhibited excellent compatibility with the substrate, i.e. much better than observed for conventional MCrAlY or aluminide coatings [18]. Furthermore, the Ti–50Al–2Ag coating seems to possess suitable mechanical properties with the result that no major cracks were found after long-term oxidation and around microhardness indents after annealing.

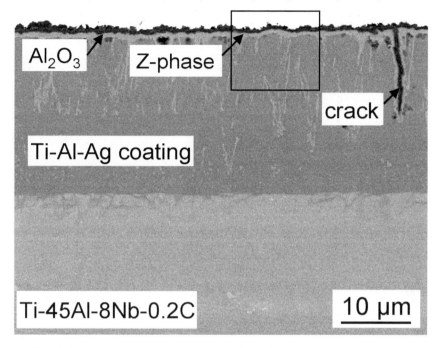

Fig. 5 *SEM/BSE image showing microstructure of Ti–Al–Ag coating on Ti–45Al–8Nb–0.2C alloy after 1000 h exposure in Ar-20%O_2 at 800°C.*

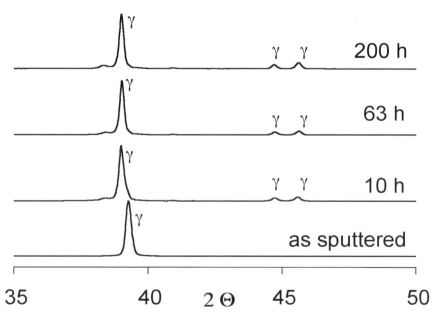

Fig. 6 *XRD patterns of the Ti–Al–Ag coating on Ti-substrate after different annealing times in air at 600°C.*

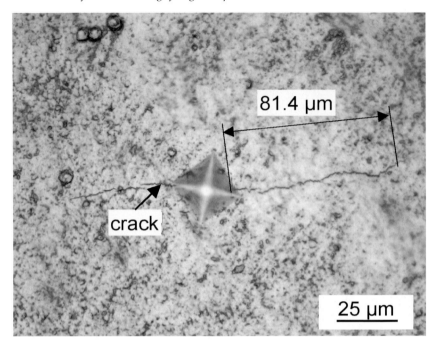

Fig. 7 *OM image showing crack formation in the Ti–Al–Ag coating (as sputtered) caused by penetration of indentor during hardness measurements (load = 0.3 kG).*

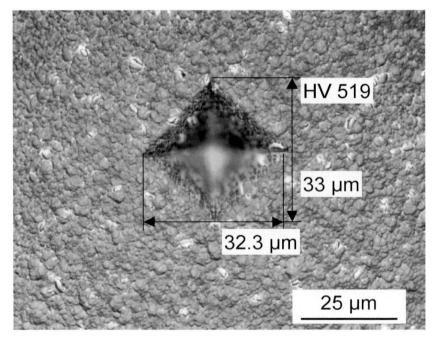

Fig. 8 *OM image showing hardness indentation in the Ti–Al–Ag coating after 63 h annealing at 600°C in air (load = 0.3kG).*

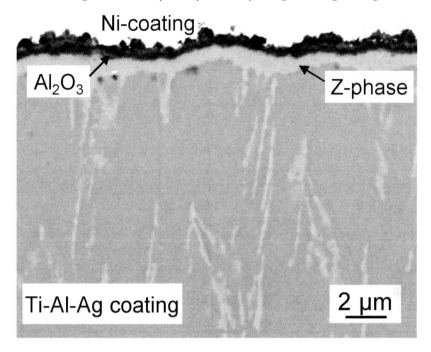

Fig. 9 SEM/BSE image showing higher magnification of the Al₂O₃ scale from the marked region in Fig. 5.

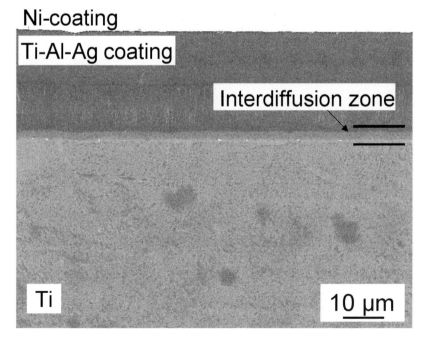

Fig. 10 SEM/BSE cross-section image of Ti–Al–Ag coating on Ti after 200 h oxidation in air at 600°C.

3.3. Interdiffusion Properties of Alumina-forming Ti–Al–Ag Coatings

The interdiffusion processes between coating and core material during exposure appeared to depend strongly on the chemical composition of the substrate. The interdiffusion observed in the Ti/Ti–Al–Ag system was much more substantial than that in case of the Ti–45Al–8Nb–0.2C/Ti–Al–Ag system. Figure 11 shows SEM/BSE cross-section images of the Ti alloy coated with Ti–50Al–2Ag after exposure at 600°C. After 200 h the interdiffusion zone was approximately 5μm in width and consisted of three layers. The first layer, from the coating side, had a thickness of 3 μm and was depleted in Al and Ag. The second consisted of α_2-Ti$_3$Al which was depleted in silver due to its low solubility for Ag (approx. 0.6 at.%). The third layer was a solid solution of Al and Ag in Ti. Figure 12(a–c) shows elemental concentration profiles measured near the alloy/coating interface after different exposure times (10, 63 and 200 h) at 600°C. EDX analyses indicated that the interdiffusion zone was wider than that shown in Fig. 11. This difference appeared to be due to the fact that the EDX signal reveals an average value of the chemical composition from a finite volume. As shown in Fig. 12, the interdiffusion zone increased with increasing exposure time.

In the third layer of this zone the concentration of Ag appeared to be higher (~1.5 at.%) than in the second layer (~0.6 at.%). This might be explained by the fact that titanium diffusion from the base alloy into the coating occurred faster than

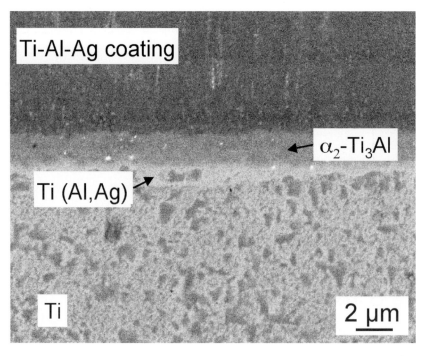

Fig. 11 *SEM/BSE image showing higher magnification of the interdiffusion zone between Ti and Ti–Al–Ag coating from Fig. 10.*

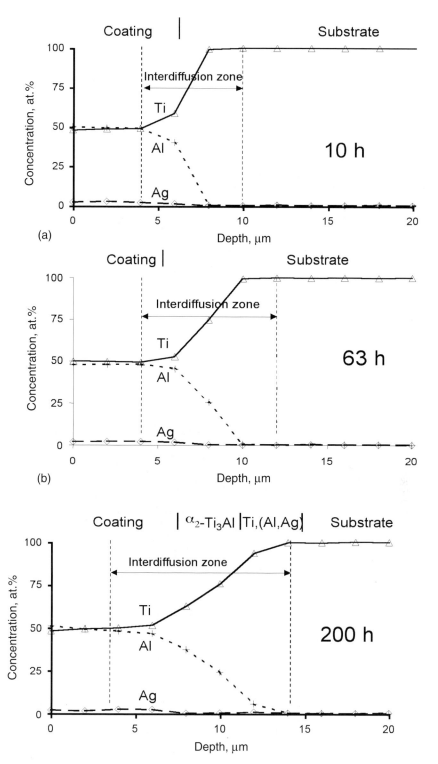

Fig. 12 *Elemental concentration profiles measured near the Ti/Ti–Al–Ag interface during exposure at 600°C, for (a) 10 h, (b) 63 h, (c) 200 h.*

diffusion of aluminium and/or silver in the opposite direction. As a consequence, Ti-rich phases will initially be formed in the coating and will move towards the alloy. Figure 12(a) shows that a very thin layer of α_2-Ti$_3$Al phase formed during short term annealing at 600°C. After 200 h annealing the thickness of the α_2-layer was approximately 4 μm and the adjacent layer contained about 6~8 at.% Al (Fig. 12c).The layer of Ti(Al,Ag) solid solution gradually formed near the coating/alloy interface. However, in spite of the increased titanium diffusion, the Ti–50Al–2Ag coating remained very adherent to the Ti-substrate and no Kirkendall-voids appeared in the coating/substrate interface.

In the case of the Ti–50Al–2Ag coating on the Ti–45Al–8Nb–0.2C substrate a smaller elemental gradient exists between coating and substrate than in case of the Ti/Ti–Al–Ag system. Because both coating and substrate mainly consist of the γ-TiAl phase, the extent of interdiffusion was substantially smaller than in case of the Ti substrate. EDX analyses after 1000 h of exposure did not give any indication for formation of secondary phases near the coating/substrate interface (Fig. 13). The Ti profile suggested that the activity of Ti in the Ti–45Al–8Nb–0.2C alloy is probably higher than that in the Ti–50Al–2Ag coating resulting in diffusion of Ti against its concentration gradient.

4. Summary and Conclusions

1. Ti–50Al–2Ag possesses excellent oxidation resistance in air and in Ar-4%H$_2$O at temperatures up to at least 800°C due to formation of a protective alumina surface scale. This scale protects the alloy against the detrimental effect of water vapour and nitrogen, commonly observed in other TiAl alloys.

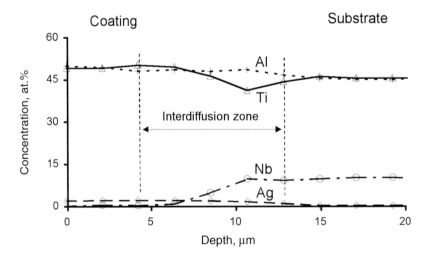

Fig. 13 *Elemental concentration profiles measured near the Ti–Al–Nb–C/Ti–Al–Ag interface after 1000 h exposure at 800°C.*

2. In the as-received state, the Ti–Al–Ag coatings obtained by magnetron sputtering are homogeneous and free from defects such as pores or cracks.

3. Microcracks, which may form during annealing, can be prevented from propagating into the base material by applying a multilayered coating. The cracks tend to stop at the interface between the two coating sub-layers.

4. The recrystallisation of the Ti–Al–Ag coating during high temperature annealing leads to an increase in ductility and a decrease in hardness.

5. The Ti–50Al–2Ag overlay coating seems to be potentially suitable for protecting Ti and γ-TiAl-based alloys at temperatures as high as 800°C. Interdiffusion did not result in a deterioration of the protective properties of the coating up to the maximum test times of 1000 h.

References

1. D. M. Dimiduk, Gamma titanium aluminide alloys — an assessment within the competition of aerospace structural materials, *Mater. Sci. Engng*, 1999, **263**, 281–288.

2. F. Appel, *et al.*, Recent Progress in the Development of Gamma Titanium Aluminide Alloys, *Adv. Engng Mater.*, 2000, **2**, 699–720.

3. R. Rahmel, W. J. Quadakkers and M. Schütze, Fundamentals of TiAl oxidation – A critical review, *Mater. Corros.*, 1995, **46**, 271–285.

4. J. Mendez, On the effects of temperature and environment on fatigue damage processes in Ti alloys and in stainless steel, *Mater. Sci. Engng*, 1999, **263**, 187–192.

5. Y. Shida and H. Anada, The effect of various ternary additives on the oxidation behaviour of TiAl in high-temperature air, *Oxid. Metals*, 1996, **2**, 197–219.

6. M. P. Brady, J. L. Smialek, J. Smith and D. L. Humphrey, The role of Cr in promoting protective alumina scale formation by γ-based Ti–Al–Cr alloys — I. Compatibility with alumina and oxidation behaviour in oxygen, *Acta Mater.*, 1997,**45**, 2357–2369.

7. M. P. Brady, J. L. Smialek, J. Smith and D. L. Humphrey, The role of Cr in promoting protective alumina scale formation by γ-based Ti–Al–Cr alloys — II. Oxidation behaviour in air, *Acta Mater.*, 1997, **45**, 2371–2382.

8. V. Shemet, A. K. Tyagi, J. S. Becker, P. Lersch, L. Singheiser and W. J. Quadakkers, The formation of protective alumina-based scales during high-temperature air oxidation of γ-TiAl alloys, *Oxid. Metals*, 2000, **53**, 211–235.

9. V. Shemet, L. Niewolak, P. J. Ennis, W. J. Quadakkers and L. Singheiser, Oxidation resistant γ-TiAl based alloys for advanced gas turbine components and coating systems, *Parsons 2000*, (A. Strang ,W. M. Banks, R. D. Conroy, G. M. McColvin, J. C. Neal and S. Simpson, eds). Cambridge University Press, Cambridge, 2000, p.874–884.

10. H. Clemens and H. Kestler, Processing and applications of intermetallic γ-TiAl-based alloys, *Adv. Engng Mater.*, 2000, **2**, 551–570.

11. N. Zheng, W. Fischer, H. Grübmeier, V. Shemet and W. J. Quadakkers, The significance of sub-surface depletion layer composition for the oxidation behavior of γ-titanium aluminides, *Script. Metall. Mater.*, 1995, **33**, 47–53

12. E. Copland, B. Gleeson and D. J. Young, Formation of Z-$Ti_{50}Al_{30}O_{20}$ in the sub-oxide zones of γ-TiAl-based alloys during oxidation at 1000°C, *Acta Mater.*, 1999, **47**, 2937–2949.

13. V. Shemet, P. Karduck, H. Hoven, B. Grushko, W. Fischer and W. J. Quadakkers, Synthesis of the cubic Z-phase in the Ti–Al–O system by a powder metallurgical method, *Intermetallics*, 1997, **5**, 271–280.

14. C. Lang and M. Schütze, TEM studies of the mechanism of the early stages of TiAl oxidation, in *Microscopy of Oxidation — 3* (S. B. Newcomb and J. A. Little, eds). The Institute of Materials, London, 1997, p.265–276.

15. B. A. Pint, J. R. Martin and L. W. Hobbs, The oxidation mechanism of θ-Al_2O_3 scales, *Solid States Ionics*, 1997, **250**, 99–107.

16. M. W. Brumm and H. J. Grabke, The oxidation behavior of NiAl-I. phase transformation in the alumina scale during oxidation of NiAl and NiAl–Cr alloys, *Corros. Sci.*, 1992, **33**, 1667–1690

17. S. Taniguchi, N. Hongawara and T. Shibata, Influence of gaseous species on the oxidation behavior of TiAl at high temperatures, in *Gamma Titanium Aluminides 1999* (Young-Won Kim, D. M. Dimiduk and M. H. Loretto, eds). The Minerals, Metals & Materials Society, Warrendale, PA, 1999, p.811–816.

18. H-P. Martinz, H. Clemens and W. Knabl, Oxidation protective coatings for γ-TiAl based alloys, in *Gamma Titanium Aluminides 1999* (Young-Won Kim, D. M. Dimiduk and M. H. Loretto, eds). The Minerals, Metals & Materials Society, Warrendale, PA, 1999, p.829–836.

Investigation and Modelling of Specific Degradation Processes

1.4 Thermal barrier systems

22

Interaction of Corrosion and Fatigue in Thermal Barrier Coatings — An Experimental Approach Towards Lifetime Assessment

M. BARTSCH, C. LEYENS and W. A. KAYSSER

Institute of Materials Research, DLR - German Aerospace Center, D-51170 Cologne, Germany

ABSTRACT

For practical reasons, lifetime assessment for ceramic thermal barrier coatings (TBCs) of gas turbine components in service requires short laboratory testing times that have to simulate the long service times experienced in practice. However, accelerated testing conditions require that damage mechanisms in laboratory tests, which supply the database for lifetime assessment, are identical to those in service. In the lifetime assessment methodology described here, fatigue and kinetic damages are treated separately. Since fatigue damages depend on the number of cycles rather than on the dwell time on a certain load level, the duration of one test cycle can be reduced drastically provided that the load alternations at the failure location are in a realistic range. The test rig used in this study simulated the cyclic loads of an internally cooled turbine blade. Initial results on as-deposited electron beam-physical vapour deposited (EB-PVD) TBCs indicated that the coatings survived the required number of load alternations, suggesting that time- and temperature-dependent mechanisms contribute significantly when TBCs fail in service. Additional work on the effect of aging on TBC spallation in bending tests revealed rapid degradation of the strain to failure, supporting the conclusion that kinetic effects play a major role in TBC lifetime. The effect of kinetically induced damages on TBC lifetime in service will be considered by fatigue testing of pre-aged specimens in future work. To determine interaction effects between fatigue and kinetic damage mechanisms fatigue tests with extended dwell times are also necessary.

1. Introduction

A typical thermal barrier coating (TBC) system consists of a ceramic top coating (usually yttria partially stabilised zirconia) deposited on to a nickel superalloy substrate coated with either an MCrAlY-type overlay or a Pt-Al diffusion bond coat. Although an increasing number of aeroengine and industrial gas turbines for power generation have specified TBCs for highly loaded airfoils in the high pressure turbine, their application today is essentially limited to lifetime improvement of the airfoil due to reduced metal surface temperatures and thus increased oxidation and creep lifetimes of the structural base material. Future thermal barrier coatings are required to exhibit prime reliance such that thermal barrier coatings can be reliably integrated into airfoil design. At least two requirements can be deduced from this ambitious design goal; firstly, significant performance improvements of TBCs are required, and there is currently substantial work under way to provide a basis for these improvements [1–4]; secondly, durability of the coatings under demanding operating conditions must be improved; in this context, reliable lifetime prediction of TBCs is a major issue. Ideally, TBCs should have lifetimes as long as those of the airfoil material or at least they should be fully utilised for the complete time to overhaul. However, in service, TBCs are subjected to various influences that initiate and/or promote failure by spallation of the ceramic top coating. The damage mechanisms that finally lead to failure can be classified into fatigue damages (number of cycles dependent only) and kinetic damages (time- and temperature-dependent).

2. Damage Evolution and Failure of TBCs

The mentioned damage mechanisms usually dominate in different sequences of the service load cycles of a gas turbine. On considering the flight cycle of an aeroengine (Fig. 1), it is seen that the airfoils are exposed to simultaneous thermal and mechanical loading, including extreme thermal shock during start-up and shut-down of the engine. These sections of the flight cycles predominantly impose mechanical and thermal loads and thus promote fatigue damage, while damages due to kinetic mechanisms at high temperatures accumulate during the dwell times at high temperatures, mainly during cruising. Indications exist that moisture-enhanced crack growth might contribute to crack propagation in pre-damaged TBCs [5]; therefore, the time interval between engine shut-down and re-start might also contribute to damage accumulation and reduction in TBC lifetime.

Thermal and mechanical load levels imposed on the airfoil and on the TBC depend on the specific location on the components. For example, the thermal load varies from low at the blade-fillet near the cooling air inlet to high at those parts of the

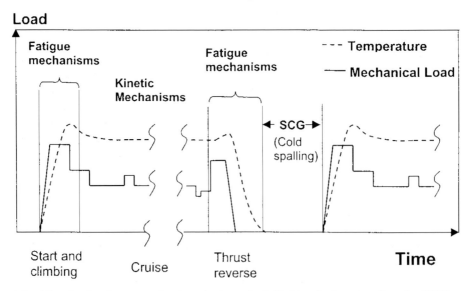

Fig. 1 *Dominating damage mechanisms during a single flight cycle of an aircraft engine (SCG = slow crack growth leading to cold spalling of TBCs).*

airfoil which are exposed directly to the hot gas and radiation from the combustion chamber. Additionally, the effect of the thermal gradient from the heated outer surface to the cooled inner surface has to be considered since the stresses due to the thermal gradient exceed the stresses caused by mechanical loading.

The thermal and mechanical cyclic loading and the mismatch of several physical properties of the substrate and the coating materials cause locally alternating stresses in and at the interfaces between the different layers of the TBC system. These stress alternations are the driving force for the evolution of fatigue damages of the TBC. Simultaneously, kinetic mechanisms at high temperature change the physical properties of the layer materials causing the change of local loading conditions as well as the resistance against damage and failure. For example, sintering of the ceramic top coat increases the Young's modulus thus reducing strain tolerance. Thermal cycling above 1200°C causes phase transformations in state of the art yttria stabilised zirconia from partially stabilised tetragonal zirconia to the monoclinic phase resulting in volume changes and hence increase of local stresses [6]. A significant part of residual stresses perpendicular to the coated surface develops due to continued growth of the oxide scale (TGO) on the bond coat. Numerous changes occur during high temperature exposure at or near the bond coat/TBC interface that contribute to damage accumulation and finally to failure of the TBC. For example, diffusion processes within the oxide scale can lead to pore formation [7] which may act as initial cracks for fatigue crack growth. Also bond coat heterogeneities are considered to play an important role in the formation of initial separations and further delamination between bond coat and TGO [8, 9]. Furthermore, diffusion of alloy elements and impurities from the substrate to the interface between bond coat and ceramic can reduce the adhesion of the TBC [10].

3. Lifetime Assessment of TBC Systems

The dominating damage mechanisms and interaction effects depend on the materials parameters of the specific TBC system and the service conditions. Consequently, practical lifetime assessment methodologies for TBC systems are required which are independent of a single mechanism-based description of damage evolution. Instead of accumulating the contributions of separate damage mechanisms determined in simplified tests (e.g. [11,12]) the approach for TBC lifetime assessment proposed here is based on damage accumulation under realistic testing conditions. The basic idea is to ensure that both damage mechanisms, fatigue and kinetic mechanisms and their interaction effects, are identical to those observed in service. However, it is impractical to simulate a service cycle in real time, since turbine blades in modern gas turbines in aircraft engines are designed for about 20 000 h equivalent to 5000–10 000 flights. The time for one test cycle can be decreased drastically if number-of-cycle-dependent fatigue damages are determined separately from damages caused by kinetic mechanisms because fatigue damages are induced by changes of the load rather than by the dwell time on a certain load level. Damages induced by time-at-high-temperature-dependent kinetic mechanisms can be considered by testing heat-treated specimens. Since several specimens can be heat-treated in a furnace simultaneously the overall testing time can be limited. However, when accelerating the kinetically controlled damages by increasing the heat-treatment temperature it must be ensured that the damage mechanisms do not change due to the higher temperature. Basically as a last step, interaction effects between fatigue and kinetic damages must be determined in extended flight cycle simulations.

4. Experimental

The test facility used simulated the cyclic loading conditions for a gas turbine blade, i.e. cycles with simultaneous mechanical and thermal loading, including thermal shock and temperature gradients over the cross section of the specimens. Hollow cylindrical specimens of directionally solidified nickel-base superalloy IN 100 with a metal substrate wall thickness of 2 mm were tested; the geometry was similar to the convex side of a turbine blade near the leading edge. Specimens were coated with an EB-PVD Ni–20Co–18Cr–13Al–0.15Y bond coat and an 7% yttria stabilised zirconia EB-PVD TBC was deposited onto selected specimens.

Mechanical loading was imposed on the specimens by a servo-hydraulic testing machine. For simultaneous heating, radiation of four quartz lamps was focused onto the specimen by elliptical mirrors. A temperature gradient was generated by internal air cooling of the specimens. Mass flow and temperature of the cooling air were controlled during the test. Thermal shock during shut down of the gas turbine was simulated by rapid cooling of the specimens with cold air blown onto the surface by 36 vents which were integrated in two sliders enclosing the specimen during cooling. High heating rates simulating the thermal shock during start and take off of an aircraft engine is achieved by switching on the quartz lamps for heating before opening the sliders again. A detailed description of this thermal gradient mechanical fatigue (TGMF) test rig is given in [13].

Each test cycle simulated an entire flight cycle of an aircraft engine, including take off, climbing, cruise, thrust reverse and shut down (Fig. 2). The duration of the test cycle was about 90 s for specimens with bond coat only and about 3.5 min for specimens with an additional ceramic top coat, representing about 1% of the real flight time. For the latter, a longer time was needed to achieve the adjusted quasi-stationary cycle temperature. Mechanical load and temperature can be adjusted independently to simulate the loading conditions for several blade locations (e.g. blade-tip or fillet).

The temperature field of the specimens was measured with thermocouples over the cross-section and along the axis parallel to the direction of the mechanical load. Calibration specimens with an integrated thermocouple were used with bond coat only and with additional ceramic top coat. The mass flow of the cooling air was adjusted to different temperatures of the control thermocouple on the outer surface of the specimens. A reference bond coat temperature of 920°C (quasi-stationary cycle temperature) was selected for all specimens (bond coat only as well as with additional ceramic top coat). For this reference temperature the stationary thermal gradient was about 35°C over the ceramic top coat and about 85°C over the metallic substrate.

Complementary four-point bending tests were performed on single-side EB-PVD NiCoCrAlY bond coated IN 617 beams overlaid with an EB-PVD TBC. Specimens were aged between 1 and 1013 h in air at 1000 and 1050°C providing oxide scale

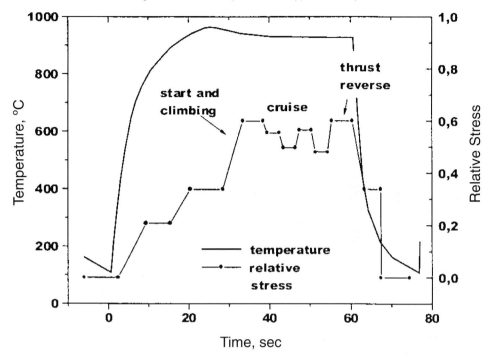

Fig. 2 *Load and temperature spectrum of a test cycle. 'Relative stress' is related to the yield stress at stationary cycle temperature.*

thicknesses between 0.5 to 8 µm. Strain-controlled four-point-bending tests were performed at room temperature. Damage evolution and the onset of TBC spallation was observed with an optical microscope during bending.

5. Results and Discussion

The first step of the experimental program was to determine a realistic load level for the TGMF experiments. For this purpose, fatigue experiments on bond coat only specimens were performed with a decreasing load level until a number of test cycles to failure of about 3000 cycles was reached which was considered a reasonable number of flights for which an airfoil is designed (Fig. 3). Subsequently, TBC-coated specimens were tested at identical load levels and bond coat temperatures of 920°C.

For the specimens coated with bond coat only the correlation between the number of cycles to failure and the maximum of the mechanical load spectrum of the cycle can be described by a power law (Fig. 3), suggesting that failure is dominated by fatigue mechanisms. Notably, the lifetime of the specimens was always significantly higher with TBC than without TBC even though bond coat temperature and mechanical load were identical. All experiments on TBC-coated specimens were terminated without failure of the specimens substrate. Visual inspection after testing revealed that neither spallation nor crack formation were evident on the surface of the TBCs.

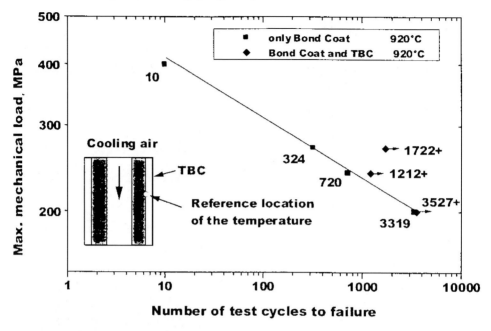

Fig. 3 *Measured number of test-cycles to failure during TGMF testing. TBC-coated specimens did not fail before termination of the test.*

Cross-section micrographs of the fractured specimens with bond coat only displayed fatigue cracks starting at the (non-coated) internally cooled surface (Fig. 4). The cracks were open and in the case of the long-term cycled specimen (3319 cycles to failure) the crack surfaces were clearly covered with an oxide scale revealing that the cracks have been open during TGMF testing. Only in the vicinity of the lethal crack were additional cracks close to the heated surface found. These cracks started in the substrate but did not propagate into the bond coat. In all cases the cracks stopped at the substrate/bond coat interface which was plastically deformed but not fractured (Fig. 5). No oxide scales were found on the crack surfaces, suggesting that these cracks were formed as a result of the high strain accumulating within the last cycles before failure.

Microstructural investigation of specimens tested with ceramic top coat revealed no crack formation or delamination either in the ceramic topcoat or in the metallic substrate or bond coat (Fig. 6a). Although the bond coat-TGO interface roughened significantly after 3527 cycles at 920°C (equal to 200 h at 920°C) the TBC was still perfectly adherent. Void formation in the TGO of TGMF-tested specimens appeared more pronounced (Fig. 6b) than that found on identical bond coats when tested at even higher temperatures and/or longer times but without superimposed mechanical load [14]. The mechanism by which these voids form is not yet understood, however, formation of a large number of voids is clearly of concern if extended exposure times are envisaged. Moreover, rumpling of the bond coat-TGO interface is more pronounced in the TGMF-tested specimens than in similar specimens not exposed to mechanical loads. This again might promote flaw formation at the bond coat-TGO interface. However, at present, results clearly indicate that EB-PVD TBCs survive

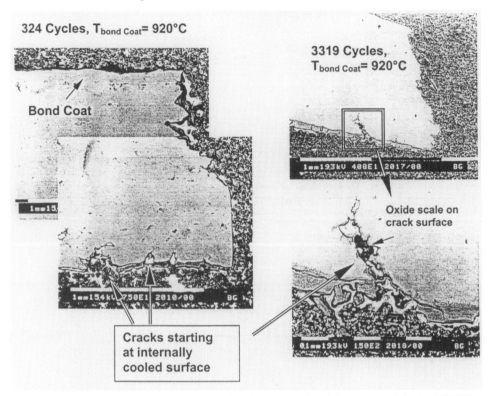

Fig. 4 *Cross sections of the specimens with bond coat only after failure occurred during TGMF test. Figures showing open fatigue cracks, starting from non coated internal surface.*

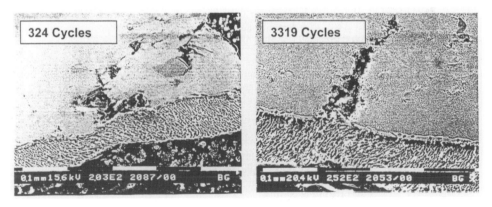

Fig. 5 *Cross sections of specimens with bond coat only after failure during TGMF tests. Cracks are due to high deformation in the tailored section beneath the lethal crack.*

the required number of mechanical load cycles without failure under the loading conditions and dwell times used in this study. Since TBCs in engines typically fail before the lifetime of the aerofoil base material is exceeded, the conclusion is that time- and temperature-dependent processes play a major role in TBC failure rather

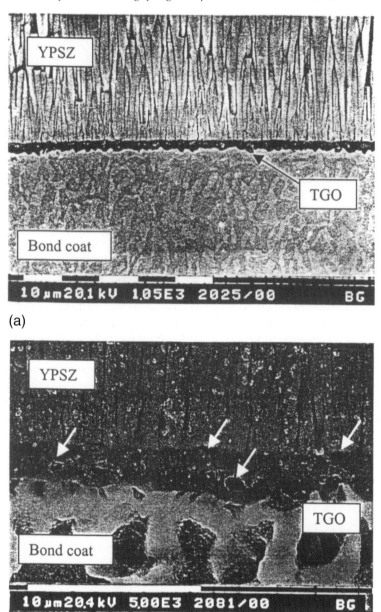

(a)

(b)

Fig. 6(a) *Microstructure of an EB-PVD TBC system after 150 h TGMF testing at 920°C in air. The TGO thickness was about 3 µm. **(b)** Higher magnification of the TGO. Despite formation of large voids in the TGO, adhesion of the top coating is still excellent.*

than fatigue damages alone. However, it appears quite plausible that fatigue damage mechanisms might accelerate failure and therefore must be taken into account in life time assessment.

Literature data on bond strength vs number of thermal cycles and complementary experiments on strain to failure vs exposure time confirmed a significant effect of kinetic damage mechanisms on adhesion of TBCs (Fig. 7). In both cases, bond strength and strain to failure as a measure of coating adhesion decreased with increasing time at high temperature. Notably, four-point bending tests on TBC coated specimens revealed that degradation occurred fairly rapidly, depending on the temperature used (Fig. 7b); as expected, with increasing exposure temperature degradation occurred more rapidly. Growth of the oxide scale appeared to be a major contributor to a weaker TGO-bond coat interface, however, the mechanism by which interfacial strength was reduced is still the subject of ongoing research efforts [3]. Although ageing temperatures of 1000 and 1050°C were somewhat higher than in the TGMF tests (920°C), these results clearly indicate that kinetic damages are indeed very important contributors to TBC failure.

The next step to validate the concept of lifetime assessment proposed here is to perform similar TGMF tests on pre-aged specimens. A reasonable assumption is that the number of cycles to failure is significantly reduced with increasing cycle and ageing times, both simulating the dwell time at cruising conditions in a real flight cycle. Experimental work is under way to generate data to quantify the model. A major issue in this context is to select appropriate ageing conditions that neither lead to failure after a few cycles or lead to no failure when a given number of cycles is exceeded (for example 3000 cycles in this study). As shown above (Fig. 7b), accelerated ageing conditions in terms of temperature appear to be critical, since degradation of adhesion occurs within a narrow time frame, thus complicating interpretation of TGMF test results under these conditions. Finally, direct interaction effects of fatigue and kinetic damages have to be determined by TGMF testing with extended dwell times. Stress effects on oxidation kinetics [15] may play a role as well as crack healing processes during extended dwell times at cruising conditions.

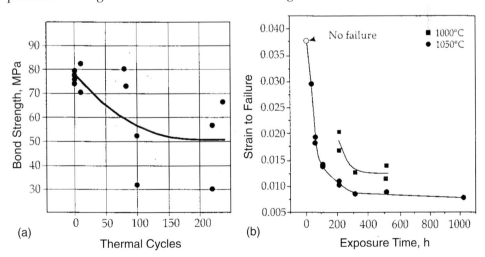

Fig. 7(a) Bond strength of an EB-PVD TBC on Pt–Al coated René N5 after thermal cycling at 1121°C (2050°F) [16]. Data were determined using a modified ASTM pull test. (b) Strain to failure vs. Exposure time at 1000 and 1050°C for EB-PVD TBCs on Ni–Co–Cr–Al–Y coated IN 617. Notably, strain to failure drops strongly after short annealing times.

6. Conclusions

Thermal gradient fatigue (TGMF) testing of EB-PVD thermal barrier coatings appears to be a promising tool for lifetime assessment. Although superimposed in-service, fatigue damages and kinetic damages, i.e. time- and temperature-dependent processes, can be studied separately and under accelerated conditions. In the present work, an aircraft engine flight cycle was simulated in about 1% of the time required for a real-time test. Initial results indicate that TBCs in service fail predominantly by kinetic rather than by fatigue damages; the as-deposited EB-PVD TBCs tested under TGMF conditions survived the required number of cycles without failure. Strain to failure vs ageing time experiments revealed that, depending on the temperature, rapid degradation of TBC occurred confirming that kinetic damages play a major role in TBC lifetime. Future work is focused on TGMF testing of pre-aged specimens as well as on tests with extended dwell times to determine interaction effects between fatigue and kinetic damages.

7. Acknowledgements

The authors are grateful to K. Mull and C. Sick at DLR for technical support and mechanical testing. M. Peters provided comments on the manuscript which are gratefully acknowledged.

References

1. M. J. Stiger, N. M. Yanar, M. G. Topping, F. S. Pettit and G. H. Meier, *Z. Metallkd.*, 1999, **90**, 1069–1078.

2. C. Leyens, U. Schulz, K. Fritscher, M. Bartsch, M. Peters and W. A. Kaysser, *Z. Metallkd.*, 2001, **92** (7), 762–772.

3. C. Leyens, U. Schulz, M. Bartsch and M. Peters. R&D status and needs for improved EB-PVD thermal barrier coating performance, in *Proc. Met. Res. Soc. Symp. Proc. Vol. 645E*, 2001, 10.1.1–10.1.12.

4. C. Leyens, U. Schulz and M. Peters, Advanced thermal barrier coating systems: research and develoment trends, in *High Temperature Coatings—Science and Technology IV* (N. B. Dahotre, J. M. Hampikian and J. E. Morrall, eds) 2001. TMS, Warrendale, PA, pp.61–76.

5. V. Sergo and D. R. Clarke, *J. Am. Ceram. Soc.*, 1998, **81**, 3247–3252.

6. U. Schulz, *J. Am. Ceram. Soc.*, 2000, **83**, 904–910.

7. M. Schütze, *Protective Oxide Scales and Their Breakdown*. 1997, John Wiley & Sons, Chichester, UK.

8. K. Vaidyanathan, M. Gell and E. Jordan, *Surf. Coat. Techn.*, 2000, **133/134**, 28–34.

9. M. Y. He, A.G. Evans and J. W. Hutchinson, *Mater. Sci. Eng.*, 1998, **A245**, 168–181.

10. U. Kaden, C. Leyens, M. Peters and W. A. Kaysser, Thermal Stability of an EB–PVD Thermal Barrier Coating System on a Single Crystal Nickel-Base Superalloy, in *Elevated Temperature Coatings: Science and Technology III* (J. M. Hampikian and N. B. Dahotre, eds). TMS, Warrendale, PA, 1999, p.27–38.

11. R. A. Miller, *J. Am. Ceram. Soc.*, 1984, **67**, 517–521.

12. S. M. Meier, D. M. Nissley, K. D. Sheffler and T. A. Cruse, *J. Eng. for Gas Turbines Power*, (*Trans. ASME*), 1992, **114**, 258–263.

13. M. Bartsch, G. Marci, K. Mull and C. Sick, *Adv. Eng. Mater.*, 1999, **2**, 127–129.
14. C. Leyens, U. Schulz, B. A. Pint and I. G. Wright, *Surf. Coat. Technol.*, 1999, **120–121**, 68–76.
15. H. E. Evans, *Int. Mat. Rev.*, 1995, **40**, 1–40.
16. M. Gell, E. Jordan, K. Vaidnanathan, K. McCarron, B. Barber, Y.-H. Sohn and V. K. Tolpygo, *Surf. Coat. Technol.*, 1999, **120–121**, 53–60.

23

New Approaches to the Understanding of Failure and Lifetime Prediction of Thermal Barrier Coating Systems

D. RENUSCH, H. ECHSLER and M. SCHÜTZE

Karl-Winnacker-Institut der DECHEMA e.V., D-60486 Frankfurt am Main, Germany

ABSTRACT

Thermal barrier coating (TBC) failure and lifetime prediction is considered as a two-step process. It is shown that micro-structural information such as defect length and porosity can be incorporated into a life time prediction model. A failure mode dependent example calculation is made based on measurement data. The example calculation is compared to recent literature data from a low cycle thermomechanical fatigue test. An additional comparison is made with an older model for lifetime prediction.

1. Introduction

The usefulness of the thermal barrier coating TBC composite system is limited by the mechanical reliability of the ceramic layers (both zirconia top coat and thermally grown oxide TGO) under service conditions. The superposition of several kinds of strains leads to the mechanical failure of either the top coat or TGO. Such applied strains can result from external loading (ε_{ext}) by the operation of the component, from temperature changes due to the different thermal expansion coefficients of the different materials (ε_{therm}) and from the formation of the TGO (ε_{int}). When the applied strains exceed a critical value (ε_{cr}), damage (i.e. macrocracking, buckling, and spallation) of the ceramic layers can occur as a strain relief mechanism [1].

$$\varepsilon_{therm} + \varepsilon_{int} + \varepsilon_{ext} \geq \varepsilon_{cr} \tag{1}$$

The primary issue in eqn (1) with regard to the TBC life expectancy is the time dependence of the strains. Under operational conditions, the strains the ceramic layers experience, due to external loading and temperature changes, have a cyclic nature which causes thermomechanical fatigue. Consequently, a common starting point for lifetime modelling is the Manson-Coffin fatigue equation [2],

$$N_f = \left[\frac{\Delta\varepsilon_{f,0}}{\Delta\varepsilon_{inel}} \right]^b \tag{2}$$

where N_f is the predicted number of cycles for failure and $\Delta\varepsilon_{inel}$ is the inelastic strain range that the material endures. $\Delta\varepsilon_{f,o}$ is the failure strain range of the undamaged system (i.e. failure in one cycle), which is the difference between the tensile $(+\varepsilon_{f,o})$ and compressive $(-\varepsilon_{f,o})$ failure strain. Consequently the failure strains for the TBC top coat and TGO in tension and compression have to be investigated.

The top coat and TGO also have to withstand internal strains of which the most commonly considered are the growth strains resulting from TGO formation within a limited volume. Other possibilities are phase transitions in the top coat, TGO, bond coat or substrate. NASA has modified the Manson-Coffin equation to include TGO growth [3].

$$N_f = \left[\left(\frac{\Delta\varepsilon_{f,0}}{\Delta\varepsilon_{inel}} \right) \cdot \left(1 - \frac{d}{d_c} \right)^c + \left(\frac{d}{d_c} \right)^f \right]^b \tag{3}$$

where d is the time-dependent TGO thickness which can be determined from the bond coat oxidation kinetics and d_c is a critical TGO thickness. More recently, Herzog *et al.* modified the fatigue equation to account for bond coat oxidation and top coat sintering [4], and the resulting eqn (4) works well for describing low cycle thermomechanical fatigue test data. Herzog also used the inelastic strain range of the base material that his specimens were made from.

$$N_f = \left[\left(\frac{\Delta\varepsilon_{f,0}}{\Delta\varepsilon_{inel}} \right) \cdot \left(1 - \frac{d}{d_c} \right)^c \cdot \left(\frac{\Delta\varepsilon_{f,\infty}}{\Delta\varepsilon_{f,0}} \cdot \left(1 - e^{-t/\tau} \right) + e^{-\frac{t}{\tau}} \right) \right]^b \tag{4}$$

However, the above models only consider TBC failure as a one-step process, whereas microstructural based coating failure theory [5,6] considers coating failure as a two step process (Fig. 1) and there exist primarily two modes.

Mode I: Assumes a strong interface and a weak coating, where the coating first develops pre-failure, macroscopic through-cracks which are perpendicular to the interface. The observed detachment of the coating occurs only after macroscopic delamination cracks form at the interface.

Mode II: Assumes a weak interface and a strong coating in which first pre-failure, macroscopic delamination cracks form at the interface and the observed detachment of the coating occurs only after the formation of macroscopic through-cracks. What is currently observed in the (electron beam-physical vapour deposited, EB-PVD) TBC system is long pre-failure delamination cracks in the TGO of oxidation samples. For the atmosphere plasma spraying, APS TBC system the current observation is long pre-failure delamination cracks in the top coat of both oxidation and mechanical testing samples. Consequently the focus of this paper is on mode II failure.

In order to develop a deeper understanding of TBC failure an approach to lifetime prediction is under consideration, an approach that combines the two-step coating failure and the Manson-Coffin fatigue equation:

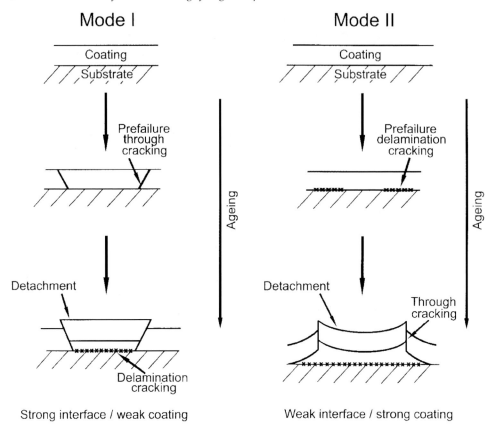

Fig. 1 *Schematic of two possible two-step coating failure routes.*

$$N_d = \left[\frac{\Delta \varepsilon_{d,0}}{\Delta \varepsilon_T} \right] \cdot F_d(t) \tag{5}$$

$$N_t = \left[\frac{\Delta \varepsilon_{t,0}}{\Delta \varepsilon_T} \right] \cdot F_t(t) \tag{6}$$

where N_d and N_t are the number of cycles for the onset of delamination and through-cracking respectively. $\Delta \varepsilon_{d,o}$ and $\Delta \varepsilon_{t,o}$ are the critical strain ranges associated with each crack type and $F_d(t)$ and $F_t(t)$ are functions that account for the time at temperature effects. This approach not only considers two-step failure processes but also, through $F_d(t)$ and $F_t(t)$, allows the effects that time at temperature has on each crack type to be considered separately. The exponent b has been removed. For the last 10 years the exponent b has been included in lifetime models for TBCs and in all applied cases it has been either set to a value of 1 or measured as being close to 1, so here it is assumed to be equal to 1 from the start. $\Delta \varepsilon_T$ is the total applied strain range in the ceramic layer that fails. The elastic strain is incorporated into applied strain range because brittle materials under the influence of elastic deformation can fail.

In Fig. 2 we have given a schematic representation of the time at temperature dependencies of eqns (5) and (6), where we assume $F_d(t)$ and $F_t(t)$ are decreasing functions in time, which is to say time at temperature decreases TBC lifetime. Also in Fig. 2 there is an ideal strain range which is assumed to be constant in time. For real systems the strain range should be calculated from the strains in eqn (1). The points of interest are the times t_1 and t_2 where the strain range intersects the critical strain functions $+\varepsilon_{d,0}F_d(t)$ and $+\varepsilon_{t,0}F_t(t)$, respectively. Prior to t_1 is the predelamination period of the TBC's life during which the top coat and/or TGO experiences microstructural changes originating from microcracking, sintering, creep and early stage oxidation of the bond coat. At time t_1 a macroscopic delamination crack forms in the top coat of the APS system or in the TGO of the PVD system. From time t_1 to time t_2 the delamination crack steadily increases in length. Time t_2 is when through cracks begin to form. Shortly after t_2 top coat detachment is observed (i.e. final failure).

The approach outlined in eqns (5) and (6) and Fig. 2 also allows the incorporation of micro-structural information into the lifetime prediction model. A large amount of literature exists [6,7] that relates critical stresses and strains to the physical defects (such as pores and microcracks) that exist in a ceramic coating. Equations (7), (8) and (9), for example, are modelling equations that have been used to describe oxide scale delamination, segmentation and shear macrocracking, respectively:

$$\varepsilon_d = \frac{K_{Ic}}{f\sqrt{\pi c}} \cdot \frac{(1+r/d)\cdot(1+\nu)}{2E_{Ox}} \tag{7}$$

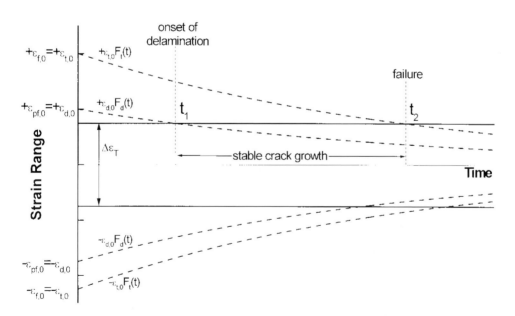

Fig. 2 *Schematic of the time at temperature-dependent strain range limit. The subscripts* pf *and* f *are for the general description of pre-failure macrocracking and final failure macrocracking. This figure considers the special case of pre-failure macrocracking as being delamination* (pf = d) *and final failure macrocracking as being through cracking* (f = t).

$$\varepsilon_s = \frac{K_{Ic}}{f \cdot E_{Ox} \sqrt{\pi c}} \tag{8}$$

$$\varepsilon_{sh} = \frac{K_{IIc}}{f \cdot E_{Ox} \sqrt{\pi c}} \tag{9}$$

where K_{Ic} and K_{IIc} are fracture toughnesses of the TBC or TGO, f is a geometric factor that equals 1.0 for a buried defect and c is half of the length of the buried defect, a defect being a microscopic object such as pores, small voids, or microcracks. The r is the interfacial roughness and d is the coating thickness. Equations similar to (7–8) and (9) can be readily incorporated into eqns (5) and (6) by placing the appropriate fracture equations into the $\Delta\varepsilon_{d/t,o}$ terms. This produces a lifetime prediction model that is dependent on the physical defects that are present in the TBC top coat or in the TGO.

The interest in physical defects comes from the fact that top coat porosity in the APS system is an adjustable spray parameter used by producers of APS TBCs to optimise the thermal barrier effect. However, changing the porosity of the top coat will have an effect on its mechanical reliability. Consequently, a physical defects-based lifetime model could prove helpful to APS coating producers. For the EB-PVD system the advantages of the physical defects-based lifetime model are a little less obvious. Recently an article [8] about the formation of thermally grown NiO on Ni has shown that the size of the physical defects in the NiO scale is heavily dependent on the impurities in the Ni substrate. A similar effect may also be present in the TGOs that form on MCrAlY and PtAl bond coats, so that data from detailed oxidation studies of bond coats and the defects that form in the TGOs can be directly connected to TBC lifetime prediction.

2. Experimental Details

The experimental data presented here are measurements of critical strains for macroscopic delamination and segmentation cracking of the TBC top coat under tensile strain and critical strains for macroscopic delamination and shear cracking of the TBC top coat under compression. Also measurements of the top coat porosity as a function of oxidation time are described.

The samples were 300 μm thick APS coating with an as-sprayed porosity of about 13%. The substrates were CSMX-4 with a 150 μm MCrAlY bond coat. Testing involved both isothermal oxidation and four point bending with acoustic emission, AE. The substrates of the oxidation samples were cylinders with a 10 mm dia and 20 mm tall. The substrates of the four point bend samples were $80 \times 6 \times 4$ mm with a 20×2 mm gauge length notch removed from the centre thus producing a substrate thickness in the centre of the sample of 2 mm (Fig. 3).

In order to pursue the critical strains in the APS TBC system a newly constructed four point bending with acoustic emission testing machine was used. This machine was constructed from components that existed on site as well as new components,

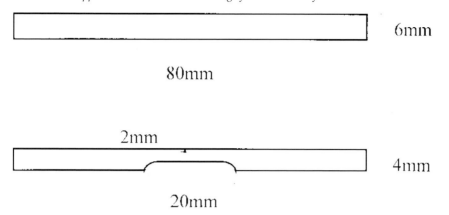

Fig. 3 *Schematic of the four point bend substrate used for determining the critical strains of TBCs.*

such as, a 45 kN ball screw jack, low voltage directive transmitter (LVDT) roller point displacement measurement system, ceramic sample grips and acoustic emission system. A two channel with linear location acoustic emission technique was used to detect TBC critical strains. The details have been described elsewhere [9]. The bending samples were unoxidised and bent at room temperature in both tensile and compression geometries. The total roller point displacement was 2 mm at a displacement rate of 35 μm/min. An alumina sample grip with roller spacings of 64 mm and 36 mm was used. During the course of the bend the roller point displacement and roller point force were measured on a 1 s interval.

The reported TBC/bond coat interface strain values were calculated from the roller point displacement measurements by using an iterative technique. The major factors in the strain calculations are the sample thickness, the length of the notch and the roller spacing. The calculation showed a minor sensitivity to material properties in this displacement driven test. To start the iterative calculation the system was first solved for elastic bending and using 130 GPa, 200 GPa and 17 GPa for the values of the Young's moduli of the substrate, bond coat and APS top coat respectively. The iterative solution corrects the strain values for the observed inelastic behaviour of the sample and is based on the elastic and inelastic stress strain distribution through the sample thickness, where the inelastic stress distribution of thin bond coat was assumed to be ideal with a flow stress of 700 MPa. The stress distribution in the superalloy was assumed to follow power law strain hardening behaviour with a flow stress of 1000 MPa and strain hardening exponent of 1/3.

3. Results and Discussion

In Fig. 4(a) the measured bending moment and the acoustic emission energy are plotted as a function of the TBC/bond coat interface strain for two samples that have been bent so that the TBC experiencies tensile strains. The acoustic emission is caused by the slow cracking of the TBC. Figures 4(b) and 4(c) are edgewise photographs of the resulting damage. Figure 4(a) shows that for strain values less

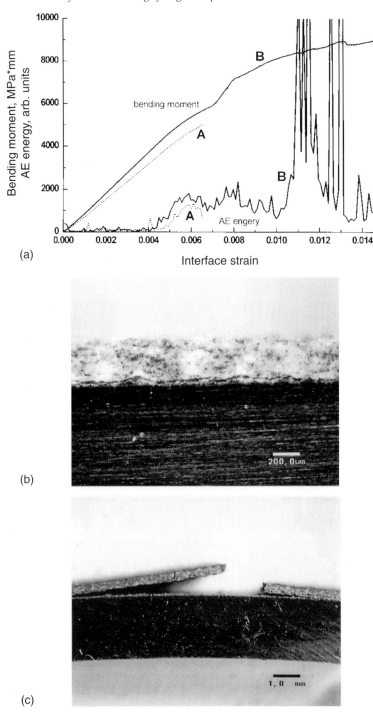

(a)

(b)

(c)

Fig. 4 *(a) Plot of the tensile bending moment and acoustic emission energy as a function of the TBC/bond coat interface strain for two samples: (b) and (c); Edgewise photos of the damage in samples A and B respectively.*

than 0.0045 there is a very small amount of acoustic emission which indicates that the amount of cracking in the TBC is low. At a strain of 0.0045 there is a sharp rise in the acoustic emission energy which is caused by the onset of macrocracking. For sample A the test was stopped at a strain level of ~0.0065 and the sample was photographed (Fig. 4(b)). In the photograph a large delamination crack can be seen near the TBC/bond coat interface. Subsequent more detailed metallographic investigations of sample A have found only delamination cracking with some crack deflection toward the air interface. Consequently, the sharp rise in AE energy at the strain level of 0.0045 is due to the interfacial strain exceeding the critical strain for delamination of the TBC in tension. Sample B was bent for a longer time and what is seen in Fig. 4(a) at interface strain levels between 0.010 and 0.012 is a second large rise in AE energy which indicates that a second type of crack has begun to form. In the photograph of the sample (Fig. 4(c)) it can be seen that the sample has suffered delamination cracking which began forming at a strain of 0.0045 and segmentation cracking which began forming at a strain just above 0.01.

In Fig. 5(a) the bending moment and acoustic emission energy are plotted for two TBC coatings that have been subjected to compressive strains and Fig. 5(b) and 5(c) are edgewise photographs of the resulting damage. Here the first dramatic rise in AE energy is at a strain level of 0.004 and from the photograph of sample C (Fig. 5(b)) it is seen that the increase in AE energy is again caused by the onset of delamination cracking. Sample D was bent for a longer time than sample C and in Fig. 5(a) at interface strain levels between 0.011 and 0.013 a second large rise in AE energy is seen which indicates the second type of cracking for the TBC in compression. From Fig. 5(c) it can be seen that the sample has suffered delamination cracking which began forming at a strain of 0.004 and a large shear crack which began forming at a strain just above 0.011.

From Fig. 4 and 5 and the above measurements the critical strain ranges for delamination and through-cracking are $\Delta\varepsilon_{d,o} = 0.0085$ and $\Delta\varepsilon_{t,o} = 0.021$ which can be used with the lifetime prediction modelling eqns (2)–(6).

Additional measurement data are available for the development of the above model. Metallographic investigations of isothermal oxidation samples have been able to track the change in porosity as a function of exposure time, where oxidation specimens have been oxidised in laboratory air at 1050 and 1200°C for times up to 1000 h and 100 h respectively. To measure the porosity the as-oxidised samples are first cross-sectioned and then investigated with back scatter scanning electron microscopy. The SEM images are examined in more detail by using image enhancing software where the SEM image is binarised (Fig. 6). The black pixels are the pores, voids and cracks. The porosity is determined by measuring the ratio of white to black pixels.

Figure 7 shows the results of the porosity measurements of the APS TBC top coat for oxidation temperatures of 1050 and 1200°C for the indicated oxidation times. The general time at temperature effect is that the porosity decreases with exposure time and this decrease is accelerated with increasing temperature. Because of the general trends in the porosity data it is believed that the porosity measurement is dominated by the sintering behaviour of the APS zirconia. The measurements are made at room temperature where they can be influenced by the microcracking that occurs in the top coat during cooling. However, the porosity decreases with exposure

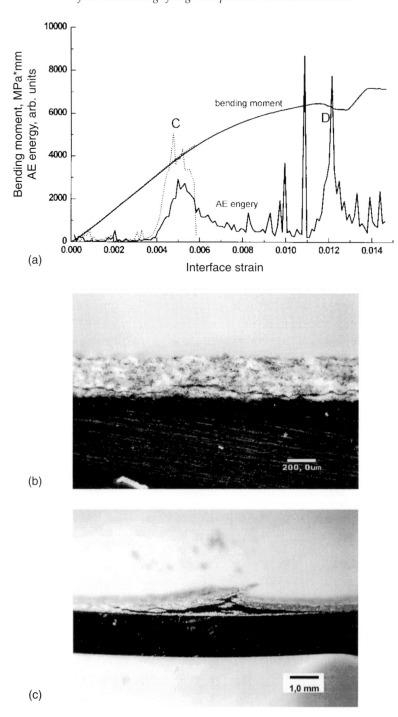

(a)

(b)

(c)

Fig. 5 (a) Plot of the compressive bending moment and acoustic emission energy as a function of the TBC/bond coat interface strain for two samples: (b) and (c); Edgewise photographs of the damage in samples C and D respectively.

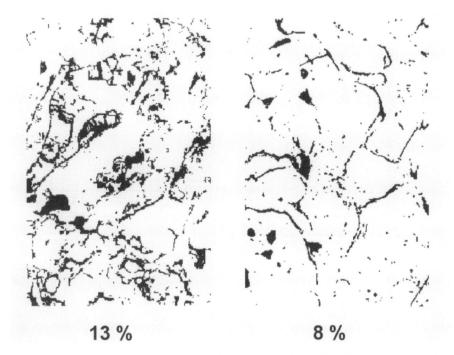

Fig. 6 *Binary images of SEM photo used to determine porosity. Right side image is as-received TBC with ~13% porosity. Left side image is after 100 h at 1200 °C with ~8%.*

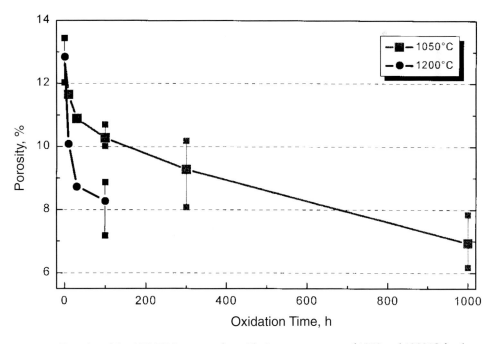

Fig. 7 *Porosity of the APS TBC top coat for oxidation temperatures of 1050 and 1200 °C for the indicated oxidation times.*

time. Consequently, the microcrack growth is assumed to be small for these isothermally oxidised samples and the observed decrease in porosity is caused by a disappearance of the smaller pores. The porosity data for the 1050°C oxidation samples can be described by eqn (10).

$$P_{1050}(t) = 4.74 + 1.95 \cdot \exp\left(-t/11.6\right) + 6.15 \cdot \exp\left(-t/972.4\right) \tag{10}$$

With respect to the above lifetime prediction model the critical strain and porosity data can be linked together by means of eqns (7), (8) and (9) and by using the relation $E = E_o(1 + 2\pi P(t))^{-1}$ and incorporating everything into eqns (5) and (6):

$$N_d = \left[\frac{\Delta\varepsilon_{d,0}}{\Delta\varepsilon_T}\right] \cdot \sqrt{\frac{c_0}{c(t)} \frac{(1+2\pi P(t))}{(1+2\pi P_o)}} \tag{11}$$

$$N_t = \left[\frac{\Delta\varepsilon_{t,0}}{\Delta\varepsilon_T}\right] \cdot \sqrt{\frac{c_0}{c(t)} \frac{(1+2\pi P(t))}{(1+2\pi P_o)}} + \theta_{d \bullet t} \tag{12}$$

where N_d, N_t, and $\Delta\varepsilon_T$ are the same as above. $\Delta\varepsilon_{d,o}$ and $\Delta\varepsilon_{t,o}$ are determined from the four point bend data. $P(t)$ and $c(t)$ are the time-dependent porosity and composite defect length. P_o and c_o are the porosity and composite defect length of the as-sprayed TBC. The $\theta_{d \bullet t}$ is a function that describes the interaction of the delamination with the segmentation or shear macrocracks. Equations (11) and (12) are simply saying that the observed decrease in porosity causes an increases in the elastic modulus of the top coat which in turn decreases the lifetime. The second insight is that as the composite defect length increases lifetime decreases. From the relationship between porosity and composite defect length, the time dependent damage situation may develop into steady state condition, i.e. stable crack growth [6]. The similarities between eqns (11) and (12) stem from the similarities between eqns (7), (8) and (9). When more complicated failures are considered, such as edge delamination and edge buckling, eqns (11) and (12) can look very different from each other.

In Fig. 8 the strain ranges vs N are plotted as calculated from eqns (11) and (12) and using the porosity and critical strain range data from above. However, the determination of the time at temperature dependent composite defect length $c(t)$ is incomplete at this time, so $c(t) = c_o$ is used. The assumption that the composite defect length remains more or less constant is born from the observed decrease in the porosity as a function of time at temperature data. An additional assumption is that there is no macrocrack interaction (i.e. $\theta_{d \bullet t} = 0$). Fig. 8 the time at temperature is assumed to be 1 hour per cycle. For purposes of comparison in Fig. 8 also includes two data points from reported failure from a low cycle TMF test of a 300 µm thick APS-TBC obtained from ref. [4], where the reported failure was mode I and the TMF test temperature was 950°C. An additional comparison is made with a calculation using eqn (3) the data for the sintering term in eqn (3) was taken from ref. [4]. The value of d_c was assumed to be 10 µm, and the oxidation kinetics were found by measuring the TGO thickness of the oxidation samples.

Figure 8 is a strain tolerance plot, showing the expected number of cycles for an applied external strain. The strains for the two literature data points are taken from the reported strain amplitude of the TMF test and the corresponding reported values for N_f. For about the first 20 cycles (eqn (3), eqn (12)), and the literature values produce predicted and measured lifetimes that are equivalent within experimental error. However, near 1000 cycles eqn (3) predicts failure to occur much sooner than eqn (12). This is because eqn (3) is dominated by the $(1-d/d_c)$ term for long times at temperature, whereas the value chosen for d_c is of critical importance. From eqn (11) a prediction for the onset of delamination cracking is obtained and is also plotted in Fig. 8. Consequently, for a strain range of 0.003 delamination starts in 3 cycles and the observed TBC detachment is at 7 cycles. Equations (11) and (12) are based on the damage and porosity measured in the top coat of the isothermally oxidised samples. Consequently they may overestimate the number of cycles to the onset of delamination and failure at high values of N. This depends on the difference in the damage accumulation rates between isothermal oxidation and cyclic oxidation.

4. Conclusions

In order to develop a deeper understanding of TBC failure an approach to lifetime prediction is under consideration, an approach that combines two-step coating failure and the Manson-Coffin fatigue equation. The approach outlined above allows the incorporation of coating defect, microcrack and porosity information into the lifetime

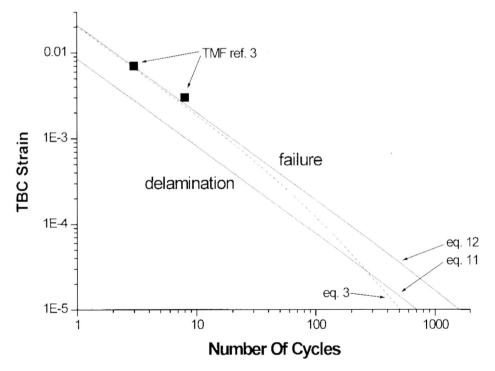

Fig. 8 Strain tolerance plot, showing the expected number of cycles for an applied external strain.

prediction model. Fracture equations, taken from the literature, that relate critical strains to the physical defects that exist in a ceramic coating, have been used to produce an example of a failure mode specific lifetime prediction model. An example calculation has been produced by using measured critical strain and porosity data from a 300 μm thick APS TBC. The model predicted the onset of TBC delamination and final failure. The results are compared to an existing model and TMF measurement data taken from the literature.

Currently there is a battery of tests running that will generate a base of data, which should provide insight into the details of the above model.

5. Acknowledgement

In the frame of the work one of the authors (D.R.) received finanical support by Deutsche Forschungsgemeinschaft under contract no. SCHU 729/11, which is gratefully acknowledged.

References

1. I. Küppenbender and M. Schütze, *Oxid. Met.*, 1994, **42**, 109.
2. G. E. Dieter, *Mechanical Metallurgy* 2 ed. McGraw-Hill, 1976 ISBN 0-07-016891-1.
3. J. T. DeMasi, K. D. Sheffler and M. Ortiz, NASA-Report 182230, 1989.
4. R. Herzog, F. Schubert and L. Singheiser, *Proc. THERMEC 2000*, 4–8. Dec. 2000 Las Vegas, USA.
5. H. E. Evans, G. P. Mitchell, R. C. Lobb and D. R. J. Owen, *Proc. R. Soc. (Lond.), A.* 1993, **440**, 1.
6. M. Schütze, *Protective Oxide Scales and their Breakdown*. John Wiley, 1997, ISBN 0-471-95904 9.
7. S. R. Choi, J. W. Hutchinson and A. G. Evans, *Mech. Mater.* 1999, **31**, 431–447.
8. W. Przybilla and M. Schütze , *In preparation*.
9. C. Bruns and M. Schütze, *Oxid. Met.*, 2001, **55** 1/2, 35–68.

Part 2

Investigation and
Modelling of Specific
Degradation Processes

24

Oxidation Lifetimes: Experimental Results and Modelling

I. G. WRIGHT, B. A. PINT, L. M. HALL and P. F. TORTORELLI

Oak Ridge National Laboratory, Oak Ridge, TN 37831, USA

ABSTRACT

Experimental results for high-temperature alloys with excellent oxidation resistance strongly suggest that their oxidation behaviour involves several distinct stages that are not well addressed in current models. An approach is suggested that relies on breaking the 'steady-state' oxidation stage into multiple stages. In higher temperature experiments, two such stages are clearly observed, whereas at more relevant, lower temperatures, these stages are more difficult to observe within reasonable experimental times. Because this approach attempts to more closely model the observed oxidation behaviour, it requires input on the rate of oxidation and the rate law exponent for each stage, as well as a criterion for changing from one stage to the next. Thus, it is necessary to improve the mechanism-based description of the oxidation behaviour in each stage. Initial results show reasonable predictions using this methodology for one alloy (MA956HT). However, extensive work will be required to determine the real potential for this approach, particularly for lower temperatures. While it is recognised that other complicating phenomena have been observed, such as effects of specimen thickness and cycle frequency on oxidation lifetime, these have not been incorporated into this modelling effort.

1. Introduction

Failure of a component exposed to a high-temperature corrosive environment will occur if the corrosion rate is sufficient to reduce the thickness of the load-bearing section below that needed to sustain the imposed mechanical load. The main aim of any model of oxidation behaviour should be to predict the maximum rate of section thinning as a function of alloy properties, temperature, time, and mode of oxidation. However, the monotonic rate of loss of section thickness from high-temperature oxidation is not usually the life-determining factor in applications where the alloys function as intended. Under such conditions, rapid metal loss and section thinning will occur only when the alloy surface becomes depleted in the protective oxide-forming elements below the minimum requirement by, for instance, consumption by the scale-forming process. This process is known as breakaway oxidation and can lead to rapid component failure. How available understanding of alloy oxidation behaviour is translated into reliable information for practical use is the concern addressed in this paper.

There exists an urgent need for oxidation lifetime modelling to provide guidance for the quantification of the high-temperature oxidation behaviour of alloys in a form that is useful for the practical application of the alloys, and to remedy the poor state of technology transfer in this area among scientists, designers, and engineers.

At the very least, a model for the high-temperature oxidation behaviour of an alloy should be capable of predicting its useful service life as a function of alloy type, component thickness, and temperature. Extension to include the effects of temperature fluctuation and changes in oxidising environment also would be desirable.

Values of oxidation lifetimes serve the purpose of quantifying the overall high-temperature environmental resistance of alloys that form protective oxide scales. Using defined criteria for end of life, oxidation lifetimes can be measured but, for certain alloys and service conditions, these times can be extraordinarily long. Therefore, the ability to model lifetimes accurately and robustly is needed so as to

(1) qualify materials for service at high temperatures based on a limited set of measurements and knowledge of the starting materials (composition, dimensions), and

(2) serve as a means to guide alloy selection and component design for extended high-temperature service.

Systematic modelling of high-temperature oxidation behaviour is obviously needed for lifetime prediction. However, even for simple alloys, this type of modelling is hindered by the need to describe a relatively complex sequence of processes in terms that can be readily quantified, yet, which relate to the observed behaviour under various conditions. Several approaches for modelling oxidation lifetime beyond the use of simple rate laws have been put forward [1–5] and have successfully described performance in certain cases. However, no one approach to lifetime modelling seems *a priori* to be wholly appropriate; a modest-to-moderate amount of experimentation for a particular material system is needed to provide the input parameters for predictions of oxidation lifetime. The goal of a single robust model for predicting oxidation lifetime that incorporates all of the most important factors and is applicable to a wide range of materials and conditions has not yet been achieved. In this paper, some of these factors are reviewed and related to long-term oxidation data for an alumina-forming alloy, Special Metals (Huntington, West VA, USA) Alloy MA956HT (Table 1 for composition), in air at 1000–1300°C, in terms of kinetics and time to breakaway. Following Quadakkers, Bennett *et al.* [3–5] emphasis is on Al reservoir and power law consumption by oxidation. The data sets involve total mass gain (specimen mass gain plus mass of any spalled oxide) instead of specimen mass gain alone, and time to breakaway oxidation vs specimen thickness.

2. Important Factors for an Oxidation Model and Related Experimental Results

2.1. Calculating the Alloy Service Lifetime

Modelling of the oxidation-governed lifetime of a high-temperature alloy requires the definition of two main parameters:

(i) a criterion that signals the end of useful life, and

(ii) a readily measured or calculated parameter that tracks the input data used by the criterion.

2.1.1. End of life criterion

As indicated earlier, the end of life criterion for many components is the point where sufficient cross section has been lost that they can no longer sustain the required mechanical load. This may be particularly true for conditions that give rise to rapid oxidation for components with thin cross sections, or where internal oxidation occurs and reduces the strength of the component. Typically, the alloys used for high-temperature service are designed to form protective Cr_2O_3 or Al_2O_3 scales by the addition of Cr or Al, or combinations of these and other elements. However, when the needs for high-temperature strength and corrosion resistance are combined, the content of protective scale-forming elements is often close to the minimum requirement in order to maximise alloy strength. In such cases, a moderate amount of oxidation can sufficiently deplete the alloy in Cr or Al so that breakaway oxidation can ensue, resulting in rapid metal consumption. In this case, a typical criterion for breakaway oxidation for a simple binary alloy AB designed to form a protective oxide B_2O_{ZB} (where valence of B is Z) is the point when the activity/concentration of B at the alloy–oxide interface (C_{B_i}) falls below the kinetically-determined minimum required to maintain the protective oxide ($C_{B_i^*}$).

2.1.2. Tracking Parameter

For the two alternative end of life criteria described above, the tracking parameters would be very different. For the critical cross section criterion, the specimen thickness would be the tracking parameter; this is the parameter of concern in, for instance, the ASSET model [2]. Specimen thickness change could possibly be inferred from mass change data but, particularly for the case of internal oxidation, microstructural information would be required. For the case of depleting B to a critical level, the key parameter would be the rate of consumption of B. This value could be calculated from mass change data or, if the relevant oxidation behaviour parameters for the alloy were known, could be calculated solely from a knowledge of the temperature history. However, for the most oxidation-resistant alloys of nominal cross section (1–5 mm), measuring the thickness change will yield little information until the alloy goes into breakaway oxidation (which would signal end of service life). For example, Fig. 1 shows a cross-section of the scale formed on MA956HT after 10 000 h at 1100°C in air (see also [6]). No internal oxidation was observed and very little thickness change could be measured. However, the mass change data could be used to calculate the amount of B (Al) consumed after this exposure, since the scale formed was predominantly Al_2O_3 at this stage. Because the current focus is on modelling the performance of the most oxidation-resistant alloys, the following discussion will deal mainly with predicting the consumption rate of B until the breakaway condition (i.e. $C_{B_i} = C_{B_i^*}$) is reached.

2.2. Calculating the Available Reservoir of B

One essential parameter required for this approach is the absolute reservoir of B available. Ideally, the amount of B available to take part in the oxidation process is

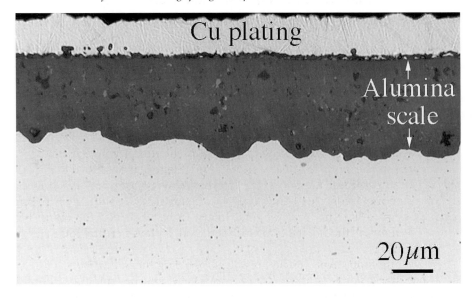

Fig. 1 *Polished cross section of alloy MA956HT after 10 000 h (100 × 100-h cycles) at 1100 °C. Optical microscopy. Only small amounts of spallation were observed and the cross sectional thickness was virtually unchanged within measuring accuracy.*

given by the density and thickness of the component and the difference between the initial and final B contents ($C_{B_0} - C_{B_i^*}$). However, this assumes that the rate of transport of B in the alloy is faster than its rate of consumption at the alloy–oxide interface. This condition may be true for Al in most alumina-forming alloys at high temperatures (e.g. refs [3–5]), especially during steady-state oxidation. However, in many alloys (particularly chromia-formers) at temperatures typically encountered in service, this condition may not be fulfilled. This represents a critical issue for modelling the performance of chromia-forming alloys, and is beyond the current scope of work and of this paper. Instead, the focus here is on the issues involved in describing and modelling the consumption of Al in alumina-forming alloys at high temperatures.

2.3. Calculating the Rate of Consumption of B

The consumption of B (W_B/A, mass per unit area) by an idealised oxidation process can be described by:

$$W_B/A = kt^n \tag{1}$$

where k is the rate constant for the oxidation process, t is the time that the oxidation process is operating, and n is the rate law exponent. However, the process suggested by eqn (1) does not fully account for the oxidation behaviour exhibited by many practical alloys, which may be represented as a series of events:

(1) a transient oxidation stage,

(2) a period of steady state oxidation,

(3) a period of modified steady-state oxidation, often observed during thermal-cycling where a process such as oxide spallation-regrowth occurs with faster oxidation rates than the normal steady-state period, and finally

(4) breakaway oxidation.

Pieraggi has pointed out the need to isolate transient effects from the steady state rate constant [7]. The modified steady state period (Stage 3) is described in the COSP model [1] as a near-linear mass loss, but in other alloy systems may result in mass gains or other, more complex behaviour [8]. Regardless of exactly what happens during Stage 3, this behaviour is most likely to be observed for many high-temperature alloys. Although it is poorly understood, it plays a critical role in lifetime predictions.

There are numerous experimental results supporting the importance of a second steady-state stage. For example, as shown in refs [3] and [4], solving a simplified equation of supply of B vs consumption rate as described solely by eqn (1) leads to the conclusion that the time to breakaway should be proportional to the component thickness to the power $(1/n)$ which, for parabolic behaviour ($n = 0.5$) would be 2. However, as illustrated in Figs 2–4 for several alumina-formers, the $1/n$ values

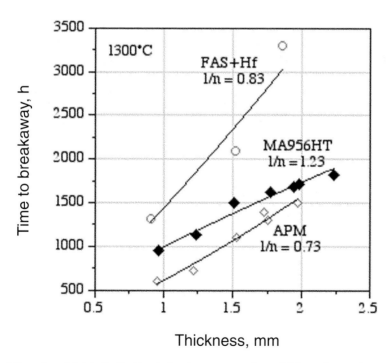

Fig. 2 *Experimental data for the time to breakaway vs specimen thickness for specimens of alloys APM, MA956HT, and FAS+Hf exposed in 100-h cycles in air at 1300°C.*

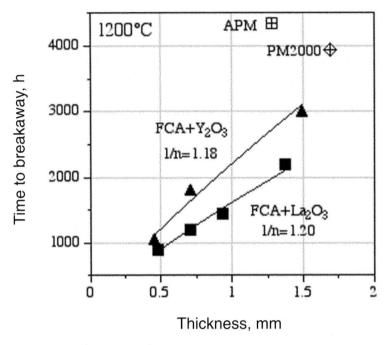

Fig. 3 *Experimental data for the time to breakaway vs specimen thickness for specimens of ORNL-made ODS-FeCrAl specimens exposed in 1-h cycles in air at 1200°C.*

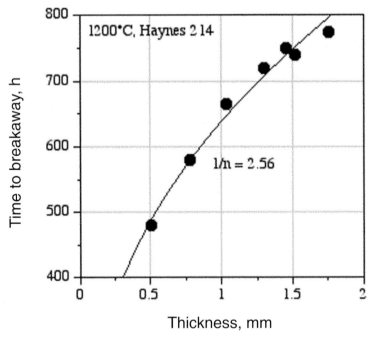

Fig. 4 *Experimental data for the time to breakaway vs specimen thickness for specimens of Haynes 214 exposed in 1-h cycles in air at 1200°C.*

Table 1. Chemical compositions of alloys studied (mass%)

Alloy	Fe	Ni	Cr	Al	Ti	Si	S*	O	Other
MA956HT	Bal.	0.11	21.6	5.9	0.4	0.05	50	0.21	0.38Y as Y_2O_3
Kanthal APM	Bal.	—	20.4	5.5	0.03	0.23	10	0.05	0.10Zr as ZrO_2
PM2000	Bal.	0.1	18.9	5.1	0.5	0.04	21	0.25	0.37Y as Y_2O_3
FCA+Y_2O_3	Bal.	0.1	19.7	4.9	<0.01	0.1	28	0.53	0.17Y as Y_2O_3
FCA+La_2O_3	Bal.	0.1	19.7	4.9	<0.01	0.1	31	0.36	0.06La as La_2O_3
FAS+Hf	Bal.	<0.01	2.1	15.6	<0.01	<0.01	24	0.004	0.38Hf (0.10 at%)
Haynes 214	3.5	Bal.	15.6	4.3	0.01	0.1	3	0.001	0.02Zr, 0.004Y

*ppm mass.

obtained experimentally are far from this value, ranging from 0.73 to 2.56. These lifetime data were obtained under cyclic oxidation conditions: MA956HT, Kanthal APM, and Fe–28Al–2Cr + Hf during 100 h cycles at 1300°C, and for Haynes 214 and ORNL-fabricated FeCrAl + La_2O_3 and FeCrAl + Y_2O_3 during 1 h cycles at 1200°C (see Table 1 for compositions). Any cracking or spallation of scale resulting from the cyclic nature of these tests might be expected to modify the assumptions implicit in parabolic oxidation behaviour. For the alloys studied, since no scale spallation was observed during Stage 2, the simplifying assumption was made that any modification of diffusion transport in the scales resulting from cracking was short-lived, so that any perturbation from parabolic behaviour was negligible. This assumption will have to be revisited for alloys that exhibit more obvious effects from thermal cycling during Stage 2. The experimental data discussed here were obtained using a technique of exposing samples in individual alumina crucibles in a cyclic test which yielded information on the total metal consumption, so that experimental problems with measuring the amount of spallation were avoided [8]. The reason for the use of such high temperatures was to generate failure in a reasonable time; testing at lower temperatures would require at least an order of magnitude increase in the test times.

2.4. Complicating Factors

While in most cases the overall oxidation processes of like alloys are assumed to be similar, variables such as the physical dimensions of the specimen, initial surface finish, microstructure of the alloy/strengthening phases present, and the temperature dependence of the mechanical properties of the alloy substrate can significantly modify the duration of the different stages of oxidation, the morphology of the oxide formed, or the time to onset of oxide spallation. While not intended to be a comprehensive catalogue of such complicating factors, Table 2 lists some of the more

Table 2. Summary of factors that complicate modelling, listed by oxidation stage

Transient oxidation	Steady-state oxidation	Modified steady-state oxidation	Breakaway oxidation
Simplest assumptions			
•shortest duration •sequence determined by *thermodynamic* considerations	•oxide thickens with time at a characteristic parabolic rate •*T*-dependence according to Arrhenius relationship •oxide is uniform in thickness •oxide is adherent	•scale spallation initiates due to growth stresses •scale spalls to substrate–oxide interface •oxide regrowth follows same parabolic rate law •spallation initiates at a given oxide thickness	•initiates when $C_{B_i} < C_{B_i^*}$ •$C_{B_i} = C_{B_i}$
Complicating issues			
•surface finish •substrate stress •spinel or M_2O_3-formation? •$C_{B_i} < C_{B_i^*}$ •desired crystal structure is critical	•constant rate law dependence •edge effects •pO_2 effects within range of stability of 'normal' oxide •water vapour effects	•substrate thickness effects + corners, loss=onset of sharp points •substrate strength •CTE mismatch, substrate vs oxide •substrate phase change •interface roughness •oxide thickness	•rapid weight breakaway oxidation •slope of concentration gradient of B •possibility of 'healing' inner layer of B_2O_{Zb}

common assumptions, and factors that can complicate or confound these assumptions.

3. A Multistage Model

Analysis of a data set of total and specimen mass gain for MA956HT oxidised in 100 h cycles in air at 1300°C clearly indicated that several mass-gain stages occurred during the oxidation lifetime (Fig. 5). As suggested in Fig. 6, if any initial period of transient oxidation is ignored, two major stages can be defined. An initial attempt to describe such multistage oxidation behaviour involves replacing the single *k* and *n* value in eqn (1) by expressions to account for the consumption of B in each stage, such as:

$$W_B / A = k_1 t_1^{n_1} + k_2 t_2^{n_2} + k_3 t_3^{n_3} \qquad (2)$$

where t_1, t_2, and t_3 are the times spent in each of the stages of oxidation: (1) transient, (2) steady state and (3) modified steady-state. Additional stages could be included if

(a)

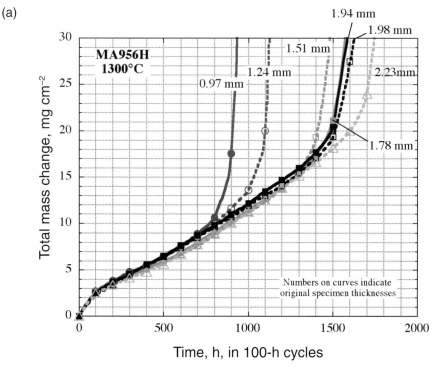

(b)

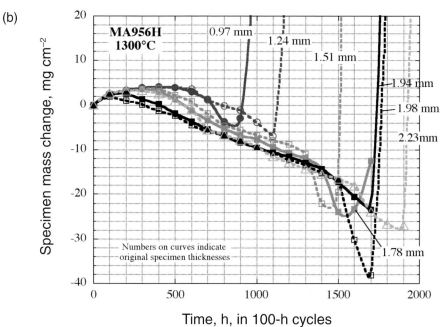

Fig. 5 *(a) Total mass gain data for specimens of alloy MA956HT exposed in 100-h cycles in air at 1300 °C. (b) Specimen mass gain data for specimens of alloy MA956HT exposed in 100-h cycles in air at 1300 °C.*

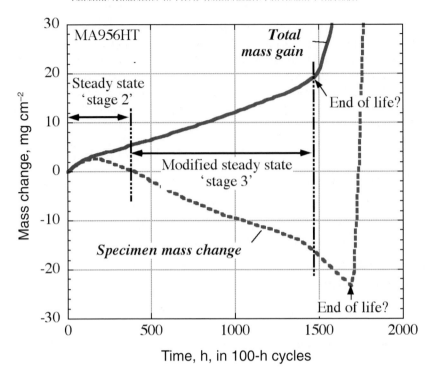

Fig. 6 *Schematic diagram of oxidation kinetics of alloy MA956HT exposed in 100-h cycles in air at 1300 °C.*

necessary (and justified mechanistically), but even this expression rapidly becomes difficult to manipulate. For any particular alloy and oxidation conditions, the time in each stage may be widely different. For the alloys considered here, the transient stage appears short-lived, thus Stages 2 and 3 represent the dominant contributions to service lifetime. Of these parameters, the value of ($k_2 t_2^{n_2}$) is the most easily obtained (for example, from the experimental data shown in Fig. 5), even at lower temperatures.

Depending on the alloy and oxidation conditions, Stage 3 ($k_3 t_3^{n_3}$) may be experimentally observed or it may have to be predicted because it only initiates after extremely long exposures at lower temperatures, so that it is not feasible to conduct experiments of appropriate duration.

3.1. Modified Calculation for Oxidation-Governed Lifetime

The time spent in oxidation according to a simple parabolic rate law (Stage 2) is readily calculated from the parabolic rate constant:

$$t_2 = (S\tau_{2-3}\rho_A / k_2)^{1/n_2} \tag{3}$$

where ρ_A is the density of alumina = 3965 mg cm^{-3} (used in the absence of an actual value for a thermally-grown oxide), τ_{2-3} is the oxide thickness at the transition from

Stage 2 to Stage 3 (µm), as determined from the experimental data (cf Fig. 5) when possible, and S converts mass gain to mass of Al_2O_3 formed = 0.4707.

The time spent in the third stage of oxidation is assumed to be governed by the amount of available Al in the alloy. This requires a knowledge of the amount of Al consumed during the parabolic oxidation stage, which is obtained by equating the duration of the parabolic stage to the change in total Al content in the alloy, after Quadakkers *et al.* [3,4]:

$$t_2 = (10^{-4} S \tau_{2-3} \rho_A / k_2)^{1/n_2} = \left[\left(C_{B_o} - C_{B_2} \right) d\rho_M / 200 M k_2 \right]^{1/n_2} \quad (4)$$

where C_{B_o} = the initial Al content of the alloy = 5.9 mass%, C_{B_2} = the Al content of the alloy at the end of the parabolic oxidation stage (mass%), C_{B_b} = the Al level at which a protective Al_2O_3 can no longer form (and, for the class of alloys studied here, = $C_{B_{i'}}$) (mass%), d = the thickness of the section being oxidised (cm), ρ_M = the density of the alloy = 7200 mg cm^{-3}, and M = converts mass gain to Al consumed = 1.1246.

Solving for C_{B_2} yields

$$C_{B_2} = C_{B_o} - 0.02 M S \tau_{2-3} \rho_A / d\rho_M \quad (5)$$

Hence, the oxidation lifetime (time to breakaway, t_b), assuming a section thickness sufficient to allow oxidation to proceed through Stage 3, is:

$$t_b = t_2 + t_3 = \left(10^{-4} S \tau_{2-3} \rho_A / k_2 \right)^{1/n_2} + \left[\left(C_{B_2} - C_{B_b} \right) d\rho_M / 200 M k_2 \right]^{1/n_3} \quad (6)$$

$$= \left(10^{-4} S \tau_{2-3} \rho_A / k_2 \right)^{1/n_2} + \left\{ \left[\left(C_{B_o} - C_{B_b} \right) d\rho_M - 0.02 M S \tau_{2-3} \rho_A \right] / (200 M k_3) \right\}^{1/n_3} \quad (7)$$

Additional stages could be added in a similar manner.

4. Applying a Multi-Stage Model

As a first step in testing the proposed approach and in examining the underlying assumptions to determine what modifications are needed to broaden its applicability, the multistage consumption model was applied to the MA956HT data set for cyclic oxidation at 1300°C, shown in Figs 2 and 5, from which actual values of oxidation-limited lifetime as a function of specimen thickness were obtained (Table 3). Any transient oxidation stage was ignored, and the consumption was based on the two stages suggested in Fig. 6. Values of n_2 and n_3 were obtained by log-log slope analysis of the data for each of the seven specimens. The average values for the data set of n_2 and n_3 were 0.52 and 1.03, respectively (see Table 4), suggesting essentially parabolic and linear oxidation behavior. Further, Stage 2 terminated at an average mass gain equivalent to an oxide thickness (τ_{2-3}, assumed all Al_2O_3) of 29 µm. Values of k_2 and k_3 were obtained from parabolic and linear plots of the respective segments of each

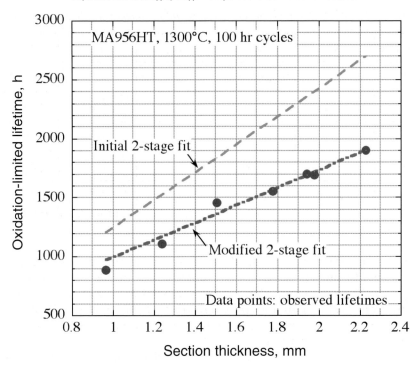

Fig. 7 *Comparison of observed and predicted oxidation lifetimes for alloy MA956HT exposed in 100-h cycles in air at 1300 °C. (Initial 2-stage fit used: $n_2 = 0.5$; $n_3 = 1.0$ ($f_{n_3} = 0$); k_2, $k_3 =$ average of data set; $\tau_{2-3} = 29$ μm; $C_{B_b} = 1.2\%$. Modified 2-stage fits used same basis, with indicated changes in f_{n_3}, or C_{B_b}).*

Table 3. *Experimentally-observed lifetimes for MA956HT*

Time to breakaway, h	Specimen thickness, mm							
	0.97	1.04	1.24	1.51	1.78	1.94	1.98	2.23
1300°C	880	—	1100	1450	1550	1700	1700	1900
1200°C	—	4700	—	—	—	—	—	—

data set, and the average values determined (Table 4). These rate constants were input to eqn (7), along with $n_2 = 0.5$, and $n_3 = 1.0$, and used to calculate the oxidation-limited lifetime as a function of specimen thickness. The result is shown in Fig. 7, and was found to over-predict significantly the specimen lifetimes.

Since a parabolic Stage 2 appeared entirely consistent with the experimental kinetic and morphological data, it was considered that the more likely source of the underestimation of the rate of Al consumption was the assumption of a strictly linear

Table 4. *Summary of experimental data over the temperature range 1000–1300 °C*

Temperature	n_2	k_2	τ_{2-3}	n_3	k_3
(°C)		$(\text{mgcm}^{-2}\text{h}^{-0.5})$	(μm)		$(\text{mgcm}^{-2}\text{h}^{-1})$
1300	0.52	0.236	29	1.03	12.49×10^{-3}
1200	0.40	0.113	28	1.07	2.60×10^{-3}
1100	0.52	0.066	25	0.89	0.70×10^{-3}
1000	0.37	0.015	23*	—	$0.13 \times 10^{-3*}$

*Extrapolated.

third stage. Examination was made of the sensitivity of eqn (7) to changes in the value of n_3 (and the linked values of k_3), using values of k_3 calculated from plots of mass gain vs $t^{1/n3}$. As shown in Table 5, for all combinations of n_3 and k_3 for values of n_3 ranging from 1.00 to 1.10, eqn (6) overestimated life by approximately 41%, indicating that the use a single process to describe Stage 3 resulted in a poor match to the experimental observations.

The sensitivity of the two-stage fit to the main variables was then re-examined by using the base values obtained from the experimental data (see Table 4), and allowing one of the variables n_3, k_3, τ_{2-3}, and C_{B_b} to vary, with the results shown in Fig. 8. This indicated that an excellent match to the experimentally observed lifetimes could be obtained if n_3 was increased to 1.06, or if k_3 was increased by a factor of 1.5, or if the value of C_{B_b} was increased to 2.5% (see Table 5). Note that the use of the experimentally determined average value of k_3 (1.03) led to an over-prediction of lifetimes. Of these changes, the increases required in k_3 and C_{B_b} to obtain a fit to the experimental data were considered to be unrealistic, based on the range of consistency of the experimentally observed values of k_3 among the seven specimens studied and the general acceptance that C_{B_b} for alloys of this type is less than 2.0% [3–5]. Since it is possible that Al consumption in Stage 3 is driven by more than one process occurring simultaneously (or sequentially), such as scale spallation at specimen ends and scale cracking and rehealing on the parallel sides, its description by a single, linear process probably is overly simplistic. A process involving localised loss of scale followed by re-oxidation can lead to an essentially linear overall oxidation rate for the area of surface affected but, as discussed in the COSP model [1] and in models for erosion-augmented oxidation [9], knowledge is needed of the area fraction over which this process occurs, the thickness of oxide lost in each event, and the frequency of the spallation events. At present, there is insufficient experimental evidence to provide such information. Hence, it was considered that augmentation of the Stage 3 rate exponent (to suggest a more complex process than scale spallation alone) was a reasonable modification of the model until an acceptable mechanism-based description of Stage 3 is available. The rate exponent in Stage 3 then becomes (n_3 +

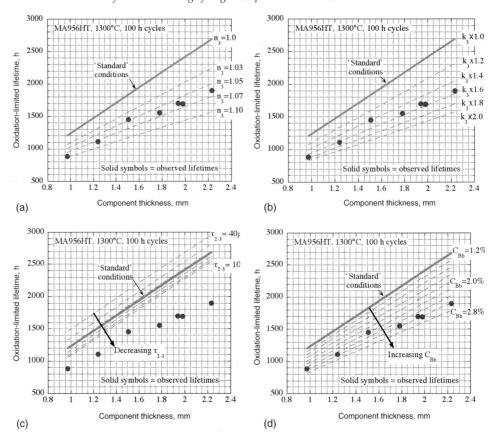

Fig. 8 *Sensitivity of two-stage model to changes in the values of the single variables (a) n_3 (b) k_3 (c) τ_{2-3} (d) C_{B_b} with all other parameters at the 'standard' values: $n_2 = 0.5$; $n_3 = 1.0$; k_2, $k_3 =$ average for standard data sets for standard n values; $\tau_{2-3} = 29 \ \mu m$; $C_{B_b} = 1.2\%$.*

f_{n_3}); in the case of MA956HT at 1300°C, $n_3 = 1.00$ and $f_{n_3} = 0.06$. The lifetimes calculated at 1300°C as a function of specimen thickness using the 2-stage model modified by an augmentation factor are shown in Table 6 and Fig. 7.

The next step was to examine the ability of the modified 2-stage model to predict the performance of the same alloy at lower temperatures, using the experimental results shown in Fig. 9. At 1200°C, the experimental total mass change data appeared to fit similar trends as at 1300°C, although the form of the specimen mass change curve was different from that at 1300°C; since only one data set was available where a specimen had accumulated sufficient exposure time at 1200°C, this trend will be further examined as more specimens are exposed. Nevertheless, there was a clear transition between Stages 2 and 3 (Fig. 9), so that values could be calculated for n_2, k_2, n_3, and k_3 (Table 4). The experimentally-determined value of n_2 for this specimen was lower (0.40) than observed at 1300°C, while the observed transition point (τ_{2-3}) was 28 μm, essentially the same as at 1300°C. In fact, the lifetimes calculated by the modified two-stage model using $n_2 = 0.4$ and the corresponding value of k_2, and by

Table 5. *Sensitivity of calculated lifetime at 1300°C to major variables*

n_3	k_3, mgcm^{-2}h$^{-n_3}$	τ_{2-3}, μm	C_{B_b}, %	Δt_b, %*
1.00	0.012490	29	1.2	+39
1.03	0.009912	29	1.2	+40
1.05	0.008484	29	1.2	+41
1.07	0.007273	29	1.2	+42
1.10	0.005776	29	1.2	+42
Changes in individual variables to give best fit:				
1.00	$k_3 \times 1.5$	29	1.2	+3
$n_3 + 0.06$	0.012490	29	1.2	+2
1.00	0.012490	29	2.5	+2
1.00	0.012490	0–40	1.2	No closure

* t_b calculated for the specimen thicknesses shown in Table 2, and compared to the observed lifetimes.

Table 6. *Oxidation lifetimes calculated using a multistage approach*

Temp., °C	Model inputs used						2–Stage model predictions, h		
	n_2	k_2*	τ_{2-3}	n_3	f_{n_3}	k_3	1.0 mm	1.5 mm	2.0 mm
		mgcm^{-2}h$^{-0.5}$	μm			mgcm^{-2}h^{-1}			
1300	0.50	0.236	29	1.0	0	12.49×10^{-3}	1258	1860	2462
1300	0.50	0.236	29	1.0	0.06	12.49×10^{-3}	1005	1388	1761
1200	0.40	0.241	28	1.0	0.06	2.60×10^{-3}	4286	5966	7605
1200	0.50	0.113	28	1.0	0.06	2.60×10^{-3}	4312	5992	7631
1100	0.50	0.066	25	1.0	0.06	0.70×10^{-3}	13 265	19 046	24 689
1000	0.37	0.046	23	1.0	0.06	0.13×10^{-3}	218 635	245 988	272 703
1000	0.50	0.015	23	1.0	0.06	0.13×10^{-3}	118 768	146 121	172 836

*Corresponding to indicated n_2.

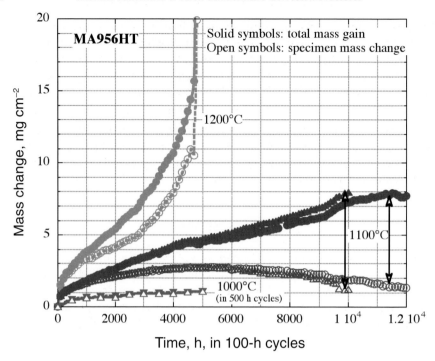

Fig. 9 *Experimental data for MA956HT oxidised at 1000 °C in 500-h cycles, and at 1100 and 1200 °C in 100-h cycles. (Basis: $<\tau_{2-3}$, oxidation is parabolic, $n_2 = 0.5$; $> \tau_{2-3}$, oxidation 'linear,' $n_3 = 1.0$; k_2, k_3 from averages of data set for $n_2 = 0.5$, $n_3 = 1.0$; $f_{n_3} = 0.06$; $C_{B_b} = 1.2$).*

using $n_2 = 0.5$ and the corresponding k_2, were very similar (Table 6). Further, the lifetime of 4455 h for the 1.04 mm-thick specimen predicted by the modified two-stage approach was close to the observed lifetime, 4700 h, which was encouraging.

Data were available for two specimens exposed to 100 h cycles at 1100°C for 10 000 h or longer, as shown in Fig. 9, allowing the calculation of values for n_2, k_2, n_3, k_3, and τ_{2-3} (Table 4). However, in this case $n_3 = 0.89$, suggesting that there may not yet be sufficient experimental data to obtain an overall value for k_3. If the model is correct, these specimens may not be fully into Stage 3; spallation of oxide was observed from only the specimen edges after 10 000 h.

For a specimen exposed for 5 000 h (in 500 h cycles) at 1000°C, a single, near-parabolic oxidation rate was followed throughout the exposure (Fig. 9), with no scale spallation observed. As a result, only values of n_2 and k_2 could be obtained experimentally. As shown in Table 4, the experimental value of n_2 was 0.37. A value of k_3 was obtained from an Arrhenius extrapolation using the higher-temperature k_3 values. Also, since there was an apparent trend for the values of τ_{2-3} to decrease with temperature (Table 4), the higher-temperature values were used to extrapolate a value of τ_{2-3} of 23 μm for 1000°C. At this point there is no indication that τ_{2-3} should be independent of temperature, since factors contributing to loss of scale protectiveness probably change with temperature. For instance, relaxation of stresses developed in the scale may become easier with increasing temperature, so that an argument could be made that τ_{2-3} would increase with temperature. The use of these data in the modified two-stage calculation with the experimental value of $n_2 = 0.37$ and the corresponding value of k_2 resulted in lifetimes approximately double those obtained when a value of $n_2 = 0.5$ and the corresponding k_2 was used (Table 6). Again, there was obviously no way to verify these values from the experimental data available.

Figure 10 summarises the predictions of lifetime as a function of temperature using eqn (7) (with $(n_3 + f_{n_3}) = 1.06$). The current status of long exposure specimens at 1100°C is indicated. It is clear that it is vital to obtain more experimental data at 1100–1200°C for MA956HT to verify the range of applicability of the modified multistage consumption methodology to this one alloy. Significantly more work is needed to examine its extension to other alloys.

5. Discussion

The initial choice in modelling is whether to use an existing model or to develop a new approach. Considering the models currently available, none appears to be inherently superior. The COSP model [1] is a two-stage model based on specimen mass change; the alloys for which it was developed exhibited an initial period of mass gain, followed by a period of uniform mass loss. The ASSET model [2] does not

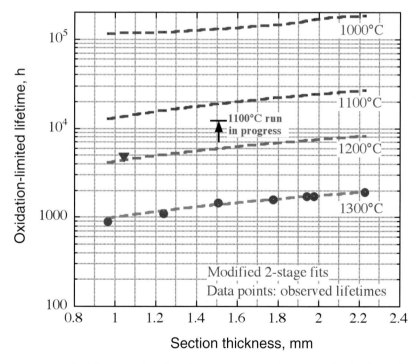

Fig. 10 *Predicted lifetimes as a function of temperature. (Basis: $<\tau_{2-3}$, oxidation is parabolic, $n_2 = 0.5$; $> \tau_{2-3}$, oxidation 'linear,' $n_3 = 1.0$; k_2, k_3 from averages of data set for $n_2 = 0.5$, $n_3 = 1.0$; $f_{n_3} = 0.06$; $C_{B_b} = 1.2$).*

attempt to reproduce the mode of oxidation or the sequence of events in the oxidation process. Instead, it uses thermodynamic calculations to determine the overall mode of corrosion (oxidation, sulfidation, etc.), and then calculates the rate of metal loss from a database of experimental metal loss data, interpolating to the temperature, oxygen or sulfur partial pressure of interest through the use of appropriate algorithms. This approach provides data that are of immediate use to engineers, and appears to be appropriate for lower temperatures and alloys that exhibit internal oxidation and/or uniform metal loss. The Quadakkers/Bennett approach [3–5] was closest to the issues considered in the present study, and served as a basis for the application of the present multistage consumption model.

As mentioned above, the main reason for incorporating a multi-stage Al consumption model into life prediction was to reflect mass change in a manner that actually reflected what is measured (cf. Fig. 5). However, one of the underlying problems with this strategy is attaching physical significance to the third stage. This modified steady-state behaviour is poorly understood and infrequently studied, judging by the sparcity of data in the oxidation literature. In order to develop confidence in the quantification of this stage, it would be useful to have a better understanding of the transition of the metal-scale system from Stage 2 to Stage 3, and more information on subsequent scale spallation and regrowth. For spallation, information is needed on the proportion of the alloy surface affected, the relative influences of specimen corners and edges, the frequency of individual events, the type and amount of damage or defects necessary to cause spallation, and whether the damaged oxide remains attached to the specimen (i.e. non-protective scale [10]) or spalls. For the regrown oxide, little information is available on its microstructure. It has been suggested that such oxide is more porous and less protective than virgin grown scale [5]. The formation of a more defective scale could result in an inherently faster or temporarily increased rate of growth, and lend physical significance to a Stage 2–Stage 3 transition scale thickness on the order of that observed here (20–30 μm). From a modelling standpoint, it also must be recognised that these types of effects likely will vary considerably among alloy systems.

In addition to the new inputs for a multi-stage model, the standard inputs for a single-stage model present some problems. In particular, values for $C_{B_{i*}}$ (C_{B_b}) are difficult to obtain and can vary considerably depending on the type of alloy and oxidation temperature. For example, for very thin specimens, values for reactive element-alloyed FeCrAl appear to approach $C_{B_{i*}} = 0$, with chromia layers observed forming below the alumina layer [11] Recent work has even suggested that $C_{B_{i*}}$ may be a function of specimen thickness for certain alloys [12].

The experimental data for breakaway times is another area of concern. For the data in Figs. 2 and 3 there is surprising agreement for the ODS FeCrAl alloys with overall n values (lifetime vs specimen thickness) of 0.63 to 0.85. However, recent work suggests that stronger ODS alloys may have lower n values than weaker cast FeCrAl alloys [12]. For another class of alloy, the Ni-based Haynes 214, the overall n value was significantly different, 0.39, which may relate to the two-phase structure of the alloy. During oxidation involving 1 h cycles at 1200°C, the alloy may have become depleted in the Al-rich phase, which could have resulted in rapid oxidation of the depleted zone, independent of specimen thickness. In that case, the lifetime would be relatively insensitive to specimen thickness, particularly for thicker specimens.

From the above discussion and other numerous examples, it is clear that the oxidation behaviour of high-temperature alloys often does not conform to the simplifying assumptions normally used. While it is thought that in most cases the overall oxidation processes are similar, variables such as the physical dimensions of the specimen and the temperature dependence of the mechanical properties of the alloy substrate can significantly modify the duration of the different stages of oxidation, the morphology of the oxide formed, or the time to onset of oxide spallation. Any realistic model for high-temperature oxidation must be able to accommodate many of these features, so that it becomes important to develop mechanistic descriptions that are amenable to some form of quantification.

6. Summary

Based on experimental results for high-temperature FeCrAl alloys with excellent oxidation resistance, currently available models are not sufficient to describe the observed behaviour. A method that attempts to link the mathematical description of the consumption rate of the scale-forming element with observed phenomena is needed. An approach is suggested that relies on breaking the steady-state consumption rate into multiple stages. In higher-temperature experiments these stages are clearly observed, whereas at more relevant, lower temperatures these stages are more difficult to observe within reasonable experimental times. Thus, it is necessary to develop criteria for the transition between the various stages of oxidation. Initial results showed reasonable predictions using this methodology for one alloy (MA956HT) at a very high temperature. However, in order to match exactly the observed lifetimes it was necessary to postulate a higher rate exponent or a higher oxidation rate than measured experimentally in the modified steady-state regime, or to postulate a significantly higher value for the minimum Al level for maintenance of the protective scale. These observations suggest that the oxidation mechanism in the modified steady state regime is more complicated than normally assumed. Clearly, extensive work will be required to determine the real potential for this approach, particularly for lower temperatures.

7. Acknowledgements

This research was sponsored by the Fossil Energy Advanced Research and Technology Development (ARM) Materials Program, U.S. Department of Energy, under contract DEA0596OR22725 with UT-Battelle, LLC. Thanks are due to Dr G. Smith of INCO, Dr D. Spörer of Metallwerk Plansee, and Mr Q. J. Mabbutt of British Gas plc. for supplying samples of ODS-FeCrAl alloys. We would like to also acknowledge the contributions of colleagues at the Oak Ridge National Laboratory: L. D. Chitwood, G. W. Garner, and M. Howell for the oxidation testing, and M. P. Brady and D. F. Wilson for critically reviewing the manuscript.

References

1. C. E. Lowell, C. A. Barrett, R. W. Palmer, J. V. Auping and H. B. Probst, *Oxid. Met.*, 1991, **36**, 81–112.

2. R. C. John, W. T. Thompson, and I. Karakaya, "Alloy corrosion data bases combined with thermochemical analysis," *CORROSION '88*, Paper 136, NACE, Houston, TX, 1988.

3. W. J. Quadakkers and M. J. Bennett, *Mat. Sci. Technol.*, 1994, **10**, 126–131.

4. W. J. Quadakkers and K. Bongartz, *Werkst. Korros.*, 1994, 45, 232–41.

5. M. J. Bennett, H. Romary and J. B. Price, "The oxidation behavior of alumina forming oxide dispersion strengthened ferritic alloys at 1200–1400°C," in *Heat Resistant Materials* (K. Natesan and D. J. Tillack, eds). ASM, Materials Park, OH, 1991, pp.95–103.

6. B. A. Pint, P. F. Tortorelli and I. G. Wright, ORNL Laboratory Report (2001), in press.

7. B. Pieraggi, *Oxid. Met.*, 1987, **27**, 177–85.

8. B. A. Pint, P. F. Tortorelli and I. G. Wright, Effect of cycle frequency on high-temperature oxidation behavior of alumina- and chromia-forming alloys (M. Schütze and W. J. Quadakkers eds). *Cyclic Oxidation of High-Temperature Materials*, Publication No. 27 in the EFC series, published by The Institute of Materials, London, 1999, pp.111–132.

9. See, for instance, I. G. Wright, V. K. Sethi and A. J. Markworth, *Wear*, 1995, **186–187**, 230–237.

10. F. H. Stott, F. A. Golightly and G. C. Wood, *Corros. Sci.*, 1979, **19**, 889–906.

11. N. Hiramatsu and F. H. Stott, *Oxid. Met.*, 1999, **51**, 479–94.

12. J. P. Wilber, M. J. Bennett and J. R. Nicholls, *Mater. High Temp.*, 2000, **17**, 125–32.

25

COSIM — A Finite-Difference Computer Model to Predict Ternary Concentration Profiles Associated With Oxidation and Interdiffusion of Overlay-Coated Substrates

J. A. NESBITT

NASA Glenn Research Center, Cleveland, OH 44135, USA

ABSTRACT

A finite-difference computer program (COSIM) has been written which models the one-dimensional, diffusional transport associated with high-temperature oxidation and interdiffusion of overlay-coated substrates. The program predicts concentration profiles for up to three elements in the coating and substrate after various oxidation exposures. Surface recession due to solute loss is also predicted. Ternary cross terms and concentration-dependent diffusion coefficients are taken into account. The program also incorporates a previously developed oxide growth and spalling model to simulate either isothermal or cyclic oxidation exposures. In addition to predicting concentration profiles after various oxidation exposures, the program can also be used to predict coating life based on a concentration dependent failure criterion (e.g. surface solute content drops to 2%). The computer code is written in FORTRAN and employs numerous subroutines to make the program flexible and easily modifiable to other coating oxidation problems.

1. Introduction

Many blades and vanes in gas turbine engines are coated with an aluminide or overlay coating to impart additional oxidation and corrosion protection to the component. These coatings provide protection by the selective oxidation of Al to form a compact and adherent Al_2O_3 scale. Under isothermal conditions, the Al_2O_3 scale thickens at a parabolic rate (scale thickness proportional to the square root of time). Diffusional transport within the coating supplies Al to the growing oxide scale at a rate consistent with the parabolic growth of the scale. Although Al is continually consumed from the coating, the rate is acceptably low such that the coating typically contains sufficient Al to easily provide protection during isothermal oxidation conditions. Although land-based turbines often operate in a nearly isothermal mode for long periods, aero gas turbines undergo thermal cycling with each flight. This thermal cycling, primarily to ambient temperatures when the engine is shut down, can cause cracking and partial spalling of the protective Al_2O_3 scale. This spallation generally occurs randomly across flat component surfaces but is higher at edges and corners. The spalling may occur to the metal surface but more commonly occurs only in the outer layers of the oxide scale. However, loss of the oxide is not catastrophic since selective oxidation of Al continues when the component again reaches high temperatures such that the damaged scale will 'heal' and continue to grow. Whenever scale spallation occurs due to thermal cycling during oxidation (i.e. cyclic oxidation), the oxide scale on the surface will, on average, be thinner than a scale grown isothermally for equivalent hot exposure times. Since the rate of scale growth is inversely proportional to the scale thickness, a consequence of this thinner oxide scale is that the average rate of Al consumption is greater during cyclic oxidation accompanied by scale spallation than during isothermal oxidation. Hence, Al is depleted from the coating at a higher rate during cyclic oxidation, the exact rate depending on the growth rate of the alumina scale and the amount of oxide spalling. This paper will focus specifically on degradation of MCrAlY overlay coatings, where M stands for either Ni, Co or Fe. These coatings are often used on components in marine environments where hot corrosion protection is required, and in military applications where higher temperatures are often encountered.

In addition to oxidation, overlay coatings are further degraded by interdiffusion with the substrate. Since the coating is, by nature, higher in Al content than the substrate, Al diffuses from the coating into the substrate and generally becomes unavailable to support the growth of the protective alumina scale. Likewise, the Cr concentration in the coating is also typically greater than that in the substrate resulting in the diffusion of Cr from the coating into the substrate. In contrast, Ni and other elements in the substrate diffuse into the coating. Although the Y in the alloy strongly affects adherence of the Al_2O_3 scale during cyclic oxidation, it is at such a low concentration in the coating (e.g. <0.2 at.%) that dilution by interdiffusion with the substrate is generally not considered. Schematic concentration profiles in the coating and substrate after a short exposure in an oxidising environment are shown in Fig. 1. When a coating is substantially depleted of Al during cyclic oxidation, it can no longer supply sufficient Al to re-heal the protective scale. Less protective oxides, such as NiO, FeO or CoO, can form signalling the end of the protective life of the coating. A critical Al content in the coating can be defined to indicate the useful life of the coating. This critical Al content could indicate the time at which less protective oxides form on the surface, or an earlier time at which the coating could be stripped from the component and a new coating applied. Modelling the Al transport in the coating and substrate during cyclic oxidation allows the coating life to be predicted. The purpose of the present work was to develop a one-dimensional ternary diffusion model to predict the concentration profiles associated with the oxidation and interdiffusion of coated superalloys undergoing cyclic oxidation. This model, given the name COSIM for coating oxidation and substrate interdiffusion model, employs finite-difference techniques embodied in a FORTRAN computer program to provide numerical solutions to the appropriate diffusion equations. The computer code employs numerous subroutines to make the program flexible and easily modifiable to other high temperature coating oxidation problems. Although the computer program and discussion below are primarily in terms of a NiCrAl coating on a Ni-based substrate, other elements could be substituted for the Al and Cr.

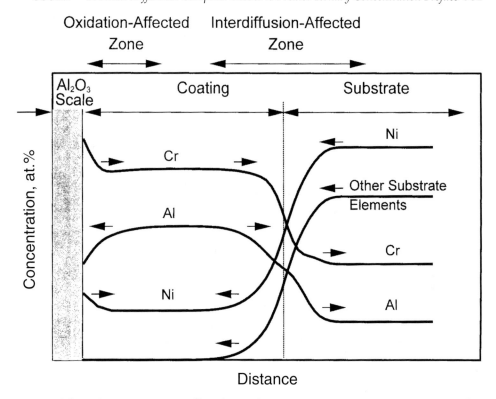

Fig. 1 *Schematic concentration profiles after oxidation exposure. Arrows indicate direction of atomic transport.*

2. Ternary Diffusion Equations

Because of the low Y content, MCrAlY overlay coatings can be approximated as ternary alloys with Al and Cr in a matrix of either Ni, Co or Fe. Ternary diffusion equations can be employed to simulate the Al and Cr transport associated with coating oxidation and coating/substrate interdiffusion. The choice of simulating Cr rather than Ni diffusion is inconsequential. In the ternary system, the concentration of the third component, being a dependent variable, is always determined by difference (i.e. $C_{Ni} = 100 - C_{Al} - C_{Cr}$). Since most superalloys are complex multicomponent, multiphase alloys, fully accounting for diffusional interactions of the various superalloy components, as well as that of Y in the coating, on the Al and Cr transport is beyond current capabilities. Hence, these potential interactions, other than those encountered in the ternary system, are ignored. In addition, NiCrAlY and CoCrAlY overlay coatings generally consist of two phases, a high-Al NiAl or CoAl β phase embedded in a Ni or Co solid solution γ phase. This β phase is depleted as Al is consumed as oxide and as Al diffuses into the substrate. It has previously been shown that this β phase is often completely dissolved well before the useful life of the coating has been reached [1]. It was also shown in this same reference that good agreement was achieved between measured and predicted concentration profiles in the coating at extended times after the β phase had been dissolved by assuming a single-phase coating for the entire oxidation exposure. Hence, in the current model development, the two-phase coating will be represented as a single phase.

Fick's laws describing diffusion in a ternary alloy are:

$$J_j = -D_{j,j}\frac{\partial C_j}{\partial X} - D_{j,k}\frac{\partial C_k}{\partial X} \quad j,k = Al,Cr \qquad \text{1st Law (1)}$$

and

$$\frac{\partial C_j}{\partial t} = \frac{\partial\left(D_{j,j}\left(\frac{\partial C_j}{\partial X}\right)\right)}{\partial X} + \frac{\partial\left(D_{j,k}\left(\frac{\partial C_k}{\partial X}\right)\right)}{\partial X} \quad j,k = Al,Cr \qquad \text{2nd Law (2)}$$

where J_j is the flux of component j, C is the concentration of either Al or Cr, D is one of the four ternary interdiffusion coefficients, or diffusivities, and X and t refer to distance and time, respectively. The first term on the right hand side (RHS) of eqn (2) can be rewritten as:

$$\frac{\partial\left(D_{j,j}\left(\frac{\partial C_j}{\partial X}\right)\right)}{\partial X} = D_{j,j}\frac{\partial^2 C_j}{\partial X^2} + \frac{\partial D_{j,j}}{\partial X}\frac{\partial C_j}{\partial X} \quad j,k = Al,Cr \qquad (3a)$$

Equation (3a) can be further expanded as:

$$\frac{\partial\left(D_{j,j}\left(\frac{\partial C_j}{\partial X}\right)\right)}{\partial X} = D_{j,j}\frac{\partial^2 C_j}{\partial X^2} + \left(\frac{\partial D_{j,j}}{\partial C_j}\frac{\partial C_j}{\partial X} + \frac{\partial D_{j,j}}{\partial C_k}\frac{\partial C_k}{\partial X}\right)\frac{\partial C_j}{\partial X} \quad j,k = Al,Cr \qquad (3b)$$

A similar equation exists for the second term on the RHS of eqn (2), namely:

$$\frac{\partial\left(D_{j,k}\left(\frac{\partial C_k}{\partial X}\right)\right)}{\partial X} = D_{j,k}\frac{\partial^2 C_k}{\partial X^2} + \left(\frac{\partial D_{j,k}}{\partial C_j}\frac{\partial C_j}{\partial X} + \frac{\partial D_{j,k}}{\partial C_k}\frac{\partial C_k}{\partial X}\right)\frac{\partial C_k}{\partial X} \quad j,k = Al,Cr \qquad (3c)$$

Substituting eqns (3b) and (3c) into the two terms on the right hand side of eqn (2) yields:

$$\frac{\partial C_j}{\partial t} = D_{j,j}\frac{\partial^2 C_j}{\partial X^2} + \left(\frac{\partial D_{j,j}}{\partial C_j}\frac{\partial C_j}{\partial X} + \frac{\partial D_{j,j}}{\partial C_k}\frac{\partial C_k}{\partial X}\right)\frac{\partial C_j}{\partial X} +$$
$$D_{j,k}\frac{\partial^2 C_k}{\partial X^2} + \left(\frac{\partial D_{j,k}}{\partial C_j}\frac{\partial C_j}{\partial X} + \frac{\partial D_{j,k}}{\partial C_k}\frac{\partial C_k}{\partial X}\right)\frac{\partial C_k}{\partial X} \quad j,k = Al,Cr \qquad (4)$$

Initially, transport associated with oxidation and the interdiffusion of the coating and substrate may be viewed as independent problems. Hence, at short times, coating/substrate interdiffusion can be treated as interdiffusion between two semi-infinite materials at a location centred on the coating/substrate interface. Similarly, oxidation of the coating can be treated as oxidation of a semi-infinite material affecting only a thin region below the surface of the coating. These two regions, the 'inner' diffusion zone resulting from coating/substrate interdiffusion and the 'outer' diffusion zone resulting from transport associated with oxidation, are shown schematically in Fig 2. At longer times, the two diffusion zones will overlap and diffusion associated with the two regions must be considered together. A solution to Fick's second law (eqn (2)), whether analytical or numerical, requires initial and boundary conditions. The boundary condition at the oxide/coating interface is discussed in the following paragraphs whereas the initial conditions and other boundary conditions used by the COSIM model are discussed in a later section.

For isothermal oxidation, the boundary condition at the oxide/coating interface, given as the rate of Al consumption, is well defined and decreases uniformly with time (the time dependence of the rate is inversely proportional to the square root of time). However, for cyclic oxidation, the rate of Al consumption can vary in a non-uniform manner as the thickness of the oxide scale increases during the high temperature exposure but decreases on cooling as spallation occurs [2]. In the present diffusion model, a separate model simulating oxide growth and spallation during

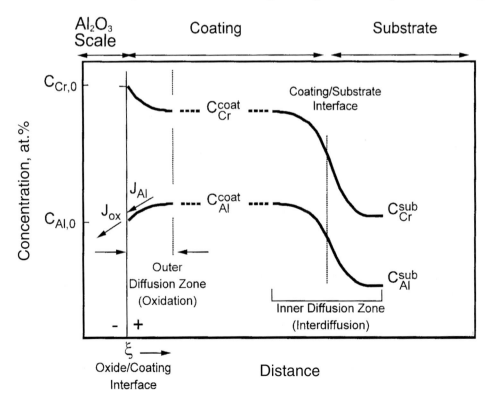

Fig. 2 *Schematic concentration profiles after a short oxidation exposure.*

cyclic oxidation has been adopted to provide the oxide/coating boundary condition. This oxide growth and spalling model [3], designated COSP by the authors, predicts the rate of Al consumption during each cycle by continuously tracking the thickness of the oxide scale, accounting for growth during high-temperature exposures, and partial oxide loss on cooling. This rate of Al consumption (J_{ox}) predicted by COSP is used as the boundary condition for the diffusion model. Hence, the supply of Al within the coating to the oxide/coating interface must equal J_{ox}. However, since Al is consumed from the coating, the coating surface recedes due to the loss of matter. This recession (ξ) is given as [4]:

$$\xi = -V_{Al} J_{ox} t \tag{5}$$

where V_{Al} is the partial molar volume of Al in the coating and J_{ox} is the rate of Al consumption discussed above (which is also the flux of Al entering the oxide). J_{ox} can also be considered as the Al flux to the left of the moving oxide/metal interface in Fig. 2, or the Al flux away from the interface. Similarly, J_{Al} is the flux in the metallic coating towards the interface. Because of the interface motion, J_{ox} is greater than J_{Al} according to the relationship:

$$J_{ox}\Big|_{X = \xi^-} = \alpha J_{Al}\Big|_{X = \xi^+} \tag{6}$$

where ξ^- and ξ^+ refer to the left hand side and right hand side of the oxide/coating interface, respectively, as shown in Fig 2. The parameter α in eqn (6) is given as:

$$\alpha = \frac{1}{(1 - V_{Al} * C_{Al,0})} \tag{7}$$

where $C_{AL,O}$ is the Al concentration in the coating at the oxide/coating interface (Fig. 2). Obviously, for the hypothetical case for $V_{Al} = 0$ (the Al atoms in the coating have no volume), $\alpha = 1$, the Al flux to the interface, J_{Al}, becomes equal to the Al flux away from the interface, J_{ox}, and the interface is stationary. The partial molar volume for Al was assumed independent of concentration due to a lack of available data for most alloy systems of interest (e.g. the γ, Ni solid solution phase in NiCrAl alloys).

Hence, the boundary condition for Al at the oxide/coating interface is given by eqn (6) whereby the rate of Al consumption due to oxide formation, predicted by the COSP oxide model, J_{ox}, is equated to the supply, or flux, of Al to the interface within the coating, J_{Al}, while taking into account the motion of the interface through the parameter α. The rate of Al transport within the coating to the oxide/coating interface (i.e. the Al flux, J_{Al}) is given by eqn (1), stated as:

$$J_{Al}\Big|_{X = \xi^+} = -D_{Al,Al} \frac{\partial C_{Al}}{\partial X} - D_{Al,Cr} \frac{\partial C_{Cr}}{\partial X} \tag{8}$$

Although no Cr is assumed to enter the Al_2O_3 scale, the surface recession requires the diffusion of Cr and Ni away from the oxide/coating interface into the coating. The boundary condition for Cr is:

$$J_{Cr}\bigg|_{X = \xi^+} = -D_{Al,Cr} \frac{\partial C_{Al}}{\partial X} - D_{Cr,Cr} \frac{\partial C_{Cr}}{\partial X} = C_{Cr,0} \frac{d\xi}{dt} \tag{9}$$

where J_{Cr} is the flux of Cr in the coating away from the interface and $C_{Cr,0}$ is the Cr concentration in the coating at this interface (Fig. 2). In the computer program, the calculations for COSP have been contained in a subroutine to facilitate the incorporation of other oxide growth and spalling models (e.g. Ref. [5]).

3. Finite-Difference (F-D) Method

The initial step in the *F-D* technique is to establish a grid of equispaced nodes across the region of the material over which diffusion will occur (i.e. the diffusion zone). Each node has a specific concentration associated with it. Fick's laws (eqns (1) and (4)) are replaced with *F-D* equivalents based on small differences in concentration, ΔC, distance, ΔX, and time, Δt. A solution yielding the concentration profile at some time t is derived by solving the appropriate *F-D* equivalents for small time increments (Δt) in an iterative manner. These iterations, or time steps, are continued until the Δt increments sum to the desired time t. A portion of a diffusion zone at time t and the new concentrations at time $t + \Delta t$ is shown schematically in Fig. 3 [6]. Because of the typically large number of repetitive and tedious calculations made each iteration,

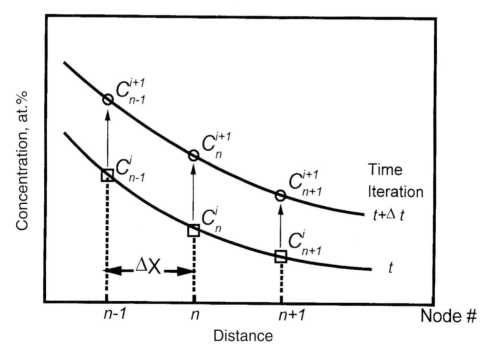

Fig. 3 Schematic diffusion zone for time t (squares) and time t +Δt (circles). The concentration subscript refers to the node number and the superscript refers to the time iteration.

F-D solutions are ideally handled by a computer.

The F-D equivalents to Fick's laws are based on Taylor series expansions [7,8]. The F-D equivalent of Fick's 2nd law (eqn (4)) can be given by either an explicit or implicit representation. The explicit form was used throughout this work. F-D expressions for both first and second order partial derivatives for concentration with respect to distance, given on the right hand side of eqn (4), were given by first central difference equations, stated as:

$$\frac{\partial C_j}{\partial X} = \frac{C_{j,n+1} - C_{j,n-1}}{2\Delta X} \tag{10}$$

$$\frac{\partial^2 C_j}{\partial X^2} = \frac{C_{j,n+1}^i - 2C_{j,n}^i + C_{j,n-1}^i}{(\Delta X)^2} \tag{11}$$

where j refers to either Al or Cr, n refers to the node number and the superscript i refers to the current iteration at time t. These equations apply to all nodes n where nodes $n-1$ and $n+1$ exist. A first forward difference expression was used for $\partial C / \partial t$ in the left hand side of eqn (4), given as:

$$\frac{\partial C_j}{\partial t} = \frac{C_{j,n}^{i+1} - C_{j,n}^i}{\Delta t} \quad j = Al, Cr \tag{12}$$

where the superscript $i+1$ refers to the next iteration at time $t+\Delta t$. The time increment Δt for the explicit F-D method is limited by a stability criterion typically given as:

$$D_{max} \frac{\Delta t}{(\Delta X)^2} \leq 0.25 \tag{13}$$

where D_{max} is the appropriate diffusion coefficient which is discussed in a later section. Although Δt is initially very small, the grid expansion scheme described below allows the time increment each iteration to increase, generally allowing long term simulation of oxidation with a reasonable number of program iterations.

The last terms in eqn (4) to be discussed are the eight derivatives of the diffusion coefficients $(\partial D / \partial C)$. Obviously, the concentration dependence of the four ternary diffusivities is required to evaluate this expression. The concentration dependence of the diffusivities is input to the computer program in polynomial form. The computer program determines an abbreviated polynomial expression for each derivative using the polynomial coefficients. If some, or all, of the ternary diffusivities are concentration independent, or if the concentration dependence is not known and cannot be input to the program, the value of the appropriate derivatives is zero. Hence, each of the terms on the right hand side of eqn (4) have known values at time t and can be evaluated. Since Δt is determined from the stability criterion in eqn (13) and values for $C_{j,n}^i$ are known, values for $C_{j,n}^{i+1}$ from the left hand side of eqn (4) for the new time $(t + \Delta t)$ can be calculated.

For the boundary condition at the oxide/coating interface (eqns (8) and (9)), no node exists in the oxide (i.e. no $n-1$ node exists, where n is located at the interface) so that the central difference formulae in eqns (10) and (11) cannot be used. Consequently, concentration gradients $(\partial C_j / \partial X)$ at the interface were determined using second order forward difference equations. Hence, the F-D equivalent for the Al concentration gradient in both eqns (8) and (9) at the interface was given as:

$$\frac{\partial C_{Al}}{\partial X}\bigg|_{X = \xi^+} = \frac{-C^i_{Al,2} + 4C^i_{Al,1} - 3C^i_{Al,0}}{2\Delta X} \tag{14}$$

where the second subscript number refers to the node which is numbered sequentially from zero at the interface and the superscript refers to the current iteration at time t (i.e. $C^i_{Al,0}$ is the Al concentration in the coating at the oxide/coating interface, $X = \xi^+$ shown in Fig. 2). A similar expression was used for the Cr gradient. The diffusivities for eqns (8) and (9) were evaluated for the Al and Cr concentrations in the coating at the oxide/coating interface (i.e. $C_{Al,0}$, $C_{Cr,0}$). Substituting eqn (14) into eqns (8) and (9) together with eqns (5–7) is sufficient to yield the interface concentrations $C^i_{Al,0}$, $C^i_{Cr,0}$.

4. COSIM Computer Program

The COSIM program initially establishes separate diffusion zones for interdiffusion and oxidation. Typically, these diffusion zones are set to be very narrow (i.e. 0.1 µm) but are allowed to expand with increasing interdiffusion. When the diffusion zones eventually overlap within the coating, the two separate zones are combined and the simulation continues with a single zone across the coating and into the substrate. This approach allows the use of a reasonable number of nodes in a zone yet with a fine node spacing during the early times when the concentration gradients are steep. For greater accuracy, all variables used in the iterative calculations were defined as double precision.

The starting width of the outer and inner zones, DXCOAT and DXSUB, and the number of nodes in each zone, NCOAT and NSUB, respectively, are parameters input to the program. The spacing between nodes, DELX1 for the outer zone and DELX2 for the inner zone (i.e. ΔX in the F-D equations), is equal to the width of the zone divided by the number of nodes minus one (i.e. NCOAT-1 and NSUB-1, the minus one since both zones are bounded by nodes). The concentrations associated with each node for both diffusion zones are stored in arrays labelled 'Al' and 'Cr'. In the outer diffusion zone, the nodes are numbered sequentially starting with zero at the oxide/coating interface through NCOATH (equal to NCOAT-1). The NSUB nodes in the inner diffusion zone are numbered from NSUBL (low) to NSUBH (high) with NSUBL starting at a value of NCOAT+2. Central difference equations, such as eqns (10) and (11), cannot operate on the endpoints of a zone since concentrations at nodes $n+1$ and $n-1$ are required. Although forward and backward difference expressions can be used at these locations, a common technique is to add ancillary nodes to the zone to allow the continued use of central difference equations. Hence, an additional ancillary node was added to the inner end of the outer diffusion zone and an ancillary node was added to each end of the inner diffusion zone (NSUBL–1 and NSUBH+1). The ancillary nodes maintain assigned constant concentration values. The node numbering scheme is schematically shown in Fig. 4 for NCOAT = 7 (e.g. 7 nodes numbered 0 to 6) and NSUB = 11 (e.g. 11 nodes numbered 9 to 19). The ancillary node for the outer diffusion zone is shown as node 7 and the ancillary nodes for the inner diffusion zone are shown as nodes 8 and 20. As shown, different values for DXCOAT or DXSUB, or for NCOAT or NSUB, may result in different values for DELX1 or DELX2.

4.1. Initial Conditions

Initially, each of the nodes in the outer zone is assigned the concentration of the coating. The ancillary node (node 7 in Fig. 4) maintains this coating concentration until the inner and outer diffusion zones overlap. These initial concentrations for the outer zone are shown in Fig. 4. An error function solution was used to assign the coating and substrate compositions to the nodes in order to provide a smooth transition between the coating and substrate compositions. The NSUBL to NSUBH nodes in the inner zone are assigned concentrations as:

$$C_j = C_j^{coat} + (C_j^{sub} - C_j^{coat})\tfrac{1}{2}(1 + erf(X_{mod})) \quad j = Al, Cr \tag{15}$$

where C_j^{coat}, C_j^{sub} refer to the initial concentration of the coating and substrate, respectively and X_{mod} is a modified distance parameter ranging from –2 to +2 across the diffusion zone of width DXCOAT. This range in X_{mod} produces a smooth concentration gradient with end compositions within 0.5% of C_j^{coat}, C_j^{sub}. The coating composition was assigned to the ancillary node at NSUBL–1 (node 8 in Fig. 4) and the substrate composition was assigned to the ancillary node at NSUBH+1 (node 20 in Fig. 4). Again, these ancillary nodes maintain these assigned concentrations until the diffusion zones overlap (i.e. the ancillary nodes are not operated upon by Fick's 2nd law, as discussed below). The initial concentrations for the inner diffusion zone are also shown in Fig. 4.

Use of the error function solution to assign initial concentration values does not have a significant effect on later concentration values because of the small initial diffusion zone width and the small initial time increments. A value on the order of 0.1 μm was typically used for the initial diffusion zone width, DXSUB. Typically, 30 to 40 nodes are assigned to the zone so that the node spacing (ΔX) is 0.0025 to 0.003 μm. Substituting this latter value into eqn (13) and using a typical value of 10^{-10} cm^2 s^{-1} for D_{max} yields a time increment Δt of only 0.000225 s. Hence, hundreds to thousands of iterations will typically be performed before the first minute of simulated exposure time allowing ample iterations for the starting concentrations to adjust to satisfy Fick's 2nd law (eqns (2)–(4)). Fortunately, when diffusion begins to affect the concentration at the ends of the zones, either at node NCOATH, NSUBL or NSUBH, the node spacings are allowed to expand (as shown in the following section)

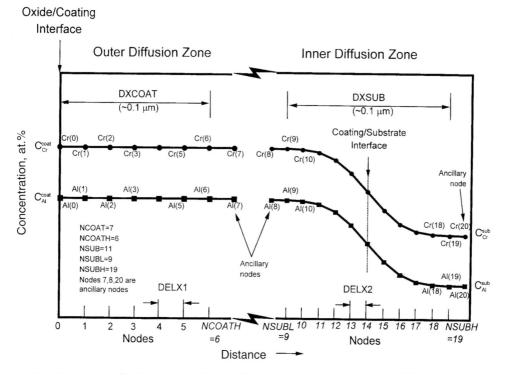

Fig. 4 *Schematic of initial concentration profiles before oxidation exposure for* NCOAT = 7 *and* NSUB = 11.

so that larger time increments can be used each iteration. Since the outer and inner zones may have different node spacings, the program determines the maximum diffusion coefficient (D_{max} in eqn (13)) for either the coating or substrate composition and uses the smaller of the two node spacings to determine the smallest time increment, DELT (i.e. Δt in the *F-D* equations), according to eqn (13).

4.2. Surface Recession, Flexible Zones and Semi-Infinite Boundary Conditions

The COSIM model utilises a flexible grid technique to account for surface recession and to simulate the semi-infinite boundary conditions. As the outer surface recedes due to Al loss, the entire outer zone is shifted and the concentrations at each node are adjusted. The technique used to accomplish this shift and adjustment is referred to as a 'Murray-Landis' (M-L) transformation [9]. This transformation shifts the nodes an amount proportional to their position from the moving boundary ($X = \xi$) to maintain a uniform node spacing. The semi-infinite boundary condition for the outer diffusion zone can be stated as:

$$ C_i \bigg|_{x = \infty} = C_i^{coat} \quad i = Al, Cr $$

This boundary condition is approximated by increasing the zone width whenever diffusion 'significantly changes' the concentration at the node NCOATH. Each iteration, the concentration at this node is changed slightly in accordance with Fick's 2nd law (eqn (4)). A 'significant change' used in the COSIM program was taken as 0.005 at.%. Varying this value from 0.002 to 0.008 at.% changed the number of iterations required to reach a fixed time but had no significant effect on the predicted concentration profiles at short or long times (i.e. 1 and 1000 h using the sample input data given in Appendix A). Hence, whenever the concentration of node NCOATH (either Al or Cr) varied from the coating composition by 0.005 at.%, the zone width was expanded by DELX1 and all node positions and concentrations were adjusted according to the M-L transformation. To illustrate further this operation, Al loss at the surface (node 0) and operation of Fick's 2nd law (eqn (4)) will eventually cause the initial Al profile in the outer diffusion zone (Fig. 4) to appear as shown schematically by the solid squares in Fig. 5. Eventually, the Al concentration at node 6 will decrease to a value 0.005 at.% below the concentration of the coating, C_{Al}^{coat}. During this iteration, the total zone width, DXCOAT, will be increased by an amount equal to the node spacing such that:

$$DXCOAT' = DXCOAT + DELX1 \qquad (16)$$

The number of nodes remains constant but the new node spacing is given as:

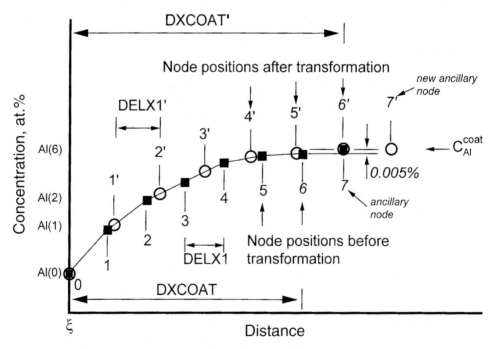

Fig. 5 *Schematic Al concentration profile in the outer diffusion zone before and after the M-L transformation.*

$$DELX1' = DXCOAT'/(NCOAT-1) \tag{17}$$

The node positions are shifted proportionately such that node 6 is shifted inward by DELX1 while node 0 at the surface ($X = \xi$) undergoes no shift (Fig. 5). The concentrations at each node are shifted in a similar manner with node 6' taking the old position of the ancillary node 7 and being reassigned the concentration of the coating, C_i^{coat} (the concentration previously held by the ancillary node 7 at the same position). The ancillary node maintains the coating composition at the new position.

The semi-infinite boundary conditions at either end of the inner diffusion zone are simulated in a like fashion. Hence, as diffusion changes the concentrations at nodes NSUBL or NSUBH in the inner diffusion zone by 0.005 at.%, the zone width is expanded by the node spacing, DELX2 and the node positions and concentrations are adjusted according to the M-L transformation. The expansion occurs at the end of the zone where the concentration change occurred such that the inner diffusion zone expands either into the coating or into the substrate for changes at node NSUBL or NSUBH, respectively. The ancillary nodes (NSUBL–1 and NSUBH+1) maintain the coating and substrate composition respectively, and are repositioned at the ends of the zones with the new node spacing. Whenever the zones are expanded or contracted, a new time increment DELT, is calculated according to eqn (13). As discussed above, the smaller of the time increments calculated for the outer and inner diffusion zones is always used. Because of the concentration dependence of the diffusivities, one end of the inner diffusion zone may expand more than the other over the course of several thousand iterations. Although the inner and outer zones may appear very different with different widths, different numbers of nodes and different node spacings, both zones always operate with the same time increment such that the total exposure time for both the inner and outer diffusion zones is always identical.

The expansion of the inner and outer zones will eventually result in their impingement, or overlap, within the coating. From this time onward, the two diffusion problems, oxidation and coating/substrate interdiffusion, become coupled and can no longer be operated independently. At the time that the diffusion zones overlap, the COSIM model sums the current width of the inner and outer zones and redefines a single zone of equal width. Equidistant node spacings are also calculated using the combined number of nodes from the two zones (i.e. NCOAT+NSUB). The COSIM model then fits a natural cubic spline curve [10,11] through the concentration profiles from both zones and generates a single profile through the coating and into the substrate for both Al and Cr at each of the new node positions. Figure 6a shows predicted Al and Cr concentration profiles in the coating and substrate at the time (time = 1.34 h) when the two diffusion zones overlap. The width of the inner diffusion zone is significantly larger with a larger number of nodes and a larger node spacing than that for the outer diffusion zone. Figure 6(b) shows the same data after the spline interpolation. Note that the nodes are equally spaced across the entire diffusion zone. The spline subroutines are based on equations and code given in [11]. Following the combination of the two diffusion zones, the COSIM model continues to simulate Al transport to the surface and interdiffusion of the coating and substrate. Both boundary conditions, at the oxide/coating interface and in the substrate remain as

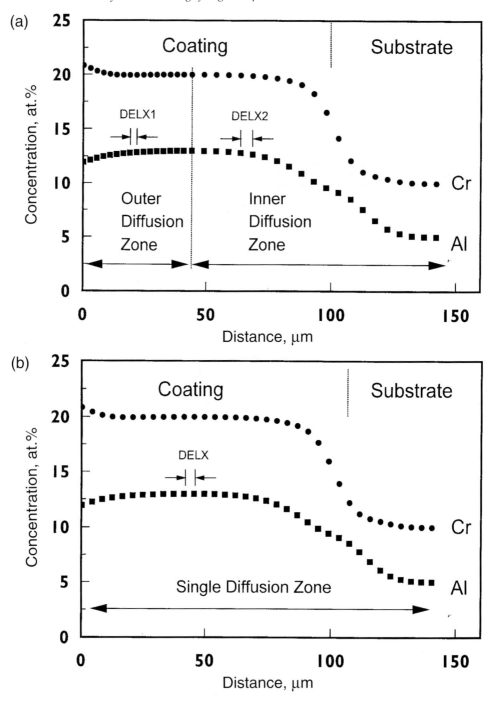

Fig. 6 (a) *Al and Cr concentration profiles at the time (time = 1.34 h) when the inner and outer diffusion zones overlap. Input data for run given in Appendix A. (Alternate nodes hidden for clarity.)* **(b)** *Al and Cr concentration profiles at time = 1.34 h following the spline operation. Input data for run given in Appendix A. (Alternate nodes hidden for clarity.)*

before the combination. Loss of Al continues to cause surface recession and a shrinking of the zone while diffusion in the substrate continues to result in zone expansion. Within the program, the transition from two zones to one is reflected in the value of the parameter ZONE. The value of ZONE changes from two to one after the diffusion zones are combined and the program operates on the single diffusion zone thereafter. Parameters are redefined so that the new, single diffusion zone utilizes parameters associated with the outer diffusion zone (e.g. DELX1, DELT1, etc.) The model continues to simulate increasing oxidation exposure with each iteration of time. The program can print out concentration data at intermediate times (e.g. 50, 100, 500, 1000, 5000 h) or at some predetermined failure condition (e.g. $C_{AL,0} = 0$ or 5 at.%). Output is also written to files to ease plotting concentration/distance profiles. Figure 7 shows predicted concentration profiles after 1, 100 and 1000 h using the data given in Appendix A. Figure 8 shows the time dependence of the surface concentrations AL(0), CR(0), rate of Al consumption and total weight of Al consumed which are also written to output files. Diffusivities for these examples were taken from [12]. The value for the partial molar volume of Al was taken from [13].

Occasionally, oscillations in the values of the Al concentration at the oxide/coating interface, AL(0), will occur in the first few cycles. These oscillations can cause the computed value for the AL(0) to become negative or to exceed 100 at.%. It has also been observed that a very high initial flux from the COSP oxide growth and spalling model can also cause the computed value for AL(0) to be negative for hundreds of cycles. To protect against this physical impossibility, the program limits the values of AL(0) to be between 0 and 100 at.% (i.e. negative values are set equal to zero and

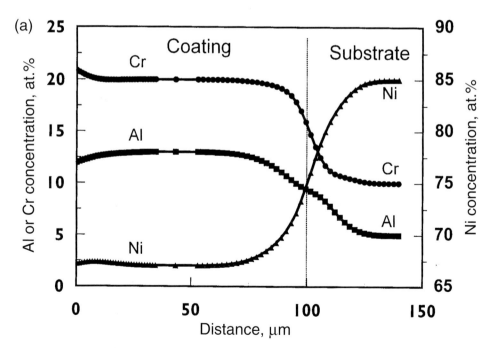

Fig. 7(a) Al, Cr and Ni concentration profiles at time = 1.0 h using the input data given in Appendix A.

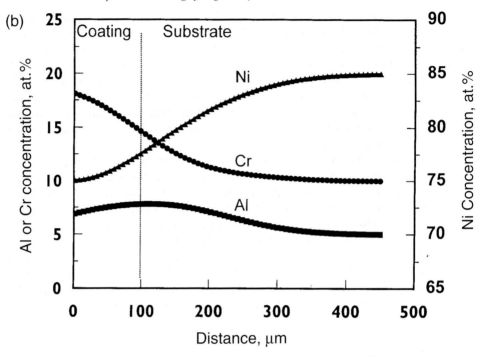

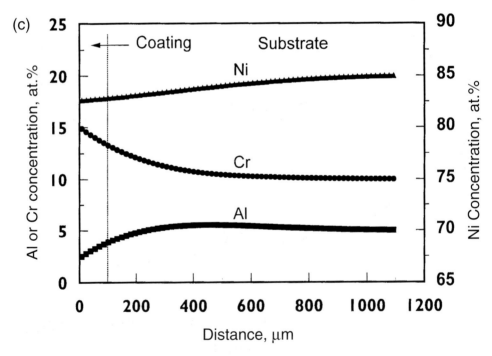

Fig. 7 (b) *Al, Cr and Ni concentration profiles at time = 100 h using the input data given in Appendix A. (c) Al, Cr and Ni concentration profiles at time =1000 h using the input data given in Appendix A.*

values greater than 100 are set equal to 100). The consequence of these limits is that eqn (6) is not satisfied during these iterations. However, monitoring the computed values for AL(0) typically show convergence to acceptable values at relatively short times (i.e. less than one hour when unreasonably high values for K_p were used with the example input data given in Appendix A). Since these early oscillations could falsely trigger the test for AL(0) less than a critical Al concentration (i.e. AL(0)≤CRITAL) this test is not initiated in the program during the first two hours of oxidation exposure. No problems have been encountered using test cases using various values for CRITAL, including zero.

Certain constraints on the time increment used for each iteration (DELT) are desirable when simulating certain oxidation conditions. For instance, a value for DELT much less than one hour might be desirable during cyclic oxidation with one hour cycles. In this case, a value of DELT of one or five minutes might be desired so that several iterations during oxide growth can be performed between periods of oxide spallation each cycle. In contrast, for isothermal oxidation for long periods, it might be preferable to have no constraint on the values for DELT in order to minimise the number of iterations to reach a solution. To provide this flexibility, a maximum value for DELT (MAXDT) may be input to the program. The default condition within the program is that no constraint be made on DELT beyond that given in eqn (13).

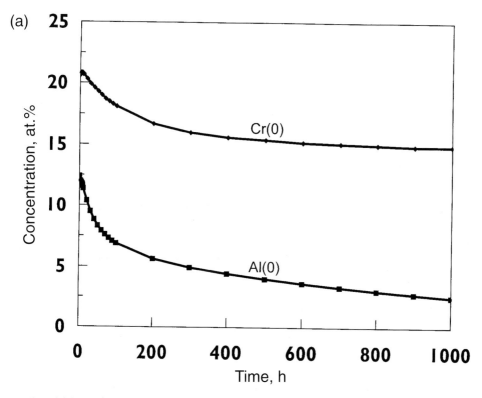

Fig. 8(a) Time dependence of Al and Cr concentrations at the surface (Al(0) and Cr(0)) from data written to unit 15 (File OUT15) for the input data given in Appendix A.

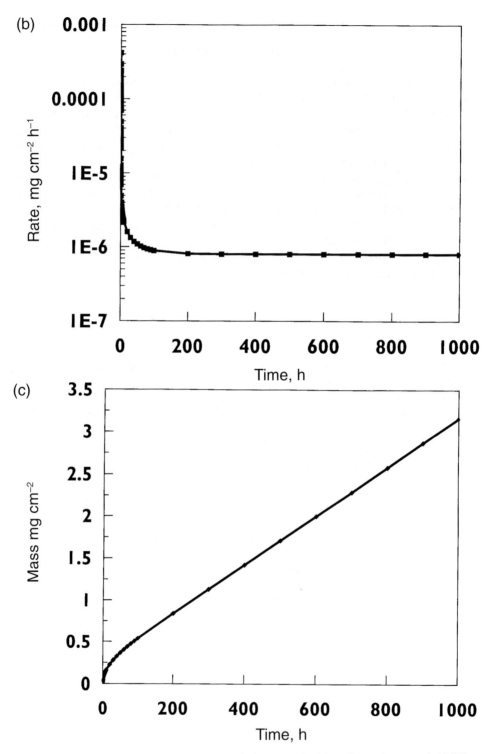

Fig. 8 (b) *Time dependence of WMDOT (rate of Al consumption) from data written to unit 15 (File OUT15) for the input data given in Appendix A.* **(c)** *Time dependence of WMINT (mass of Al consumed) from data written to unit 15 (File OUT15) for the input data given in Appendix A.*

5. Program Application

The program has been used to predict both concentration profiles and coating life after cyclic oxidation of overlay coated substrates [1,14]. The COSIM program has also been used to perform parametric studies to examine the effect of coating thickness and coating and substrate composition on coating life [1]. Recently, the program was modified to examine the coating life extension due to the presence of a perfect diffusion barrier between the coating and substrate [14].

The COSIM program was initially compiled and executed on mainframe computers and later revised for execution on a desktop PC. The current version of the program has been compiled with ANSI FORTRAN95.

A description of the main program and each of the subroutines, as well as flowcharts, is given elsewhere [15] in which a description of each of the input parameters, an alphabetical list of all program variables, a sample input file and the corresponding output files are also given.

References

1. J. A. Nesbitt and R.W. Heckel, *Thin Solid Films*, 1984, **119**, 281.
2. J. A. Nesbitt, 'Diffusional aspects of the high-temperature oxidation of protective coatings', in *Diffusion Analysis & Applications* (A. D. Romig, Jr. and M. A. Dayananda, eds). TMS, Warrendale, 1989, p. 307–324.
3. C. E. Lowell, C. A. Barrett, R. W. Palmer, J. V. Auping and H. B. Probst, *Oxid. Met.*, 1991, **36**, 81.
4. J. A. Nesbitt, *J. Electrochem. Soc.*, 1989, **136**, 1518.
5. K. W. Chan, *Met. and Mat. Trans.*, 1997, **28A**, 411.
6. J. A. Nesbitt, *Oxid. Met.*, 1995, **44**, 309.
7. R.W. Hornbeck, *Numerical Methods*. Quantum Publishers, New York, 1975.
8. M. L. James, G. M. Smith and J. C. Wolford, *Analog and Digital Computer Methods*. International Textbook, Scranton, 1964.
9. D. Murray and F. Landis, *J. Heat Transfer*, **81**, 1959, 106.
10. L.W. Johnson and R. D. Riess, *Numerical Analysis*. Addison-Wesley, Reading, MA, 1977.
11. J. H. Mathews, *Numerical Methods for Mathematics, Science and Engineering*, 2nd Edn. Prentice Hall, Englewood Cliffs, NJ, USA, 1992.
12. J. A. Nesbitt and R. W. Heckel, *Met. Trans.*, 1987, **18A**, 2075.
13. J. A. Nesbitt, NASA TM 83738, Cleveland, OH, USA, 1984.
14. J. A. Nesbitt and Jih-Fen Lei, 'Diffusion barriers to increase the oxidative life of overlay coatings', in *Elevated Temperature Coatings: Science and Technology III* (J. M. Hampikian and N. B. Dahotre, eds). TMS, Warrendale, USA, 1999, pp.131–142.
15. J. A. Nesbitt, NASA TM-2000-209271, Cleveland, OH, USA, 2000.

Appendix A

Al concentration in the coating (at.%)	13.0
Cr concentration in the coating (at.%)	20.0
Al concentration in the substrate (at.%)	5.0
Cr concentration in the substrate (at.%)	10.0
Density of the coating (g cm^{-3})	7.754
Partial molar volume of Al in the coating (cm^{3}mole^{-1})	7.1
Coating thickness (µm)	100.0
Number of nodes in the 'outer' coating diffusion zone	30
Number of nodes in the 'inner' coating/substrate diffusion zone	40
Initial width of the outer diffusion zone (µm)	0.1
Initial width of the inner diffusion zone (µm)	0.1
Parabolic oxide growth parameter due to weight of oxygen in the oxide (mg^2 cm^{-4}h^{-1})	0.002
The hot cycle duration for each thermal cycle (h)	1.0
The spall parameter for the COSP spalling model (Ref. [3])	0.008

The four ternary diffusion coefficients (cm^{2}s^{-1}):

$$D_{AlAl}(C_{Al},C_{Cr}) = [1.229 + (0.0731\ C_{Al}) + (-0.0083\ C_{Cr}) + (0.0101\ C_{Al}^2) + (0.00016\ C_{Cr}^2)]\ 10^{-10}$$

$$D_{AlCr}(C_{Al},C_{Cr}) = [0.0116 + (0.0923\ C_{Al}) + (-0.0010\ C_{Cr}) + (0.00016\ C_{Al}^2) + (0.000017\ C_{Cr}^2)]\ 10^{-10}$$

$$D_{CrAl}(C_{Al},C_{Cr}) = [0.0766 + (-0.0153\ C_{Al}) + (0.0837\ C_{Cr}) + (0.00062\ C_{Al}^2) + (-0.0015\ C\ C_{Cr}^2)]\ 10^{-10}$$

$$D_{CrCr}(C_{Al},C_{Cr}) = [0.783 + (-0.0123\ C_{Al}) + (0.0247\ C_{Cr}) + (0.00096\ C_{Al}^2) + (-0.00057\ C_{Cr}^2)]\ 10^{-10}$$

26

Corrosion and Lifetime Modelling of Components in Coal Fired Combined Cycle Power Systems

N. J. SIMMS, J. R. NICHOLLS and J. E. OAKEY

Power Generation Technology Centre, Cranfield University, Cranfield, Bedfordshire, MK43 0AL, UK

ABSTRACT

Combined cycle power generation is one approach to the challenge of generating electricity from coal much more efficiently and cleanly than conventional pulverised coal power generation. The systems that are being developed are mainly based on coal gasification and/or combustion, with associated gas cleaning technologies to meet system and emission requirements. The influence of materials issues on the development of these processes can be considerable as it is necessary that components in these new processes have adequate lifetimes in their required operational environments. In modelling the behaviour of materials within components in such systems and producing component lifetime predictions, several distinct aspects of materials behaviour and component requirements have to be taken into account: environmental degradation, mechanical factors, synergistic effects, and different component life criteria. As components tend to fail from the regions of worst corrosion damage, corrosion modelling work has been particularly targeted at the development of 'maximum' corrosion damage models based on large numbers of metal loss measurements. These statistical models of corrosion damage need to be combined with appropriate information on materials mechanical properties, synergistic effects and component life criteria to give realistic estimates of component lives. This paper describes an approach to integrating these different factors that has been developed during the course of work on components throughout the hot gas path of coal-fired combined cycle power plants. The integration of plant and component data, life criteria and corrosion models are illustrated for a gasifier heat exchanger and blades/vanes within a gas turbine utilising dirty fuel gases.

1. Introduction

Combined cycle power generation is one approach to the challenge of generating electricity from coal much more efficiently and cleanly than conventional pulverised coal power generation [1–5]. A wide range of competing combined cycle power generation processes have been devised over the past 20 years and the development of many of these systems is continuing within Europe, Japan and the USA [1–9]. The most advanced of these systems can utilise state of the art gas and steam turbine technologies to optimise their efficiency and environmental performance. However, such new systems need to be seen to be able to compete with the gradual improvement in performance of conventional power generation systems. An important aspect of this competition is that the new systems need to be shown to be as reliable as existing systems.

The combined cycle power systems that are being developed are mainly based on coal gasification and/or combustion, with associated gas cleaning technologies to meet system and emission requirements [9]. The influence of materials issues on the development of these processes can be considerable as it is necessary that components in these new processes have adequate lifetimes in their required operational environments [9–11]. The selection of a particular process, or option within a generic process, may be governed by these materials issues as a result of their effects on capital, operating and maintenance costs, which may affect the overall process viability. While some components in these systems are very similar to ones in gas or oil fired systems (e.g. gas turbines), other components are unique to coal fired systems (e.g. hot gas clean-up systems). For all components, the use of coal as a fuel provides different materials issues as a result of the different hot gas path environments produced. The presence of particles may cause erosion, abrasion or deposition, and gaseous/vapour species (e.g. sulfur oxides (SO_x) or hydrogen sulfide (H_2S), hydrogen chloride (HCl), alkalis and other trace metal species) may cause deposition and/or enhanced corrosion.

Ideally, materials would be exposed as components in operating plants and their performance monitored over realistic timescales: periods of 1–10 years depending on the component. However, most of the coal fired combined cycle technology has not been developed beyond the pilot plant, or in a few cases, demonstration plant stage. The most realistic exposures for materials are within such plants. At the pilot plant stage of development, the expense of running these plants usually limits their operation to relatively short periods, often only 500–1000 h. Much useful information on materials performance can be obtained from such periods of pilot plant operation, if suitable metrology of material damage is carried out [12]. With only limited plant data available, the prediction and assessment of materials performance has to use laboratory data obtained under realistic conditions to produce models of materials performance that show the sensitivity of the degradation rates to realistic changes in the exposure conditions. The plant data available have to be used to validate these models, and check that the same forms of degradation are observed in the plant and laboratory exposure conditions.

In modelling the behaviour of materials in components in such systems and producing component lifetime predictions, several distinct aspects of materials behaviour and component requirements have to be taken into account for each different operating environment:

- Environmental degradation (e.g. high temperature oxidation, hot corrosion, erosion);

- Mechanical factors (e.g. creep and high/low cycle fatigue behaviour)

- Synergistic effects (e.g. effect of fatigue behaviour on allowable corrosion damage, erosion/corrosion); and

- Component life criteria.

As components tend to fail from the regions of worst corrosion damage, corrosion modelling work has been particularly targeted at the development of 'maximum' corrosion damage models based on large numbers of metal loss measurements [13,14]. Thus, statistically based models have been developed for the various different types of high temperature corrosion damage that can arise in coal-fired combined cycle systems. These statistical models of corrosion damage need to be combined with appropriate information on exposure conditions (e.g. gas chemistry, gas pressure, metal temperatures, deposition flux, deposit compositions), materials mechanical properties, synergistic effects and component life criteria to give realistic estimates of component lives.

This paper describes an approach to integrating these different factors that has been developed during the course of work on components throughout the hot gas paths of coal-fired combined cycle power plants. This approach, shown as a flow diagram in Fig. 1, is described using the illustrations of a gasifier heat exchanger and blades/vanes within a gas turbine utilising coal derived gases.

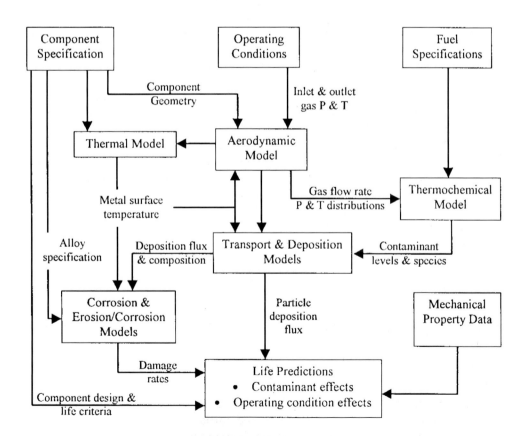

Fig. 1 *Flow diagram for component life modelling.*

2. Gasifier Heat Exchanger

2.1. Gasification Hot Gas Path Conditions

A number of coal gasification systems have been developed [2–4, 9], based on different types of gasification processes, e.g. entrained flow, fixed bed, fluidised bed. Variations on these processes may use either oxygen or air as their oxidant in the gasifier vessel and can be controlled to give varying degrees of conversion of the coal to fuel gas, from complete (>99% conversion) to partial (e.g. 75% conversion). Once generated, the fuel gases need to be cooled and cleaned before being burnt in gas turbines. It is possible to carry out the gas cleaning by water scrubbing the cooled fuel gases, but this leads to lower cycle efficiencies and requires the provision of a scrubbing and waste water treatment facility, which generates liquid waste for later disposal. The first generation gasification systems have been built to use water scrubbing for gas cleaning. Hot dry gas cleaning, using barrier filters to remove particulates and sorbents/catalysts to remove gaseous species (e.g. sulfur, chlorine, ammonia), offer higher cycle efficiencies as well as lower capital and operating/disposal costs. However, hot dry gas cleaning is still at the developmental stage and so only parts of the processes are included in the latest demonstration plants, often as parts of sidestreams.

There are many significant differences between the various gasification systems [4,9], both in terms of the operation of the actual gasification process and the requirements for different downstream components with different ranges of operating conditions. These give rise to economic and efficiency differences between the systems that are beyond the scope of this paper. However, from the perspective of materials performances and the optimum materials selection, the component operating conditions and the environments produced in each of the systems are critically important. Table 1 lists published bulk gas compositions produced by some of these gasification systems. Further differences arise from the minor and trace gas species that are very important in determining materials performances, both through direct reaction and indirectly through deposit formations and subsequent reaction. The levels of the minor and trace gas species in these gasification systems are not readily available, but limited data have been published for some systems (e.g. Table 2 [9] and [30,31]).

The majority of the coal gasification processes which have reached the pilot and demonstration plant scale are based on pressurised oxygen blown entrained flow slagging gasifiers, e.g. the Texaco, Dow, Shell (Fig. 2), Prenflo and GSP processes [3]. In addition, air blown pressurised fluidised bed gasification (PFBG) processes have been developed by, for example, British Coal and HT-Winkler. As well as IGCC cycles, air blown gasifiers can be optimised for use in partial gasification systems resulting in hybrid cycles, in which unburnt carbon from the gasifier is burnt in a combustor to raise steam for the steam cycle (Fig. 3). These cycles have many advantages in terms of efficiencies, staged construction, fuel flexibility (e.g. range of coals, co-firing with biomass) and availability of different parts of the system for power generation. British Coal's Air Blown Gasification Cycle (ABGC) [5,15] is an example of such a cycle. In this system, an air blown fluidised bed partial (~70% conversion) gasifier is used with a circulating fluidised bed as the char combustor (Fig. 3). Similar hybrid

Table 1. Gasifier gas compositions

| Gas species | Units | Oxygen blown gasifiers | | | Air blown gasifier |
| | | Entrained slagging | | Fluidised bed | Fluidised bed |
		Dry feed	Slurry feed		
CO	%	62–64	35–45	30–40	15–20
CO_2	%	2–4	10–15	10–15	5–8
H_2	%	27–30	27–30	24–28	10–15
H_2O	%	0–3	15–25	11–20	5–12
N_2	%	1–5	0–2	0–2	40–50
CH_4	%	n/a	n/a	3.5	2–4
H_2S *	vol ppm	2000–12 000	2000–12 000	2000–12 000	300–5000
NH_3	vol ppm	200–500	2000–5000	200–500	500–11500
HCl *	vol ppm	50–1000	50–1000	50–1000	50–500

*Dependent on coal S and Cl content. n/a=not available

Table 2. Measured data available on selected gas phase trace elements from an Air Blown Gasification System

Element	Concentration in gasifier fuel gases ($\mu g\ m_n^{-3}$)
Na	<15
K	<5
Pb	20–180
Zn	10–125

'partial' gasification/combustion cycles have been proposed by other companies, for example Foster Wheeler's Carboniser/Pressurised Fluidised Bed Combustor (PFBC) Cycle [16] and ABB's second generation PFBC cycle [17].

2.2. Materials Degradation Modes

The performance of materials in various simulated and real coal gasification gases have been investigated by research groups in the US, Japan and Europe for more than 25 years [e.g. 15,18–24]. The initial generic studies investigated the performance

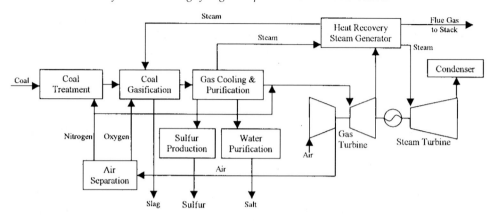

Fig. 2 *Example of an Integrated Gasification Combined Cycle (IGCC) using a pressurised oxygen blown slagging gasifier [15].*

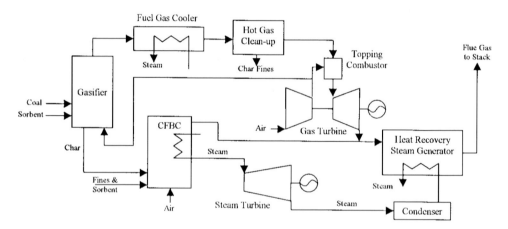

Fig. 3 *Example of an air blown gasification cycle using an air blown fluidised bed partial gasifier [15].*

of materials in a range of highly reducing atmospheres with varying levels of sulfidation at high temperatures. In view of the results obtained, later studies have tended to concentrate on higher alloyed materials and/or lower exposure temperature. Following plant experience, more recent studies have been targeted at increasingly realistic exposure simulations to match the degradation morphologies observed in practice [9,15,24,28].

There are several potential degradation processes for these environments [15–24], including:

- elevated temperature gas phase induced corrosion: including oxidation, sulphidation, carburisation/ metal dusting and chlorination;

- corrosion induced by surface deposits: either particles from the gasifier, species condensed onto those particles or species condensed onto the component surfaces;

- dewpoint corrosion: induced by high temperature deposits becoming damp during plant idling, gaseous species (e.g. HCl, H_2S, etc) reacting with condensing water during idling (e.g. forming HCl and polythionic acids) or deposits forming on cooler gas path surfaces;

- downtime corrosion: induced by high temperature deposits becoming damp during plant shut-down, gaseous species (e.g. HCl, H_2S, etc) reacting with condensed water during plant shut-down (e.g. forming HCl and polythionic acids) or hygroscopic deposits being exposed to damp air during plant shut-downs (e.g. during the course of maintenance operations);

- interaction of any of the above degradation modes with mechanical factors, e.g. creep or fatigue, to produce synergistic degradation, e.g. creep-corrosion or corrosion-fatigue; and

- spallation of corrosion products: important on the clean side of the filter unit from where spalled scale may enter the gas turbine and cause erosion damage.

In practice, several materials degradation modes will occur on each component in the hot gas path. In any one gasification system, different combinations of degradation modes will be found on components along the hot gas path due to differences in operating temperature and local plant environment (deposition, local gas composition, ie the extent of gas clean-up, gas temperature at that point in the system, component temperature, etc).

2.3. Corrosion Models

The development of corrosion models for gasification environments was an activity in the completed European COST501 programme, with investigations covering high temperature gaseous corrosion, downtime corrosion and some forms of synergistic damage (e.g. stress corrosion cracking, alternate periods of high temperature and downtime corrosion) [15]. In this work programme, British Coal Corporation, KEMA BV and ENEL-Research SpA generated materials performance data from realistic laboratory tests investigating specific degradation mechanisms [15,28]. The development of these models is continuing in the current European COST522 programme [25]. A parallel activity in developing corrosion models for IGCC gasifier environments has been carried out by Shell [26,27].

The quantity of materials performance data available to the COST501 partners (from that and previous research programmes [e.g. 15,20,21,23]), provided enough data to model reliably the performance of at least some of the candidate materials in such gasification conditions. After analysing the available data sets [15], it was found that the best model for the dependence of corrosion rate on gas composition and temperature took a logistic, or sigmoidal, form (e.g. Fig. 4 shows the form for alloy 800):

$$\text{Corrosion rate} = \frac{a-d}{1+\left(\dfrac{\log pS_2 - 1.5\log pO_2}{c}\right)^b} + d \tag{1}$$

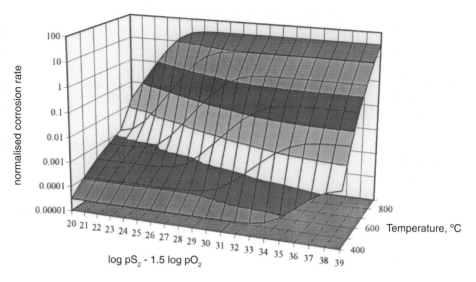

Fig. 4 *Illustration of modelling of high temperature gaseous corrosion in gasification environments for alloy 800.*

where, a, b, c and d are temperature dependent constants. In theory, the constants a and d should also be slightly dependent on the pO_2 and pS_2 levels, but the corrosion data currently available are not sufficiently detailed to show this effect.

This form of the model compares well with the data generated by Shell for the dependence of corrosion rate on gas composition for their gasification process (e.g. Fig. 11 in reference [27]). The Shell data relate to the higher pS_2/lower pO_2 part of the model, where the corrosion rate increases rapidly in the oxidation/sulfidation regime with increasingly sulfidising conditions and then increases more slowly on entering the sulfidising regime. It should be noted that the ABGC process operates under more oxidising conditions than the oxygen blown IGCC systems, and for alloy 800 (Fig. 4) this is around the transition from the oxidising regime to the rapidly increasing oxidation/sulfidation regime (the location of this transition relative to the gas parameters is alloy dependent).

Less sophisticated models of downtime corrosion under simulated gasifier 'deposits' were generated by correlating deposit compositions and exposure conditions with the damage observed [15,28]. For the more complex synergistic damage modes, candidate materials were either ranked in terms of their resistance (e.g. Fig. 5 [28]) or simple multiplication factors produced to allow for the extra damage produced [15]. With the current state of the materials degradation models for these conditions, such issues therefore have to be considered in parallel with to the more thorough life prediction modelling that can be carried out for other degradation modes, so that particularly damaging conditions are avoided. Further development of the materials degradation models is currently in progress to widen their applicability in terms of the conditions covered [25].

Validation of the materials performance models requires the prediction of damage in pilot plant hot gas paths from the known exposure conditions to be compared with the actual damage observed on materials exposed to those conditions.

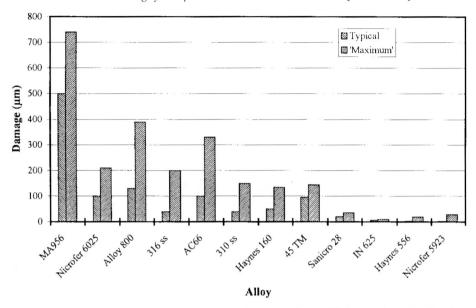

Fig. 5 *Example of the ranking of the relative performance of materials in tests investigating the synergistic effects of alternate periods of high temperature gaseous corrosion and downtime corrosion in a typical ABGC fuel gas environment (4 cycles of 500 and 100 h exposure per cycle respectively) [28].*

A selection of candidate materials was exposed during the operation of atmospheric and pressurised fluidised bed gasification pilot plants [15,28]. The performance of the materials was determined from accurate dimensional metrology before and after exposure. Predictions for typical and maximum high temperature corrosion damage levels were calculated using eqn (1), with constants evaluated for each material and both evaluations of damage. Total damage was calculated as the larger of the values of 'typical high temperature damage + maximum downtime damage' and 'maximum high temperature damage' (with the maximum downtime damage model coming from the laboratory downtime corrosion test with the most aggressive deposit [28]). Figure 6 illustrates the results of the comparison of predicted and measured damage for AISI 310. As expected there is some scatter of data in this comparison. There are several sources of this scatter [27]:

- measurement errors in all stages of materials performance monitoring in both the laboratory and pilot plant exposures;

- mixing different methods of data gathering with different degrees of accuracy and sensitivities to types of corrosion damage (e.g. mass change, optical microscopy, contact metrology);

- data analysis and curve fitting in the materials performance modelling; and

- knowledge of plant exposure conditions: gas composition, deposition, gas/metal temperatures, etc.

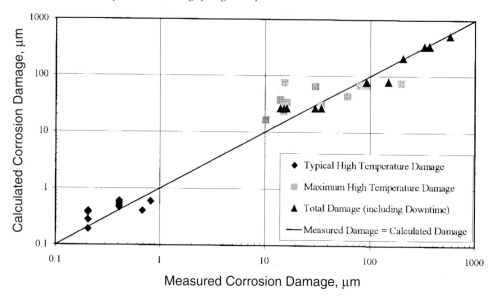

Fig. 6 *Comparison of predicted and measured gaseous corrosion damage for AISI 310 exposed in ABGC pilot plants [28].*

Of all these factors, by far the most important for model validation was the uncertainty associated with the plant exposure conditions. As the pilot plants were dedicated to process performance evaluation, materials exposure conditions have varied in terms of gas/metal temperatures, gas composition, deposition, etc. For the purposes of these validations, ranges of exposure conditions have been determined and appropriate values of the most important variables taken (e.g. metal temperature and gas H_2S content).

Examination of cross-sections during the course of this data generation allowed the morphology of the damage to be checked to ensure that models were generated and validated for the same types of corrosion behaviour.

2.4. Component Life Criteria and Predictions

Life criteria for wide range of components in gasifier hot gas paths, will be component and process specific due to the large differences in operating requirements.

The life criteria for combustor heat exchangers are normally set on the basis of the creep/fatigue lives of components with a somewhat arbitrary corrosion allowance (e.g. 2 mm). However, in coal gasifiers, the corrosive environment around the heat exchanger can be severe and in fact limit the lives of the heat exchanger if operated with too high a metal temperature [9]. Thus, more consideration needs to be given to the acceptable life criteria from a corrosion point of view, e.g. damage with a 4% probability of being exceeded [13,14] over 2 mm could be equated to a life (tube replacement) criteria.

The corrosion models available for gasifier hot gas path conditions are not yet as sophisticated as those for gas turbine conditions (section 3), but can still be used to make predictions of component lives with appropriate assumptions. For example, a requirement for a heat exchanger with a 2 mm corrosion allowance to last ~25 years and an assumption of a linear corrosion rate gives a target damage rate of ~0.01 μm/h. Comparison with Fig. 4 shows that whilst operation at 400°C will be result in a gaseous corrosion damage rate less than this target, higher temperature operation could require restrictions on the acceptable gas compositions (i.e. fuel compositions) for some types of gasifiers (especially oxygen blown IGCC systems). Alternatively, unrestricted fuel use would require limited metal temperatures. In addition, the aggressive nature of many deposits formed on gasifier heat exchangers when damp, would necessitate the avoidance of component operation below this dewpoint and very good downtime control during shutdowns.

3. Gas Turbine Hot Gas Path Components

3.1. Operating Conditions

All combined cycle power systems utilise industrial gas turbines. The durability and performance of the gas turbines underlie the increased efficiencies and economic viability of these advanced power generation systems, with periods of at least three years desired between major overhauls and longer periods between major overhauls desired in the future. However, the requirements for the gas turbines vary significantly between the different types of power systems.

In first generation Pressurised Fluidised Bed Combustion (PFBC) systems, the fuel is burnt in the combustor at ~800–850°C and then the combustion gases are passed through cyclone(s) and/or a hot gas filter before passing into the gas turbine without further heating [11]. Thus, the hot gas path is oxidising throughout its length, with the major gaseous contaminants of concern for the gas turbine being transported through this gas path being SO_x, HCl and alkali metal vapour phase species. The level of these contaminants can be controlled to varying degrees by process operation and fuel composition: e.g. sorbent additions to the fluidised bed can be used to reduce SO_x (or a lower sulfur-containing fuel), lower bed operating temperatures can be used to reduce levels of volatile alkali species.

In contrast, gasification and hybrid gasification/combustion systems (section 2.1) produce a fuel gas that has to be burnt in the gas turbine combustor. Thus, these systems can use advanced gas turbines with higher entry temperatures and efficiencies (with suitable modifications to allow the use of fuel gases with much lower calorific values than natural gas). The gas compositions that can be produced from these systems vary widely as they pend not only on the fuel composition, but also the gasification process and the gas clean-up systems used (e.g. water scrubbing or hot dry gas cleaning — section 2.1). The production of fuel gases (i.e. reducing conditions) causes many more vapour species to be generated (e.g. lead, zinc, tin, arsenic, cadmium [9,29,30]) and these have the potential to be carried through the fuel gas path to the gas turbine, as well as the H_2S and HCl. Table 2 gives examples of trace metal contaminant levels in the vapour phase of a gasifier fuel gas measured downstream of a hot gas filter.

The hot gas cleaning systems are predominantly aimed at reducing the emissions of potentially damaging gaseous species and dust to the environment [9]. At the same time these gas cleaning systems could reduce the corrosion and erosion potential of gas contaminants to gas turbine and combustor components. However, the gas cleaning requirements to meet the objectives of reduced environmental emissions and adequate gas turbine component life are not necessarily the same. Targets for permitted environmental emissions are being progressively reduced and these currently dictate the required performance of the gas cleaning systems. However, the demands on the gas cleaning systems to give adequate lives for gas turbine components have not been set for use on coal derived gases. These contaminants may be the same as those of environmental concern (e.g. SO_x, after combustion), but can also be different (e.g. alkali), and the required clean-up levels may also be different (e.g. species below emissions limits may be above sensible limits for gas turbine usage). It is also possible that the introduction of hot dry gas cleaning in gasification systems could introduce new contaminants (e.g. Zn) from the sorbents/ catalysts used, and/or change the concentrations of trace contaminants in the fuel-gas such that when combusted it becomes more aggressive to the turbine components.

Generic ranges of gas turbine operating conditions, in terms of gas pressures, gas temperatures and metal temperatures are given in Fig. 7.

3.2. Materials Degradation Modes

The requirements for materials in the hot gas paths of gas turbines are very demanding. The materials need to be capable of operating at bulk temperatures up to ~900°C under both high and fluctuating stresses (i.e. having creep and both high and low cycle fatigue resistance), whilst also withstanding the surrounding environments [31]. The environments produced can be both physically and chemically aggressive, with particles producing erosion or deposition and gaseous species

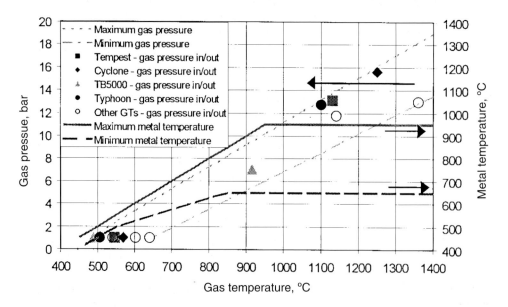

Fig. 7 *Typical gas turbine operating conditions.*

producing different forms of deposition, oxidation and hot corrosion [32,33]. During the last 40 years, these topics have been the subject of many investigations and the potential problems which many be encountered in gas and oil fired gas turbines have been well characterised [32,33]. Many similar types of materials degradation can be expected in gas turbines using coal-derived fuels, as some of the contaminant species are the same as for oil and/or gas fired systems, but the levels of contamination are different and there are additional species which can be expected to modify the materials behaviour.

Gas turbine materials will oxidise in the combustion gases produced in all gas turbine systems, but the rate of oxidation below ~900–1000°C metal temperature is sufficiently slow so as not to be life limiting. However, hot corrosion of turbine materials can occur much more rapidly and is potentially life limiting. For hot corrosion to occur, a liquid (usually sulfate) deposit is required on the surface of components. The formation of this deposit depends on trace metal species (e.g. sodium, potassium, lead, zinc compounds) in the gas streams and other reactive gas species (e.g. sulfur dioxide (SO_2), sulfur trioxide (SO_3), HCl). The rates of corrosion will depend (among other factors) on the rate of deposit formation, temperature and the surrounding environment. Two general types of hot corrosion in gas turbine environments have been identified to date: Type I hot corrosion at ~750–900°C; Type II hot corrosion at ~600–800°C. Reviews of these forms of damage are available in the literature, for example reference [32].

3.3. Corrosion Models

Studies have been carried out on the corrosive effects of contaminants expected in gas turbines using coal derived gases, but few systematic studies have been carried out to generate suitable dimensional data for modelling these forms of corrosion. However, limited corrosion models have been developed in UK and European research activities carried out in support of coal fired PFBC and solid fuel fired gasification systems [12,34–37]. These activities are on-going and the latest forms of these developing corrosion models are reported elsewhere in these conference proceedings. Figure 8 illustrates the general features of these corrosion models for IN738LC: deposition fluxes, deposit compositions and SO_X partial pressures dominate the corrosion behaviour of the materials, with less effects from HCl partial pressures. Figure 9 shows some of the validation data generated from materials exposures in plant and burner rig environments. As for the gasification environment, the largest source of errors in this validation is the uncertainty in exposure conditions in the plant environments.

3.4. Component Life Criteria

There are no set life criteria for blades and vanes in industrial gas turbines. Different life criteria are used by different gas turbine manufacturers and turbine users, depending on how conservative the view of component life is and how/whether components will be refurbished. For example, four alternative life criteria for gas turbine blades and vanes are:

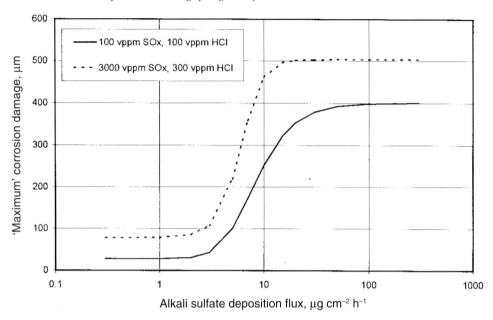

Fig. 8 *Illustration of the use of a 'maximum' hot corrosion damage model for IN738LC at 700°C (time = 500 h).*

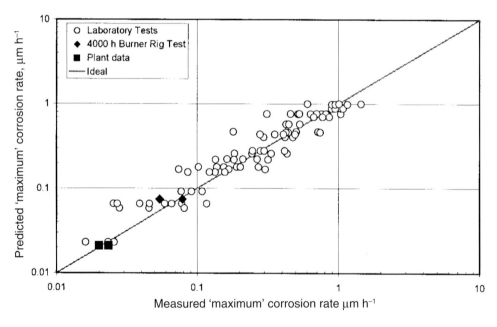

Fig. 9 *Comparison of model predictions with burner rig and plant results for IN738LC metal temperatures of ~700°C.*

1.   Penetration of 2/3 of the thickness of a corrosion resistant coating (applicable to blades and vanes);

Table 3. Typical composition of gas leaving a Pressurised Fluidised Bed Combustor [11]

Parameter	Units	Typical values
Pressure	Bar abs	16
Temperature	°C	850
Gas composition		
N_2	%	76.5
O_2	%	4.7
CO_2	%	13.7
H_2O	%	4.2
Ar	%	0.9
HCl	vol ppm	160
SO_X	vol ppm	130
Na	vol ppm	0.3
K	vol ppm	0.01

2. Total penetration through a corrosion resistant coating (applicable to blades and vanes);

3. For a blade, the minimum defect size to trigger high cycle fatigue cracking in the base alloy; and

4. For an air cooled vane, total penetration of the corrosion resistant coating and 2/3 penetration of the minimum wall thickness over the cooling channels.

Life criteria 1 and 2 represent different degrees of conservatism on the component refurbishment route, but give lower predicted lives than criteria 3 and 4. However, there would be doubt about the viability of component refurbishment using life criteria 3 or 4. These criteria vary with gas turbine blade and vane designs and coating specifications and so are gas turbine specific.

3.5. Component Life Predictions

The lives of gas turbine components in gas turbines operating on gasifier derived fuel gases will depend on many factors. From the point of view of corrosion damage to the components, important factors include the type of gasification system, fuel composition and gas cleaning process (especially the operating temperature of hot gas cleaning systems), as well as gas turbine blade/vane designs and operating conditions. Differences in such factors can have a significant effect on the corrosion damage to the gas turbine.

In order to make corrosion life predictions for gas turbines operating on a gasification system, detailed information is required about the exposure conditions within the gas turbine hot gas path, in terms of the entry gas composition, gas partial pressures, deposition fluxes, metal temperatures, etc. (i.e. the outputs of the other models in Fig. 1). Such detailed information about gas turbine and system performance is commercially sensitive. However, to illustrate the use of such modelling a set of generic conditions have been used (from Fig. 7 and Table 2), together with gas contaminant levels of 80 vol ppm SO_X and 140 vol ppm HCl. These data have been used in a vapour deposition model [38] to predict the variation of deposition flux with gas pressure. It is also necessary to assume that creep is not life limiting, but that high cycle fatigue could be and a critical defect damage size can be defined (for criterion 3). Estimated dimensional details needed to use the models are: a corrosion resistant coating of 120 μm (for use in life criterion 1 and 2), a minimum wall thickness over the vanes' cooling passages of 6 mm (for use in life criterion 4), and a minimum defect size to trigger HCF of 1.3 mm (for use in life criterion 3).

These various assumptions and estimates have been combined to show the variation in predicted component lives with reducing gas pressure (and gas temperature) but constant metal temperature. Figures 10 and 11 illustrate the results obtained when these assumptions are combined with the four life criteria outlined

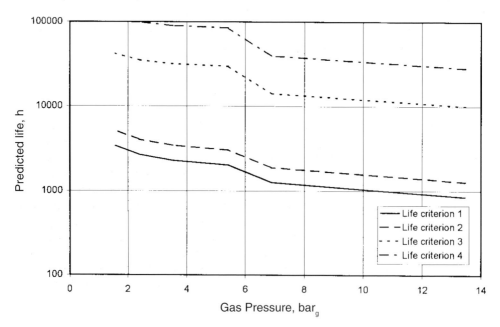

Fig. 10 *Illustration of the variation of life predictions with different life criteria (section 3.4) at 700°C (assumptions for use of the model given in section 3.5).*

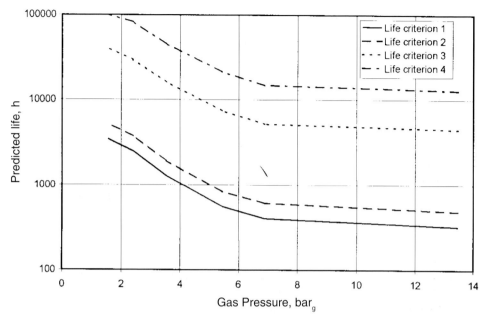

Fig. 11 *Illustration of the variation of life predictions with different life criteria (section 3.4) at 650 °C (assumptions for use of the model given in section 3.5).*

in section 3.4 to give predictions of component lives for metal temperatures of 700 and 650°C respectively. These figures show that the predicted lives increase (as expected) going from criteria 1 to 4 (the 'step' in the predicted lives in these figures is related to crossing a deposition dewpoint). Care is needed in relating such predictions to gas turbine vanes and blades, as not all conditions for which predictions can be made occur in actual gas turbines, and criteria 3 and 4 are component specific.

4. Conclusions

In modelling the degradation behaviour and potential lives of materials in new advanced coal-fired power systems, a wide range of factors have to be taken into consideration, including the detailed exposure environment, mechanical requirements and life criteria. As components tend to fail from the regions of worst corrosion damage, statistically based corrosion models derived from metal loss data that show the sensitivity of corrosion damage to exposure parameters are needed. Some corrosion models have been developed to meet these objectives for specific components within the hot gas paths of these power systems. This paper has used the examples of gasifier heat exchangers and blades/vanes in gas turbines fired using coal-derived gases to illustrate the use of these corrosion models for life predictions, together with the interactions necessary with component design data and plant hot gas path environments.

5. Acknowledgements

Funding has been provided by British Coal Corporation, UK Department of Trade and Industry (Clean(er) Coal Programme contracts C/05/00086, C/05/00135 and C/05/00266), European Commission JOULE programme (contract JOUF-0022), European Coal and Steel Community programme (contracts 7220-ED/069 and 7220-PR/053).

References

1. W. Schlachter and G. H. Gessinger, in High Temperature Materials for Power Engineering 1990, (E. Bachelet *et al.*, eds). Kluwer, 1990, pp1–24.
2. Proc. 1st International Workshop on Materials for Coal Gasification Power Plant, Petten, The Netherlands, June, 1993, in 'Materials for Coal Gasification Power Plant', Special Issue of Materials at High Temperature, 11 (1993).
3. Proc. Corrosion in 'Advanced Power Plants', Special Issue of *Mater. High Temp.*, **14** (1997).
4. T. Takematsu and C. W. Maude, 'Coal Gasification for IGCC Power Generation', IEACR/37, IEA Coal Research, London, UK (1991).
5. S. G. Dawes, C. J. Bower, C. Henderson, D. Brown and J. A. C. Hyde, in 'Power Generation and the Environment', Inst. Mech. Eng. (1990).
6. I. Mendez-Vigo, J. Chamberlain and J. Pisa, in 'Corrosion in Advanced Power Plants', Special Issue of *Mater. High Temp.*, **14**, 15–20 (1997).
7. J. van Liere and W. T. Bakker, in 'Materials for Coal Gasification Power Plant', Special Issue of *Mater. High Temp.*, **11**, 4–9 (1993).
8. P. L. Zuideveld and A. Postuma, in 'Materials for Coal Gasification Power Plant', Special Issue of *Mater High Temp.*, **11**, 19–22 (1993).
9. J. E. Oakey and N. J. Simms, in *Materials for Advanced Power Plant 1998* (J. Lecomte-Beckers *et al.*, eds). Forschungszentrum Julich GmbH, Germany, 1998, pp.651–662.
10. J. B. Marriott, M. Van De Voorde and W. Betteridge, Coal Conversion Processes and their Materials Requirements, EUR 9182 EN (1984).
11. N. J. Simms, J. E. Oakey and M. A. Smith, in *Materials for Combined Cycle Power Plant*. Publ. Inst. of Metals, 1991.
12. N. J. Simms, J. E. Oakey and J. R. Nicholls, in *Proc. Per Kofstad Memorial Symposium on High Temperature Corrosion and Materials Chemistry* (M. J. McNallan, ed.). Proc Electrochemial Society PV99–38, pp.305–317 (2000).
13. J. R. Nicholls and P. Hancock, in High Temperature Corrosion (R.A. Rapp, ed.). NACE, Houston, TX, USA, p.198 (1983).
14. *Introduction to Life Prediction of Industrial Plant Materials: Application of the Extreme Value Statistical Method for Corrosion Analysis* (M. Kowaka, ed.) Allerton Press, Inc., New York (1994)
15. N. J. Simms, F. Bregani, W. M. M. Huijbregts, E. Kokmeijer and J. E. Oakey, in *Materials for Advanced Power Plant 1998* (J. Lecomte-Beckers et al., eds). Forschungszentrum Julich GmbH, Germany, 1998, pp.663–678.
16. J. L. Blough and A. Robertson, in 'Materials for Coal Gasification Power Plant', Special Issue of *Mater. High Temp.*, **11**,10–14 (1993).
17. I. Stambler, *Gas Turbine World*, 1993, **23**, 22–27.
18. D. M. Lloyd, in *Research and Development of High Temperature Materials for Industry* (E. Bullock, ed.). Elsevier, Oxford, pp.339–359 (1989).
19. J. Stringer, in *High Temperature Materials Corrosion in Coal Gasification Atmospheres* (J. F. Norton, ed.). Elsevier, Oxford (1984).

20. F. Gesmundo, in High Temperature Materials for Power Engineering 1990 (E. Bachelet *et al.*, eds). Kluwer, pp.67–90 (1990).

21. F. Gesmundo, Advanced Materials for Power Engineering Components: The Corrosion of Metallic Materials in Coal Gasification Atmospheres – Analysis of Data from COST 501 (Round 1) Gasification Subgroup. EUCO/MCS/08/1991 (1991).

22. W. T. Bakker, in *Proc. 7th EPRI Ann. Conf. on Coal Gasification*, EPRI, Palo Alto, CA, USA (1987).

23. 'Erosion/Corrosion of Advanced Materials for Coal-Fired Combined Cycle Power Generation', Final and Summary Reports on EC contract JOUF-0022 (1994).

24. W.T. Bakker, 'Mixed Oxidant Corrosion in Nonequilibrium Syngas at 540°C', EPRI Report TR-104228, (EPRI, USA, 1995)

25. D. H. Allen, J. E. Oakey and B. Scarlin, in Materials for Advanced Power Plant 1998 (J. Lecomte-Beckers *et al.*, eds). Forschungszentrum Jülich GmbH, Germany, pp.1825–1839 (1998).

26. R. C. John, in *Proc. Symp. on Life Prediction of Corrodible Structures*, Cambridge, UK, 1991, pp.33/1–21, National Association of Corrosion Engineers, Houston, USA (1991).

27. R. C. John, W. C. Fort and R. A. Tait, in 'Materials for Coal Gasification Power Plant', Special Issue of *Mater. High Temp.*, **11**, 124–132 (1993).

28. N. J. Simms, J. R. Nicholls and J. E. Oakey, *Mater. Sci. Forum*, to be published (2001).

29. G. Reed, S. Brain, P. Cahill and I. Fantom in *Proc. 14th Int. Conf. on Fluidized Bed Combustion* (D. S. Fernando, ed.). ASME Book No. G1052B–1997, pp.1285–1293 (1997).

30. A. Pigeard and J. J. Helble, in *Coal Fired Power Systems '94 – Advances in IGCC and PFBC Meeting,* DOE/METC-94/1008 (1994)

31. J. Stringer and I. G. Wright, *Oxid. Met.*, 1995, **44** (1/2), 265–308.

32. C. T. Sims, N. S. Stoloff and W. C. Hagel, in *Superalloys II*. Wiley, New York, 1987.

33. 'Hot Corrosion Standards, Test Procedures and Performance', *High Temp. Technol.*, 1989, **7**, (4).

34. *Effects of Contaminants on Materials Performance in Industrial Gas Turbines for Advanced Combined Cycle Power Plants*, Coal R&D Report COAL R129, ETSU for UK DTI (1997)

35. N. J. Simms, J. R. Nicholls and J. E. Oakey, *Mater. Sci. Forum*, to be published (2001).

36. N. J. Simms, J. E. Oakey, J. R. Nicholls, P. J. Smith and D. J. Stephenson, in *Erosion by Liquid and Solid Impact/1984* (J. A. Little and I. M. Hutchings, eds), Wear, 1994, **247**, 186–187.

37. J. R. Nicholls, P. J. Smith and J. E. Oakey, in *Materials for Advanced Power Engineering 1994* (D. Coutsouradis *et al.*, eds). Kluwer, p.1273 (1994).

38. J. E. Fackrell, R. J. Tabberer, J. B. Young and I. R. Fantom, Paper 94-GT-177, International Gas Turbine and Aeroengine Congress and Exposition, June 1994, The Hague, Netherlands, AMSE, USA (1994).

27

The ASSET Project — A Corrosion Engineering Information System for Metals in Hot Corrosive Gases

R. C. JOHN, A. D. PELTON*, A. L. YOUNG[†], W. T. THOMPSON[§]
and I. G. WRIGHT[¶]

Shell Global Solutions (US), Houston, TX 77251-1380, USA
*CRCT, École Polytechnique de Montréal, Montreal H3R 1Z8, Canada
[†]Humberside Solutions Ltd., Toronto, Ontario M6N 4X7, Canada
[§]Royal Military College of Canada, Kingston, Ontario, K7K 5LO, Canada
[¶]Oak Ridge National Laboratory, Oak Ridge, TN 37831-6156, USA

ABSTRACT

Corrosion of metals and alloys in industrial equipment is an important consideration in terms of safety, economics, and process design. This paper reviews how corrosion of alloys in process equipment can be identified, simulated, and managed. The discussion highlights the following:

• several corrosion mechanisms prevalent in high temperature process equipment;

• laboratory corrosion testing procedures to simulate conditions in operating equipment; and

• an approach to predict corrosion-limited equipment lifetimes at high temperatures.

Corrosion of alloys by exposure to complex gases is discussed for temperatures of 200–1100°C. Many alloys are based upon combinations of iron, nickel, iron–chromium, and iron–nickel–chromium. These alloys are used in high temperature corrosive conditions even though there are often limited corrosion data available to predict lifetimes. We can correlate corrosion defined by the total metal penetration with exposure conditions such as the temperature, exposure time, and gas composition. This allows compilation of high temperature corrosion data so that engineering decisions can be made with respect to lifetime, maximum allowable temperature, and influence of gas composition, by analysis of corrosion data from well defined conditions for a variety of alloys. The likely corrosion mechanism is determined, based on alloy composition, exposure conditions, gas composition, and temperature. This approach has been incorporated into software called ASSET (Alloy Selection System for Elevated Temperatures) and is used for failure analysis, equipment design, alloy selection, and alloy design.

1. Introduction

The ability to predict corrosion of alloys in high-temperature corrosive gases in many processes could aid management of corrosion in various types of equipment. An important obstacle to predicting corrosion is the variety of alloys and corrosive environments. Another obstacle is a lack of recognition of how to generate and then use high-temperature corrosion data to assess engineering lifetimes. The technology discussed in this article offers solutions by providing a method to deal with many alloys and corrosive conditions and by providing a systematic method to predict corrosion in terms of metal loss (penetration). Metal loss is directly related to the loss of metal thickness used in equipment design and operation decision making.

Many years of tradition have led engineers to archive and report corrosion data in ways, which are cumbersome, expensive to update, and limited by the expertise of the compiler of the handbook or data collection in question. Most corrosion data are reported as mass change and cannot be used when making decisions about equipment lifetimes and maximum allowable conditions such as temperatures or gas compositions. The expertise needed to properly update these compilations is held by relatively few persons. Also, attempts to compare recent and previous corrosion data or to extrapolate measured data to predict corrosion in untested conditions are often difficult.

This paper discusses the potential of high-temperature gases to corrode metals by several different mechanisms and recommends how to predict sound metal losses for a wide range of conditions. Gaseous corrosion is possible in processes such as: petroleum refining, gas processing, fired equipment, process heaters, burners, flares, furnaces, boilers, thermocouples, instrumentation, process heaters, hydrocracking, coking, vacuum flashing, hydrotreating, coal/coke/oil gasifying, gas processing, petrochemical production and catalytic reforming.

Many aspects of equipment/process design, process operation, alloy selection, alloy design, and plant maintenance are influenced by the expected lifetimes of equipment in high-temperature, corrosive gases. These lifetimes are greatly affected by the conditions present in process heating, energy conversion, and power generation. These processes are usually implemented with equipment, which has maximum allowable temperatures, or other process conditions, which are limited by the corrosion rates expected for the equipment. This equipment can be found in heaters, steam boilers, flares, waste incinerators, hydrogen plants, hydrotreating plants, ethylene plants, other chemical plants, crude oil refineries, heat treatment ovens, various fossil-fueled equipment, instrumentation, and electric heating devices.

Most corrosion data for alloys in high-temperature gases have been reported in terms of mass change/area for relatively short exposures and inadequately defined exposure conditions. The mass change/area information does not directly relate to the thickness of corroded metal, which is often needed in assessing the strength of equipment components. Corrosion is best reported in penetration units, which indicate the sound metal loss, as discussed earlier [1–3]. Corrosion in high-temperature gases is affected by parameters of the corrosive environments such as temperature, alloy composition, time, and gas composition. Important aspects of four different corrosion mechanisms are discussed in this paper.

2. Corrosion Mechanisms

The dominant corrosion products are used to determine the names of the corrosion mechanisms discussed in this paper. For example: oxidation implies oxides, sulfidation implies sulfides, sulfidation/oxidation implies sulfides plus oxides, and carburisation implies carbides.

2.1. Oxidation

Oxidation often occurs upon exposure of metals to temperatures above 300°C in gases containing more than 1% (vol.) O_2. The first step in determining the potential for oxidation is to confirm that oxides are present. X–ray analysis by diffraction of the surface scale or analysis of the gas composition are common methods to confirm oxidation.

The dependence of corrosion upon exposure time for alloys after sufficient time has passed is commonly assumed to be proportional to $(time)^{0.5}$, which is known as parabolic time dependence. Several thousand hours may be required to establish this time dependence, suggesting that oxidation measured after hundreds of hours is unlikely to be useful in estimating long time oxidation rates. It has been shown that many alloys establish parabolic time dependence after a time of 500–1000 h in air at constant temperatures of 870–1090°C [1,2].

The gas composition influences the rate of oxidation in terms of variables such as O_2. The influence of O_2 concentration upon oxidation is specific to each alloy [1,2]. Most alloys do not show a strong influence of the O_2 concentration upon the total penetration. Alloys such as MA 956, 120, and 214 exhibit slower oxidation rates as the O_2 concentration increases. Compositions of alloys mentioned in this article are listed in Appendix 1. These alloys form surface oxides, which become more stable as the O_2 concentration increases. They are rich in Cr or Al, whose oxides are stabilised by increasing levels of O_2 concentration. Some alloys, which generally exhibit increased oxidation rates as the O_2 concentration increases, are AISI 304, AISI 410, 9Cr–1Mo, DS, 617, AISI 446, and 253MA. These alloys tend to form rapidly growing oxide scales and increasing O_2 concentrations effectively increase the growth rates of the corroding alloy components, thereby increasing the corrosion rate.

Most alloys tend to have increasing penetration rates with increasing temperature for all O_2 concentrations. Some exceptions will be alloys with 1–4% Al concentrations, such as MA 956 and 214. These alloys need high temperatures to form Al_2O_3 as the dominant surface oxide, which grows more slowly than the Cr_2O_3, which in turm dominates at the lower temperatures. Figure 1 summarises oxidation after one year for some common alloys exposed to air. The metal temperature is used in assessing the oxidation rate of metals and not the gas temperature.

Most of the commercial heat resistant alloys are based on combinations of Fe–Ni–Cr and they show about 80–95% of the total penetration as subsurface oxidation [1,2]. Some alloys change in how much of the total penetration occurs by subsurface oxidation as time passes, until long-term behavior is established, even though the corrosion product morphologies may remain constant [1,2].

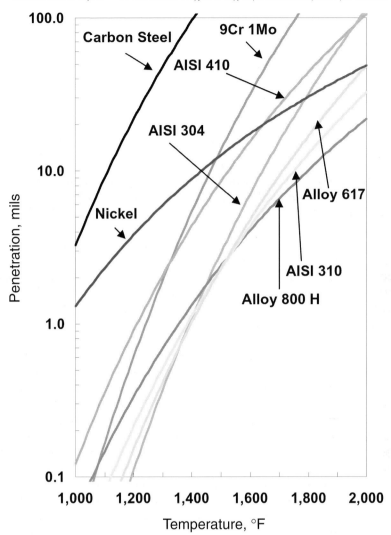

Fig. 1 *Effect of temperature upon metal penetration of some common alloys by oxidation after exposure for one year to air (21% vol. O_2); 10 ml = 0.25 mm, 1000–2000°F = 538–1093°C.*

2.2. Sulfidation

Sulfidation occurs upon exposure of metals to temperatures above approximately 200°C in gases containing H_2S at concentrations greater than 1 ppm vol. The presence of sulfides confirms sulfidation. X-ray analysis by diffraction of a scale sample or analysis of the gas are common methods to confirm the existence of sulfidation. Sulfidation occurs upon exposure of metals to gases containing CO–CO_2–COS–H_2–H_2O–H_2S. Variables which influence the sulfidation rate are the exposure time, and partial pressures of H_2 and H_2S, and temperature.

The time dependence of sulfidation is controversial [5–7] with reports of a parabolic time dependence (metal loss proportional to time$^{0.5}$), linear time dependence (metal loss proportional to time), power law dependence (metal loss proportional to time^x), and combinations of these dependencies. An undisturbed sulfide scale and the exposure time in excess of 2000 h, probably yield parabolic time dependence — although, some studies report linear time dependence after several thousand hours [5,6].

Increasing the concentration of H_2S tends to increase the sulfidation rate of alloys, as illustrated for several alloys in Fig. 2. The line for carbon steel stops for lower concentrations of H_2S because FeS is not stable and the steel can not corrode.

High Ni alloys used either as base metals or as welding filler metals are a special concern in sulfidation conditions. Sulfidation of high Ni alloys can be especially rapid and yield corrosion rates greater than 2.5 mm/year, if the temperature exceeds 630°C, which is the melting point of a potential corrosion product which forms as a mixture of Ni and nickel sulfide.

2.3. Sulfidation/oxidation

Sulfidation/oxidation occurs upon exposure of alloys containing elements such as Cr and Al to hot gases containing various combinations of CO–CO_2–COS–H_2–H_2O–H_2S gases. Sulfidation/oxidation is found in hydrocrackers, hydrotreaters, coal/coke/oil gasifiers, and Flexicokers, where Cr- containing alloys are exposed to complex gases. Sulfidation/oxidation occurs when corrosion products are mixtures of sulfides and oxides. Elements such as Cr, Al, and Si may be present in oxides, while Fe, Ni, and Co may be present in sulfides, because they typically do not form both oxides and sulfides simultaneously. Pure metals such as iron, low alloy steels, or Ni form only sulfides or oxides and rarely undergo sulfidation/oxidation. X-ray analysis by diffraction is a common method to determine the presences of oxides and sulfides.

The important variables for sulfidation/oxidation of each alloy are the alloy composition, pO_2 and pS_2, metal temperature, and time. The oxidation potential, pO_2, is the partial pressure of O_2 and the sulfidation potential is pS_2. The pO_2 can be calculated by using the partial pressure ratios of H_2O/H_2 or CO_2/CO and metal temperature [4]. The pS_2 can be calculated by using the partial pressure ratios of H_2S/H_2 or CO/COS and metal temperature [4].

The presence of oxidising gases such as H_2O or CO_2 slows the sulfidation rate below that expected if only H_2S–H_2 were present. This is important because gases, which are thought to contain only H_2S–H_2 often also contain some H_2O, because of exposure to liquid water. A gas exposed to water at room temperature (such as a water wash) may contain up to 2% water in the gas. Sulfidation rates predicted using the H_2S–H_2 concentrations might overestimate the rate, if H_2O is present. This lowered corrosion rate is sulfidation/oxidation, which is a transition between rapid corrosion of sulfidation and slow corrosion of oxidation. This is illustrated in Fig. 3 for AISI 304 at 700°C in a gas H_2S–H_2–H_2O, based upon the ASSET analysis methods discussed earlier [3] and reviewed in this article. The right-hand pO_2 corresponds to air (normal oxidation), while the left-hand pO_2 corresponds to O_2-depleted conditions (normal sulfidation). The minimum rate is rate of oxidation in O_2-containing gases and the maximum rate is rate of sulfidation in H_2S–H_2 gases.

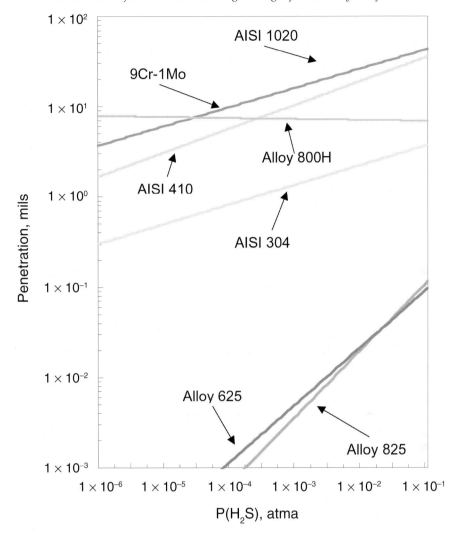

Fig. 2 *Effect of H_2S partial pressure upon sulfidation corrosion after one year in H_2–H_2S gases at 500 psia (34 atma) and 540°C. 10 mils = 0.25 mm.*

2.4. Carburisation

Carburisation forms carbide corrosion products and occurs upon exposure of metals to temperatures above approximately 760°C in gases containing CH_4, CO, hydrocarbons, and solid carbon. The first step in determining the potential for carburisation is to confirm that carbides are present. X-ray analysis by diffraction of the surface scale and analysis of the gas are common methods to confirm carburisation. Variables, which influence the carburisation rate are the temperature, exposure time, partial pressures of H_2, CH_4, H_2S, and alloy composition.

Alloys tend to have more penetration with increasing temperature for all gas conditions. Figure 4 summarises carburisation after one year for some alloys exposed to carbon and 200 ppm (vol.) H_2S.

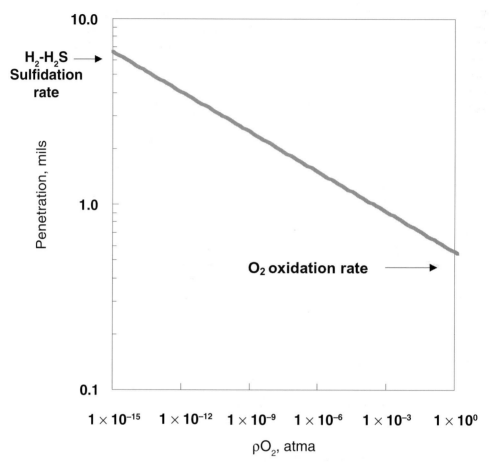

Fig. 3 *Effect of H$_2$O or CO$_2$ as indicated in the p(O$_2$) upon the extent of sulfidation/oxidation of AISI 304 after one year at 700°C in H$_2$–H$_2$S–H$_2$O gases. 10 mils = 0.25 mm.*

The time dependence of carburisation has been reported to be parabolic (metal carburisation proportional to time$^{0.5}$). One thousand hours may be required to establish the time dependence expected for long-term service, suggesting that carburisation rates measured after periods of only hundreds of hours (as is often the case for available data) may not be useful in estimating carburisation corrosion rates for long-term service.

Increasing the concentration of H$_2$S tends to slow the carburisation rate of alloys. Figure 5 shows the effect for several alloys widely used in petrochemical equipment. The H$_2$S slows decomposition of the CH$_4$, which adsorbs onto the metal surface, thus slowing the rate of carburisation. Increasing concentrations will slow carburisation, until the concentrations are high enough to sulfide the alloy. The conditions for the initiation of sulfidation depend upon the alloy and gas compositions. Approximately this means a concentration of 300 ppm (vol.) of H$_2$S for ethylene furnace conditions hydrocarbons and steam at 980–1090°C.

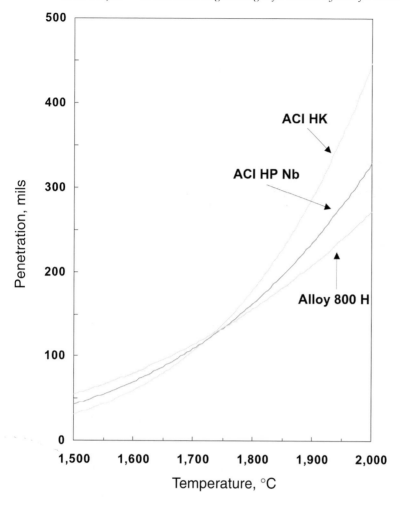

Fig. 4 *Effect of temperature upon carburisation of several alloys exposed to solid carbon and 200 ppm (vol.) of H₂S at 1 atm. 10 mils = 0.25 mm, 1500–2000°F = 816–1093°C.*

3. The ASSET Project Plan

The concepts we have just reviewed on corrosion data interpretation have been expanded and developed into a project to create an information system to provide alloy corrosion data for a wide range of conditions. ASSET (Alloy Selection System for Elevated Temperatures) is the information system. An overview of the three-year project plan, which started in early 2000, is discussed here. The project is producing corrosion prediction software that includes a database and thermochemical calculation programs that use laboratory corrosion data from well-controlled conditions to predict corrosion for alloys over a range of high-temperature corrosive environments. The project is improving corrosion predictions for alloys in gases at temperatures of 250–1150°C. ASSET operates under Windows on a PC and manages/correlates corrosion data for alloys corroding by several mechanisms in high temperature gases.

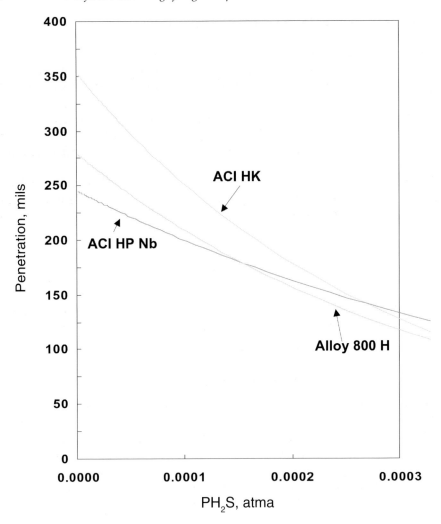

Fig. 5 *Effect of pH$_2$S upon carburisation of several alloys exposed to solid carbon and 200 ppm (vol.) of H$_2$S at 982°C. 10 mils = 0.25 mm.*

The applications of ASSET are in the following:

- equipment failure analysis to reduce maintenance costs and improve process safety;

- alloy evaluations to select cost effective alloys for equipment;

- equipment design/operation guidelines to optimise process economics;

- alloy design studies to optimise alloy properties;

- corrosion research to archive and exploit data; and

- process evaluations to look for alloy corrosion concerns.

The project involves a diverse group of organisations: Caterpillar, USA; Creusot-Loire Industrie, France; Ecole Polytechnique de Montreal, Canada; Foster Wheeler Development Corporation, USA; Haynes International, USA; Humberside Solutions Ltd, Canada; KEMA, Netherlands; Kvaerner Pulping Oy, Finland; Materials Technology Institute, USA; Oak Ridge National Laboratory, Oak Ridge, USA; Royal Military College of Canada, Canada; Shell Global Solutions (US), USA; Special Metals Corporation, USA; Texaco, USA; and US Department of Energy — Office of Industrial Technologies, USA.

The project goals for each of the three years involve four main tasks each year. The tasks are software development, thermochemical modelling, corrosion data generation, and information exchange. Details of these tasks are described below:

3.1. Software Development

Humberside Solutions Ltd. provides support to modernise ASSET, incorporate additional corrosion data, increase the number of corrosion mechanisms covered by ASSET, distribute software, and instruct participants in the operation of the ASSET information system.

3.2. Thermochemical Modelling

The Center for Research in Computational Thermochemistry of Ecole Polytechnic de Montreal, Quebec, Canada evaluates available thermochemical data to produce consistent data sets to be included into ASSET. This will improve the accuracy of predictions of corrosion mechanism, by determining the most stable corrosion products formed by alloys in contact with the gas. The data include the system of Fe–Cr–Ni–Co–C–S–O–, over the temperature range of 250–1200°C, with the data covering conditions applicable to many industrial processes.

3.3. Corrosion Testing

Corrosion rates are being determined under well-defined conditions, according to guidelines, which have been rigorously established, and the data stored in defined formats. Corrosion is evaluated via metallographic measurements of the maximum corrosion depth. The test requirements are shown in Appendix 2. Commercial alloys are used. The measured data are recorded in the formats illustrated in Appendix 3.

Corrosion data are being measured for the following exposure conditions:

- Temperature range of 250–1150°C;

- Exposure times of 500–20 000 h;

- Oxygen partial pressure (pO_2) range of 0.01–1.0 atma for oxidising conditions;

- Hydrogen sulfide partial pressures (pH_2S) of 0.001–0.3 atma, and hydrogen partial pressures (pH_2) of 0.1–10 for sulfidising conditions;

- pO_2 of 1×10^{-30}–1×10^{-20} atma and pS_2 of 1×10^{-15}–1×10^{-5} atma for sulfidising/oxidising conditions; and

- pH_2S of 0–3×10^{-4} atma, carbon activity of 0.5–1.0, and pH_2O of 0–0.5 atma for carburising conditions.

Each year has themes for the corrosion mechanism being evaluated and the type of compounds being assessed by the thermochemical modelling, as listed below:

Schematic Corrosion Testing Plan for the Project

	YEAR ONE	YEAR TWO	YEAR THREE
Mixed Gases	Sulfidation	Sulfidation/Oxidation	Carburisation
Oxidation	Low O_2 (0.01 atma)		High O_2 (1 atma)
Oxidation	Air	Air	Air

The oxidation testing will generate data in three environments: air, N_2 + 1% O_2, and O_2, at the temperatures of 300–1100°C. Examples of the alloys being tested are: Copper, 153 MA, 253 MA, AISI 347, Alloy DS, Alloy 803, Alloy 625, Alloy 600, Alloy X, HR-160, Nickel, 304, 310, 316L, 600, 800H, and MA 956.

The sulfidation testing will generate data in environments containing H_2 and H_2S. The H_2S concentration will range from 100 to 10 000 ppm (vol.) at one atma and the pH_2 will range from 0.1 to 10 atma, at temperatures of 300–900°C. Examples of the alloys are being tested are:

153 MA, 253 MA, 304, 310, 347, 446, 600, 800 H, HR-160, HR-120, Alloy DS2, 12% Cr (AISI 410), 316L, and MA 956 (the superscript refer to the year of testing).

The sulfidation/oxidation test alloys are not yet selected. The tests will be in mixtures of H_2–H_2S–H_2O or CO–CO_2–COS. The pS_2 will range from 1×10^{-15} to 1×10^{-5} atma and the pO_2 will range from 1×10^{-30} to 1×10^{-20} atma at temperatures of 300–900°C.

The carburisation test alloys are not yet selected. The tests will be in environments containing mixtures of CH_4–H_2–H_2S. The carbon activities will range from 0.5 to 1.0 and the pH_2S will range up to 3×10^{-4} atma. Exposure times will be up to 6000 h.

Year One – Oxidation in Low pO_2 Conditions and Sulfidation Conditions

- Develop thermodynamic solution phase models for Fe–Ni–Cr–Co–O systems for oxides over the temperature range of 250–1150°C.

- Complete sulfidation testing of ten alloys in gases containing 0.1–10 atma of H_2, 100–10 000 ppm vol. H_2S at temperatures of 250–900°C.

- Oxidation testing of 10 alloys in low pO_2 conditions (1%O_2 — balance N_2) and air at temperatures of 500–1100°C with times up to 20 000 h will be completed or started.

Year Two – Oxidation in Air Conditions and Sulfidation/Oxidation Conditions

- Develop thermodynamic solution phase models for Fe–Ni–Cr–Co–C systems for metals and oxides over the temperature range of 250–1150°C.

- Complete sulfidising/oxidising tests at temperatures of 300–700°C, times up to 6000 h, and gases containing H_2–H_2O–H_2S.

- Continue oxidation testing of 10 alloys in air conditions at 300–1100°C at exposure times up to 20 000 h.

Year Three – Oxidation in High pO_2 conditions and Carburisation Conditions

- Develop thermodynamic solution phase models for solids and liquids in the Fe–Ni–Cr–Co–S systems for metals and sulfides over the temperature range of 250– 1150°C.

- Complete sulfidation/oxidation testing of 10 alloys in representative conditions.

- Complete carburisation testing of at least 10 alloys in representative conditions.

- Complete oxidation testing of 10 alloys in O_2 at temperatures of 800–1100°C and times up to 20 000 h.

5. Compilations of Corrosion Rate Predictions

Comparisons of methods to present corrosion data for several different classes of corrosion, including the ASSET concepts being developed are now discussed.

5.1. Oxidation Rates

The traditional approach for predicting oxidation rates in high-temperature gases has been to use the type of information summarised in the central portion of Table 1 (including data from [7]) and in Fig. 6, based upon Ref. [8]. These sources of information are useful but they do not allow a quantitative prediction of sound metal loss, which could be used, for lifetime prediction. They provide qualitative indication of relative oxidation behaviours of the alloys indicated and perhaps temperatures which are associated with 'excessive scaling', which is not well defined. This approach does not allow a quantitative prediction of equipment lifetimes caused by different allowable penetration rates. The recommended approach for assessing oxidation metal losses (penetration) for alloys exposed to ranges of temperature is illustrated in Table 1 and Fig. 1 for a standard condition of oxidation in air. Data were analysed by ASSET and reported previously [1,2,9–14].

The right-hand column of Table 1 shows temperatures corresponding to 0.25 mm of sound penetration after one year, while Fig. 1 shows the amount of penetration occurring after one year for a range of temperatures. The temperatures in Table 1 corresponding to 0.25 mm of metal loss after one year for alloys such as carbon steel and AISI 304 (which experience most of the total metal penetration by surface scale formation) agree to ±30°C with the traditional maximum temperatures to avoid excessive scaling. The temperatures, which yield 0.25 mm of penetration after one year of exposure for the more highly alloyed metals, are lower than the temperatures to avoid 'excessive scaling' by approximately 80–110°C. The temperature difference is due to the subsurface oxidation, which affects more metal thickness than would be expected, from the mass change due to surface scale formation.

5.2. Sulfidation Rates

The first step in assessing the rate of sulfidation is to evaluate the potential for sulfide corrosion products to form. Confirmation of sulfides on existing

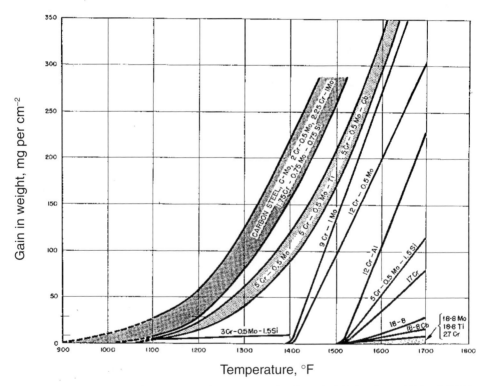

Fig. 6 *Temperature effect on gain of various steels after 1000 h in air at 1100– 1700 °F (593–927 °C).*

Table 1. *Maximum temperatures to avoid excessive scaling and temperatures yielding 0.25 mm of penetration (recommended approach) after one year in air*

UNS alloy number	Common alloy name	Maximum allowable temperature, °C	Maximum allowable temperature, °C	Temperature, °C for 10 mils/year
		Cycling	Isothermal	Isothermal
–	Carbon Steel*		565	605
K11597	1Cr–0.5Mo steel*		595	Unknown
K21590	2.25Cr–1Mo steel*		620	Unknown
K41545	5Cr–0.5Mo steel*		650	Unknown
C11000	Copper			680
N02270	Nickel			780
S50400	9Cr–1Mo steel		705	800
S41000	410		705	830
S30400	304 **	870	930	895
N06617	617			940
	803			955
N06625	625			960
N08810	800 H	1895	1150	965
–	601GC			980
–	DS			980
N06230	230			980
S31000	310	1035	1150	980
S33000	330			1000
S44600	446	1175	1095	1010
R30556	556			1010
–	120			1010
S30815	253MA			1080
–	602CA			1120
S67956	MA 956			>1150
N07214	214			>1150

* Alloys with 0–5% Cr oxidise similarly, modest increases in maximum allowable temperatures are possible with increasing Cr concentration.
** Some alloys with oxidation behaviour similar to 304 are 321, 347, and 316.

equipment or a thermochemical evaluation of the corrosive gas to produce sulfides may be done by analyses of corrosion products, use of the charts produced by Gutzeit [15] or the ASSET program. Once sulfidation is expected, sulfidation rates can be predicted by using either the traditional sulfidation curves for corrosion in $H_2S–H_2$ gases shown in Figs 7 and 8 or the information in this presentation, including the effects of temperature, gas composition, and alloy composition. Sulfidation rate data for some commercial alloys are summarised in Figs 2, 7, and 8. Figures 7 and 8 show corrosion of carbon/low alloy steels (< 5% Cr), 5% Cr, 9% Cr, 12% Cr, and 18%Cr–8% Ni steels (type 304, 321, 347, and 316 steels). They are based upon research information and operating experience gathered during the 1950s and 1960s [15–20] and have been the best available method to predict sulfidation rates, based upon temperature and H_2S concentration, for several alloys.

One important concept assumed in Fig. 7 and 8 is that corrosion cannot occur below the line shown to fall from the lower left to the upper right. This line represents the limiting conditions for formation of FeS, which is the primary corrosion product of carbon/low alloy steels (<5% Cr) during sulfidation in contact with $H_2–H_2S$ gases. H_2S concentrations above the line favour formation of FeS, which is a requirement for sulfidation of iron and low alloy steels. It should be noted that it is incorrect to indicate that alloys with significant concentrations of Cr cannot corrode at conditions representing lower H_2S concentrations and higher temperatures indicated by this line. Alloys with Cr can form CrS, which can form at lower H_2S concentrations and higher temperatures than needed to form FeS on low alloy steels. Figure 9 shows the limits of H_2S concentration and temperature corresponding to corrosion products of Cr. Alloys with greater than 5% Cr can corrode in conditions where low alloy steels cannot corrode. The corrosion rates may be low (such as 0.025 – 0.25 mm/year) but significant for these conditions where steels have been traditionally thought to be immune to sulfidation. Extrapolations of the iso-corrosion contours of Figs 7 and 8 for the alloys with Cr concentrations greater than 5% to combinations of high-temperatures and low H_2S concentrations below the Fe/FeS line have been found to give reasonable predictions of the rates expected for low H_2S concentrations.

A summary of maximum allowable temperatures which will limit the extent of metal loss by sulfidation to less than 0.25 mm/year is shown in Table 2 for several gas compositions of $H_2S–H_2$ at a pressure of 34 atma. The maximum allowable temperatures for alloys exposed to different gas pressures and compositions can be evaluated with this information.

5.3. Sulfidation/Oxidation Rates

Examples of predictions of metal losses by sulfidation/oxidation (in comparison to sulfidation) for some important commercial alloys are shown in Table 3, based upon analyses by ASSET and previously reported data [5,6,14,21–24], for an example temperature of 704°C. Metal losses by sulfidation/oxidation for low steam concentrations and sulfidation are similar for a given alloy. Increasing the H_2O (or CO_2) concentration, reducing the H_2S concentration, and lowering the temperature reduce metal losses. Decreasing values of the pO_2 and increasing values

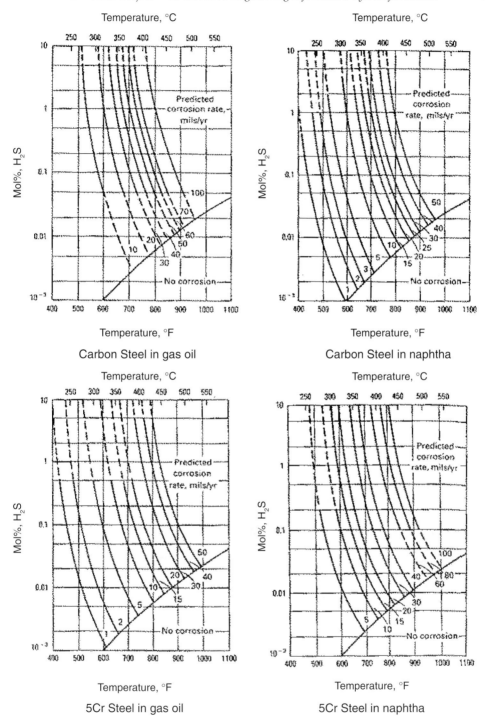

Fig. 7 *H$_2$S Concentration and temperature effects on steel sulfidation rates at 400–1100°F (204–593°C).*

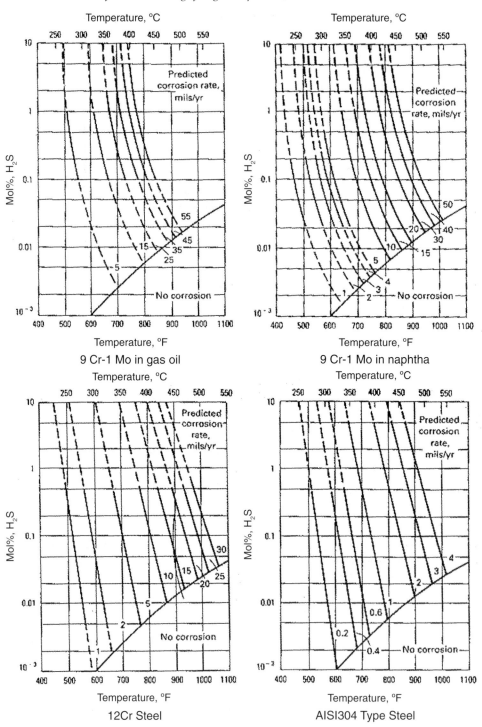

Fig. 8 *H₂S Concentration and temperature effects on 9Cr, 12Cr, and AISI 304 steel sulfidation rates at 400–1100 °F (204–593 °C).*

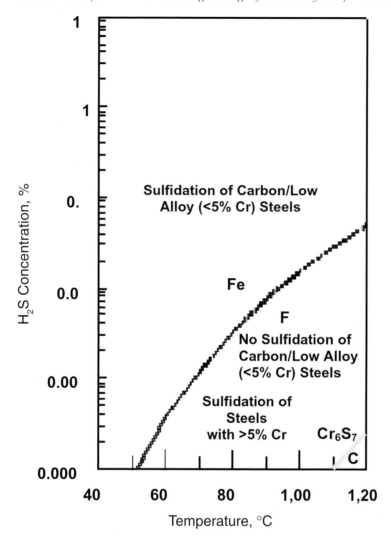

Fig. 9 *Conditions for possible sulfidation, based upon H_2S concentrations in H_2–H_2S gases and temperatures above the FeS/Fe line for carbon/low alloy (<5% Cr) steels and above the Cr_6S_7/Cr line for alloys with >5% Cr. 400–1200°F = 204–648°C.*

of the pS_2 will increase the rate of sulfidation/oxidation because they tend to increase the stability of sulfide corrosion products (corresponding to more rapid corrosion) and reduce the stability of the oxide corrosion products (corresponding to slower corrosion).

The time dependence of the instantaneous parabolic rate constant for corrosion by sulfidation/oxidation is shown in Fig. 10 for several alloys exposed to CO–CO_2–COS gases at 600°C. This figure illustrates that sulfidation/oxidation is similar to other mechanisms in that times greater than several hundred hours are needed to establish the parabolic rate constant which can be used to predict long-time corrosion losses.

Table 2. *Sulfidation corrosion temperatures corresponding to a maximum metal loss of 0.25 mm after one year in H_2S–H_2 gases at 500 psia gas pressure*

UNS alloy No.	Common alloy name	Maximum allowable temp., °C	Maximum allowable temp., °C	Maximum allowable temp., °C	Maximum allowable temp., °C	Maximum allowable temp., °C
		H_2S	H_2S	H_2S	H_2S	H_2S
		10 ppm (vol.) (0.001%)	100 ppm (vol.) (0.01%)	1000 ppm (vol.) (0.1%)	10 000 ppm (vol.) (1%)	100 000 ppm vol. (10%)
N02270	Nickel	395	360	340	310	295
G10200	Carbon Steel	425	415	405	400	390
S50400	9Cr–1Mo	505	445	395	350	310
S41000	410	570	500	440	390	345
N08810	800 H	580	580	580	580	580
S43000	430	760	680	615	555	500
S30400	304	880	790	670	625	565
N08825	825	>930	630*	630*	630*	630*
N06625	625	>760	630*	1170*	630*	630*
N07718	718	>760	630*	1170*	630*	630*

* Temperatures for nickel-base alloys are limited to below 631°C, which is the minimum value needed to form liquid Ni–NiS, which causes very rapid corrosion in gases with H_2S levels greater than 0.01% vol.

5.4. Carburisation Rates

Quantitative measurements of sound metal loss (metal penetration) at well defined values of the important variables of exposure time, carbon activity, temperature, and H_2S concentration, and alloy characteristics have been used to produce correlations of carburisation rates for standard conditions. The data used for the results in the analyses shown in this paper were reported previously [24–31].

A suggested, standard method to present carburisation data is shown in Table 4. The maximum allowable temperatures corresponding to 5.1 mm of carburisation for several alloys and H_2S concentrations are shown. The value of 5.1 mm of penetration is selected because it is a typical wall thickness of tubes used in steam methane reformers and ethylene furnaces. The common concern

Table 3. *Metal losses by sulfidation/oxidation at 704°C and several different gas compositions after one year in H₂S–H₂O–H₂ gases at 500 psia gas pressure. NB: This table continues overleaf.*

UNS alloy No.	Common alloy name	Metal loss mpy 5% H₂S - Bal. H₂	Metal loss mpy 5% H₂S - 1% H₂O - Bal. H₂	Metal loss mpy 0.5% H₂S - 1% H₂O - Bal. H₂	Metal loss mpy 0.5% H₂S - 10% H₂O - Bal. H₂
		Sulfidation	Sulfidation/ Oxidation	Sulfidation/ Oxidation	Sulfidation/ Oxidation
S30415	153MA	Unknown	–	378	23
–	803	Unknown	–	487	23
N06601	601 GC	Unknown	284	135	45
N06002	X	Unknown	160	147	122
N06617	617	Unknown	123	82	46
–	120	Unknown	94	57	28
N06625	625	20	67	46	26
N08825	825	30	58	32	6
–	MA 6000	Unknown	50	21	6
S30400	AISI 304	24	50	35	21
N07718	718	Unknown	49	31	16
N06600	600	Unknown	46	61	92
–	3220	15	39	31	23
J94224	ACI HK	Unknown	37	17	5
S31000	AISI 310	Unknown	36	21	9
N08810	800 H	23	35	23	9
–	801	Unknown	28	35	48
S67956	MA 956	Unknown	24	16	5
S31803	2205	Unknown	22	23	24

Table 3 (continued)

UNS alloy No.	Common alloy name	Metal loss mpy 5% H_2S - Bal. H_2	Metal loss mpy 5% H_2S - 1% H_2O - Bal. H_2	Metal loss mpy 0.5% H_2S - 1% H_2O - Bal. H_2	Metal loss mpy 0.5% H_2S - 10% H_2O - Bal. H_2
		Sulfidation	Sulfidation/ Oxidation	Sulfidation/ Oxidation	Sulfidation/ Oxidation
R30188	188	Unknown	21	14	8
–	6	Unknown	21	7	1
–	DS	Unknown	20	12	4
–	671	Unknown	19	11	4
R30556	556	Unknown	16	18	20
–	160	Unknown	12	12	10
S30815	253MA	Unknown	10	19	49
R30155	Multimet	Unknown	10	5	2
S33000	AISI 330	Unknown	7	6	4
S44600	AISI 446	Unknown	6	4	2

The metal losses are typically accurate to ±50% of the reported value; 39.4 mpy = 1 mm/year.

with carburisation is how much time is needed before the wall is carburised, making the tube wall not repairable by welding. Increasing the H_2S concentration increases the maximum allowable metal temperature, with respect to carburisation.

5.5. Generalised Procedure for Corrosion Predictions

ASSET (Alloy Selection System for Elevated Temperatures) [3,32,33] is publicly available. The program stores the corrosion measurements, exposure conditions, and corrosion mechanisms. It predicts alloy corrosion by accessing the stored data for that alloy and determining the parameters of the rate equation. A different equation is used for each corrosion mechanism.

The software uses the alloy composition and the corrosive environment information to calculate the stable corrosion products and the equilibrium gas composition, for a given combination of alloy and exposure conditions. The

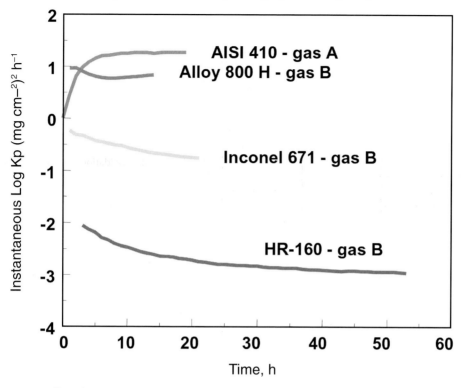

Fig. 10 *Effect of time upon the instantaneous parabolic rate constants for sulfidation/oxidation of several alloys at 600°C. Gas A — 2% CO, 2% COS, and balance CO_2; gas B — 48% CO, 48% CO_2, and 4% COS.*

computations use the ChemSage program from F*A*C*T [34]. Some of the potential solution or mixed corrosion product phases considered are:

$$MO, M_3O_4, M_2O_3, MS, M_2S_4, M_3C, M_{23}C_6, \text{ and } M_7C_3$$

Table 4. *Maximum allowable temperatures to limit carburisation to 5.1 mm of penetration after one year in contact with solid carbon, gases containing H_2S, and in the absence of oxidising gases such as H_2O and CO_2*

UNS alloy number	Common alloy name	Maximum allowable temperature, °C 0 ppm (vol.) H_2S	Maximum allowable temperature, °C 100 ppm (vol.) H_2S	Maximum allowable temperature, °C 300 ppm (vol.) H_2S
J94224	ACI HK	920	960	1040
J95705	ACI HP Nb	955	980	1045
N08810	800 H	925	980	1095

M represents a combination of Fe, Ni, and Cr in the compound. The thermodynamic solution behaviours of the solid austentic and ferritic alloys are also considered. Thermochemical characteristics such as the pO_2, pS_2, and carbon activity of the environment, which help determine corrosion product stabilities are also provided by the calculation. The software assists identification of the likely corrosion mechanism by knowing the stable corrosion products at the corrosion product/corrosive gas interface, the alloy in question, and the partial pressures of pO_2 and pS_2. Different alloys in the same exposure conditions may exhibit different stable corrosion products and different corrosion mechanisms.

In the absence of experimental data for the specific conditions of interest, predictions made by using the approach discussed here may be the best available for the corrosion mechanisms that are incorporated into the system, in comparison with those made using the traditional methods of literature review and data analysis. Corrosion predicted without familiarity with the specific environment should be experimentally confirmed if high confidence is required.

Lifetime predictions as limited by corrosion depend strongly upon the corrosion predictions. Examples of the accuracy using the ASSET system are shown in Figs 11 and 12. They show how large amounts of corrosion data can be well correlated. The correlations are quite good for three decades of variation in corrosion penetration for several alloys and corrosion mechanisms, considering the uncertainty associated with this type of data.

6. Conclusions

This paper has reviewed metal loss information for alloys in several types of corrosive environments. Several traditional standard compilations for corrosion of metals in hot gases were discussed and their limitations were mentioned in predicting losses of sound metal for wide ranges of conditions. The ASSET project and information system have been designed to go beyond traditional means with the following objectives:

- Develop an objective and consistent approach to archive and use data for alloys corroding in diverse conditions and by different mechanisms;

- Create a database to manage data on corrosion of metals and alloys corroding in complex, corrosive, and high temperature gases;

- Form a state-of-the-art capability in thermochemical calculations to determine corrosion mechanisms of alloys in industrial processes;

- Create an information system with the following uses: equipment failure analysis, alloy evaluations to select cost effective alloys for equipment, guidelines to optimise equipment design/operation process economics, alloy design studies, and corrosion research.

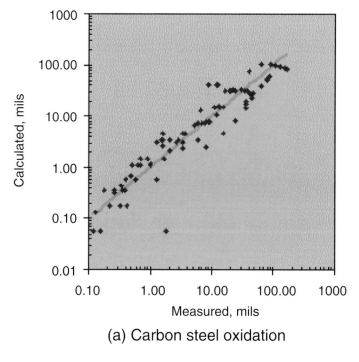

(a) Carbon steel oxidation

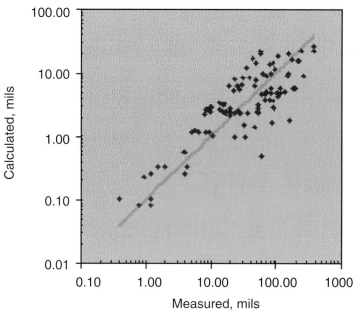

(b) AISI 310 sulfidation/oxidation

Fig. 11 (a–d) *Verification of corrosion prediction capability of ASSET (10 ml = 0.25 mm).*
(a) Carbon steel oxidation; (b) AISI 310 sulfidation/oxidation. See overleaf for (c) and (d).

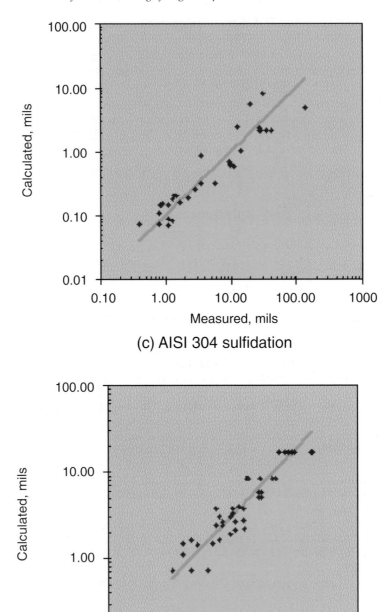

(c) AISI 304 sulfidation

(d) HK 40 carburisation

Fig. 11 (continued) *Verification of corrosion prediction capability of ASSET (10 ml =
0.25 mm). (c) AISI 304 sulfidation; (d) HK 40 carburisation.*

7. Acknowledgements

Participation and support by the following organisations are recognised and appreciated: US Department of Energy — Office of Industrial Technologies via cooperative agreement DE-FC02-00CH11020, Shell Global Solutions (US), Humberside Solutions Ltd., Centre for Research in Computational Thermochemistry in Université de Montréal, Royal Military College of Canada, Oak Ridge National Laboratory, Materials Technology Institute, Foster Wheeler Development Corporation, KEMA, Caterpillar, Special Metals Corporation, Texaco, Haynes International, Creusot-Loire Industrie, and Kvaerner Pulping Oy.

References

1. R. C. John, "Oxidation Studies of Commercial Alloys at 871–1093°C (1600–2000°F), in *Proc. 2nd Int. Conf. on Heat Resistant Materials*, Gatlinburg, TN, USA. 1995.

2. R. C. John, Engineering Assessments of Oxidation of Commercial Alloys", *Corrosion '96*, Paper No. 171, NACE International, Houston, Tx, 1996.

3. R. C. John, A. L. Young and W. T. Thompson, 'A Computer Program for Engineering Assessment of Alloy Corrosion in Complex, High Temperature Gases', *Corrosion '97*, Paper No. 142, NACE International, Houston, Texas, 1997.

4. R. C. John, W. C. Fort III and R. A. Tait, Prediction of Alloy Corrosion in the Shell Coal Gasification Process, *Mater. High Temp.*, 1993, **11**, 124.

5. E. W. Haycock, *High Temperature Metallic Corrosion of Sulfur and Its Compounds*, (ed. Z. A. Foroulis). The Electrochemical Society, Princeton, New Jersey, 1970, p. 110.

6. E. W. Haycock and W. H. Sharp, "Corrosion of Ferrous Alloys by H_2S at High Temperatures", Session from the 24th Meeting of the American Petroleum Institute, New York 1959.

7. R. S. Treseder (ed.), R. Baboian and C. G. Munger (co-eds.), *NACE Corrosion Engineer's Reference Book*. NACE International, Houston, TX, 1991, p. 128.

8. E. McGannon (ed.), *The Making, Shaping and Treating of Steel, United States Steel Corporation*. Herbick & Held, 1971.

9. J. Paidassi, Sur La Cinetique De L'Oxydation Du Fer Dans L'Air Dans L'Intervalle 700–1250 C, *Acta Metall.*, 1958, **6**, 184–194.

10. J. Paidassi, Thesis, University of Paris, France, 1954. (See *Oxydation des Metaux*, by J. Benard, pub. Gauthier-Villars-Editeur, Paris, France, Vol. 2 – Monographies (1964).)

11. J. Benard, *Oxydation des Metaux, Vol. 2 – Monographies*. Publ. Gauthier-Villars Editeur, Paris, France, 1964, p. 179.

12. J. Paidassi and R. Abarca, see p. 93 of *Oxydation des Metaux, Vol. 2 — Monographies*, by J. Benard, pub. Gauthier-Villars Editeur, Paris, France, 1964.

13. M. F. Rothman, Oxidation Resistance of Gas Turbine Combustion Materials, presented at the Gas Turbine Conference and Exhibit, Houston, TX, March, 1985. 85-GT-10.

14. G. Y. Lai, *High-Temperature Corrosion of Engineering Alloys*. ASM International, Materials Park, OH, 1990 and literature from Haynes International Marketing, 'Technical Information, Oxidation Resistance of Haynes High Temperature Alloys'.

15. J. Guzeit, High temperature sulfidic corrosion of steels, in *Process Industries Corrosion – The Theory and Practice*. NACE International, Houston, TX, 1986.

16. E. B. Backensto, R. E. Drew and C. C. Stapleford, High temperature hydrogen sulfide corrosion, *Corrosion*, 1956, **12**, (i), 22.

17. G. Sorell and W. B. Hoyt, Collection and correlation of high temperature hydrogen sulfide corrosion data, *Corrosion*, 1956, **12**, (i), 213t.

18. G. Y. Lai, Materials behavior in high temperature, sulfidizing environments, *Corrosion '84,* Paper No. 73, NACE, Houson, Tx, 1984.

19. Stoklosa and J. Stringer, Studies of the kinetics of Ni sulfidation in $H_2S–H_2$ mixtures in the temperature range 450–600°C, *Oxid. Met.,* 1977, **11**, (4), 263.

20. J. D. McCoy and F. B. Hamel, Effect of hydrodesulfurizing process variables on corrosion rates, *Mater. Perform.,* 1971, **10**, 17.

21. I. G. Wright, R. O. Dodds, R. B. Palmer, H. A. Link, W. E. Merz and G. H. Beatty, Correlation of the high-temperature corrosion behavior of structural alloys in coal conversion environments with the components of the alloys and of the corrosive environments', Battelle Columbus Laboratories, Columbus, OH, U.S. Department of Energy, Contract No. AT02-76CH92092, February, 1980.

22. H. Howes, 'High Temperature Corrosion in Coal Gasification Systems', final report 1 October 1972 – 31 December 1985 by IIT Research Institute, Chicago, IL, #IITRI-M08251-97, for Gas Research Institute (1987).

23. R. C. John, Alloy corrosion in reducing plus sulfidizing gases at 600–950°C, in *High Temperature Corrosion in Energy Systems,* AIME, Warrendale, PA, 1985.

24. P. Garguet-Moulins, P. D. Frampton, S. Canetoli and J. F. Norton, Compilation of high temperature corrosion data derived during studies in mixed gaseous atmospheres, Joint Research Centre Petten, Final Report – May 1991 (Ref. CEC/SIPM/CON/91/1).

25. E. Bullock, P. D. Frampton and J. F. Norton, Structural aspects of gaseous carburization of austenitic steels, *Microstruct. Sci.,* 1981, **9**, 215–224.

26. H. J. Grabke and A. Schnaas, Review on high-temperature gaseous corrosion and mechanical performance in carburizing and oxidizing environments, in *Alloy 800.* Publ. North Holland, pp. 195–211, 1978.

27. R. C. Hurst and J. F. Norton, The effect of high temperature carburization upon the ambient temperature ductility of alloy 800H, *High Temp. Tech.,* 1983, **1**, (6), 319–325.

28. H. J. Grabke, Materials degradation by carburization and nitriding, in *8th Int. Congr. on Metallic Corrosion, Vol. III,* West Germany 6–11, September 1981, DECHEMA, pp. 2143–2156.

29. H. J. Grabke, R. Moller and A. Schnaas, Influence of sulphur on the carburization of a CrNiFe-alloy at high temperatures, *Werkst. und Korros.,* 1979, **30**, 794–799.

30. G. M. Smith, D. J. Young and D. L. Trimm, Carburization kinetics of heat-resistant steels, *Oxid. Met.,* 1982, **18**, (5/6), 229–243.

31. R. C. John, Alloy carburization and testing in simulated process gases at 1800–1950°F, in *Corrosion' 95,* Paper No. 460, Nace International, Houston, Tx, 1996.

32. R. C. John, Compilation and use of corrosion data for alloys in various high-temperature gases, *Corrosion '99,* Paper No. 73, NACE International, Houston, Tx, 1999.

33. R. C. John, Exploitation of data of alloy corrosion in high-temperature gases, *Mater. Sci. Forum (Trans. Tech. Publ.),* 2000, in press.

34. W. T. Thompson, C. W. Bale and A. D. Pelton, "Facility for the Analysis of Chemical Thermodynamics (FACT)", McGill University Montreal, Royal Military College of Canada in Kingston, Ecole Polytechnique, Montreal, 1985.

Appendix 1

Compositions of Some Alloys in ASSET Databases

Alloy	UNS	Fe	Cr	Ni	Co	Mo	Al	Si	Ti	W	Mn	Cu	Re	C	Nb
153 MA	S30415	69.66	18.30	9.50	0.00	0.42	0.00	1.23	0.00	0.00	0.56	0.23	0.05	0.05	0.00
253 MA	S30815	65.60	20.90	11.00	0.00	0.00	0.00	1.77	0.00	0.00	0.64	0.00	0.00	0.09	0.00
9 Cr 1Mo	S50400	88.61	8.90	0.00	0.00	1.03	0.00	0.86	0.00	0.00	0.47	0.00	0.00	0.13	0.00
ACI HK40	J94224	50.65	25.10	21.20	0.00	0.00	0.44	1.35	0.37	0.00	0.60	0.00	0.00	0.29	0.00
ACI HP–Nb	J95705	31.90	24.65	38.75	0.00	1.46	0.00	1.65	0.43	0.00	1.01	0.00	0.00	0.15	1.00
AISI 1020	G10200	99.42	0.00	0.00	0.00	0.00	0.00	0.04	0.00	0.00	0.38	0.00	0.00	0.16	0.00
AISI 304	S30400	71.07	18.28	8.13	0.14	0.17	0.00	0.49	0.00	0.00	1.48	0.19	0.00	0.05	0.00
AISI 309	S30900	61.20	23.50	13.10	0.00	0.00	0.00	0.40	0.00	0.00	1.80	0.00	0.00	0.00	0.00
AISI 310	S31000	52.41	24.87	19.72	0.05	0.16	0.00	0.68	0.00	0.00	1.94	0.11	0.00	0.06	0.00
AISI 314	S31400	52.41	24.07	20.30	0.00	0.00	0.00	2.04	0.00	0.00	1.12	0.00	0.00	0.06	0.00
AISI 316	S31600	68.75	17.00	12.00	0.00	2.25	0.00	0.00	0.00	0.00	0.00	0.00	0.00	0.00	0.00
AISI 321	S32100	69.94	17.22	9.85	0.21	0.14	0.00	0.46	0.43	0.00	1.61	0.10	0.00	0.04	0.00
AISI 330	N08330	44.00	35.00	18.20	0.00	0.00	0.00	1.30	0.00	0.00	1.50	0.00	0.00	0.00	0.00
AISI 347	S34700	68.14	17.75	10.75	0.00	0.00	0.00	0.55	0.00	0.00	1.80	0.00	0.00	0.05	0.96
AISI 410	S41000	86.50	12.30	0.50	0.00	0.10	0.00	0.60	0.00	0.00	0.00	0.00	0.00	0.00	0.00
AISI 446	S44600	74.12	24.36	0.36	0.02	0.20	0.00	0.33	0.00	0.00	0.45	0.10	0.00	0.06	0.00
Alloy 188	R30188	1.32	21.98	22.82	38.00	0.00	0.00	0.37	0.00	14.55	0.82	0.00	0.04	0.10	0.00
Alloy 214	N07214	2.49	16.04	76.09	0.14	0.10	4.71	0.10	0.00	0.10	0.20	0.00	0.00	0.03	0.00
Alloy 230	N06230	1.30	21.90	59.70	0.28	1.20	0.38	0.42	0.02	14.20	0.49	0.01	0.00	0.10	0.00
Alloy 556	R30556	32.50	21.27	21.31	18.09	2.88	0.17	0.33	0.00	2.38	0.96	0.00	0.00	0.11	0.00
Alloy 600	N06600	7.66	15.40	75.81	0.00	0.00	0.32	0.16	0.00	0.00	0.29	0.32	0.00	0.04	0.00
Alloy 601	N06601	14.00	23.00	60.45	0.00	0.00	1.40	0.30	0.00	0.00	0.40	0.40	0.00	0.05	0.00
Alloy 601GC		13.53	23.48	60.00	0.06	0.16	1.26	0.50	0.27	0.00	0.31	0.38	0.00	0.05	0.00
Alloy 617	N06617	0.76	22.63	53.20	12.33	9.38	1.15	0.15	0.27	0.00	0.02	0.05	0.00	0.06	0.00
Alloy 625	N06625	2.66	21.74	62.79	0.00	8.46	0.10	0.41	0.19	0.00	0.10	0.00	0.00	0.03	3.52
Alloy 671		0.32	46.02	53.13	0.00	0.00	0.00	0.00	0.41	0.00	0.07	0.00	0.00	0.05	0.00
Alloy 718	N07718	18.05	18.51	53.40	0.14	3.03	0.49	0.15	1.05	0.00	0.11	0.14	0.00	0.04	4.89

(continued overleaf)

Appendix 1. *Compositions of some alloys in ASSET databases (continued)*

Alloy	UNS	Fe	Cr	Ni	Co	Mo	Al	Si	Ti	W	Mn	Cu	Re	C	Nb
Alloy 800 H	N08810	44.22	21.22	31.71	0.00	0.00	0.33	0.60	0.41	0.00	0.92	0.51	0.00	0.08	0.00
Alloy 803		35.94	26.19	35.04	0.21	0.01	0.58	0.63	0.33	0.00	0.98	0.00	0.00	0.09	0.00
Alloy 825	N08825	30.41	23.34	40.22	0.00	2.74	0.04	0.17	0.89	0.00	0.41	1.76	0.00	0.02	0.00
Alloy DS		44.31	16.60	34.90	0.00	0.29	0.00	2.52	0.00	0.00	1.14	0.17	0.00	0.07	0.00
Alloy X	N06002	19.53	21.76	45.97	1.99	8.73	0.00	0.39	0.00	0.55	0.68	0.00	0.00	0.08	0.32
Alloy X–754		7.00	15.50	74.30	0.00	0.00	0.70	0.00	2.50	0.00	0.00	0.00	0.00	0.00	0.00
Copper	C11000	0.00	0.00	0.00	0.00	0.00	0.00	0.00	0.00	0.00	0.00	100.00	0.00	0.00	0.00
HR–120		34.53	25.12	37.44	0.11	0.37	0.11	0.57	0.02	0.10	0.73	0.18	0.00	0.06	0.66
HR–160		8.00	28.00	34.30	27.00	0.00	0.00	2.70	0.00	0.00	0.00	0.00	0.00	0.00	0.00
Incoloy MA956	S67956	75.22	19.40	0.28	0.05	0.00	4.50	0.11	0.33	0.00	0.09	0.00	0.00	0.02	0.00
Multimet	R30155	31.31	20.97	19.59	19.24	3.00	0.00	0.65	0.00	2.49	1.37	0.10	0.00	0.12	1.16
Nickel	N02270	0.00	0.00	99.99	0.00	0.00	0.00	0.00	0.00	0.00	0.00	0.00	0.00	0.01	0.00
Stellite 6		1.66	29.89	2.31	59.28	0.74	0.00	0.33	0.00	4.05	0.73	0.00	0.00	1.01	0.00
602 CA		9.45	25.35	62.63	0.00	0.00	2.09	0.06	0.14	0.00	0.09	0.01	0.00	0.18	0.00

Appendix 2

Summary of the Characteristics of Data to be Included into ASSET

The considerations used to generate and compile the corrosion data for the corrosion mechanisms involved in oxidation, sulfidation, sulfidation/oxidation, and carburisation are listed below.

1. The corrosion mechanisms for which predictions are currently possible are:

   - Isothermal oxidation.
   - Isothermal sulfidation.
   - Isothermal sulfidation/oxidation.
   - Isothermal carburisation.

2. The corrosion mechanisms for which data can be archived, but correlations not yet available are:

   - Cyclic oxidation.
   - Condensing sulfuric/sulfurous acid corrosion.
   - Isothermal oxidation with SO_2/SO_3.

3. For the ASSET corrosion mechanisms, the absolute minimum number of data points to define a model for an alloy corroding by a given mechanism is 9–12, depending upon the mechanism. However, the correlations are more accurate as the data collection represents wider ranges of temperature, exposure time, and gas composition variables.

4. A reasonable range of temperature would be 300°C, measured to +/– 5°C.

5. A reasonable range in gas compositions will be 2–3 orders in magnitude variation in terms of partial pressures of O_2 and S_2, and carbon activity. The partial pressures should be defined to a factor of +/– 100% (e.g. pO_2 of 1×10^{-25} atm in a range of 9×10^{-26} to 2×10^{-25}), variables such as partial pressure of H_2 and H_2S should be defined to +/– 20% of the value in question.

6. Commercial homogeneous alloys only.

7. Exposure times of at least 100 h, with a strong preference of a minimum of 500 h (isothermal conditions or a well–defined temperature cycle).

8. Corrosion measured by total metal penetration (sum of thickness of metal lost by surface scale formation — mass change and thickness of surface metal containing internal corrosion products — metallography), as shown in the attached picture.

9. Gas composition either at thermochemical equilibrium or a well–defined and confirmed non–equilibrium gas composition.

10. Corrosion coupons to have a slab (flat) geometry with a total thickness and test conditions such that at least one–half of the original metal thickness remains uncorroded after the exposure.

11. Data have been individually measured and numerical values placed into table format (data will not be interpreted from graphs, photos, etc.).

12. Metal temperature is used to define the experiment temperature.

13. These conditions have been achieved for many combinations of alloys and corrosion mechanisms present in the ASSET databases. For example, if there are no other data for a given alloy, a reasonable minimum data collection would be duplicate samples at combinations of three temperatures and two variations of each important gas composition variable.

14. The data must be accompanied by a written report which will serve as the bibliographic citation for the data source and should also completely describe the procedures used to prepare corrosion coupons, operate and design the corrosion test equipment, monitor the test apparatus, and to analyse the corroded coupons after the exposure to produce the tabulated corrosion data. Adequate description of a thorough and valid test procedure are crucial to confirmation of adequate quality in the corrosion data.

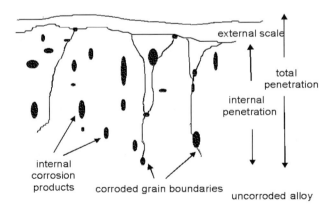

Schematic view of total penetration measurement, which is the corrosion measurement archived and used in ASSET, for a typical corrosion product morphology.

Appendix 3

Format for Data Stored in ASSET (with example data)

The data are organised according to the stable corrosion products that form on the corroding alloys. The types of corrosion products formed are used to infer the corrosion mechanism. For example, oxides suggest corrosion by oxidation, sulfides suggest corrosion by sulfidation, sulfides plus oxides suggest sulfidation/oxidation, and carbides suggest carburisation. Samples of the data and associated variables for each corrosion mechanism are shown below.

Sample Data for Sulfidation Corrosion Mechanism

Alloy name	Reference ID	Penetration, mils	Exposure, h	Temperature, °C	pH_2S, atma	pH_2, atma
9 Cr 1Mo	4	4.30	219.00	482.00	4×10^{-2}	2.04×10^1
Alloy 3220	2	1.51	2000.00	400.00	1.05×10^0	3.17×10^0

Sample Data for Oxidation Corrosion Mechanism

Alloy name	Reference ID	Penetration, mils	Exposure, h	Temperature, °C	pO_2, atma
HR–120	38	2.06	1152.00	982.22	1×10^{-1}
HR–120	38	16.69	1152.00	1093.33	1×10^{-1}

Sample Data for Sulfidation/Oxidation Corrosion Mechanism

Alloy name	Reference ID	Penetration, mils	Exposure, h	Temperature, °C	pO_2, atma	pS_2, atma
Alloy 625	3	2.14	527.00	600.00	6.31×10^{-27}	8.71×10^{-10}
Alloy 718	23	5.71	1330.00	522.00	2.57×10^{-32}	2.14×10^{-10}

Sample Data for Carburisation Corrosion Mechanism

Alloy name	Reference ID	Penetration, mils	Exposure, h	Temperature, °C	$A_c{}^*$	pH_2S, atma
Alloy 800 H	19	175.50	1400.00	1000.00	8×10^{-1}	0
ACI HP–Nb	35	39.37	500.00	900.00	8×10^{-1}	0

*carbon activity

(continued overleaf)

Appendix 3 . Format for data stored in ASSET (with example data)(continued)

Sample Data for Isothermal Oxidation with SO_2–SO_3 Corrosion Mechanism

Alloy name	Reference ID	Penetration, ml	Exposure, h	Temperature, °C	pO_2, atma	pSO_2, atma	pSO_3, atma
253 MA	29	0.12	500.00	1000.00	4×10^{-2}	1×10^{-3}	0
Alloy 617	29	0.53	2000.00	1000.00	4×10^{-2}	1×10^{-3}	0

Sample Data for Cyclic Oxidation Corrosion Mechanism

				Low	High	Cycle	
Alloy name	Reference ID	Penetration, ml	Exposure, h	Temp., °C	Temp., °C	Time, h	$p(O_2)$, atma
253 MA	15	0.22	500.00	20.00	980.00	20.00	2.1×10^{-1}
AISI 309	15	0.88	500.00	20.00	1095.0	20.00	2.1×10^{-1}

Sample Data for Sulfuric/Sulfurous Acid Dew Point Corrosion Mechanism

Alloy name	Reference ID	Penetration, ml	Exposure, h	Temperature, °C	pO_2, atma	pSO_2, atma	pSO_3, atma
AISI 1020	14	0.08	48.00	103.00	6.3×10^{-3}	1.3×10^{-3}	3×10^{-6}
AISI 1020	14	0.13	48.00	103.00	6.4×10^{-2}	1.3×10^{-3}	3.5×10^{-6}

28

The NiO/Ni(111) System in Creep at 550°C: Lifetime Prediction using Several Techniques

L. GAILLET, G. MOULIN, M. VIENNOT and P. BERGER*

Université de Technologie de Compiègne, Laboratoire Roberval, UPRESA 6066, BP 20529,
F-60205 Compiègne, France
*Laboratoire Pierre SUE, CEA/CNRS, Centre d'Etudes Nucléaires de Saclay, F-91191 Gif sur Yvette, France

ABSTRACT

Lifetime prediction, or more precisely, the cracking behaviour of oxide scales, has been studied on oxides developed on nickel single crystals during creep deformation under oxygen at 550°C. Three types of techniques have been employed: mechanical testing, oxide degradation monitoring by acoustic emission and oxygen diffusion analysis. Creep deformation of samples covered with an oxide scale is compared with vacuum tested ones. Differences in creep mechanisms between nickel with or without oxide scale have been noted. The acoustic emission technique provides a large amount of information to aid the understanding of degradation of the oxide scales during creep deformation. In addition, the oxidation rate, i.e. oxygen diffusion, depends strongly on the applied stress. These techniques enable a more accurate and realistic lifetime prediction of oxide scales grown on materials.

1. Introduction

As nickel is one of the most common constituent base materials for high temperature uses, its oxidation behaviour has been the subject of many studies [1]. The oxidation resistance of metallic materials depends on their abilities to form protective oxide scales on metal. This protective behaviour can be greatly modified when a mechanical stress is applied during oxidation [2,3]. This applied stress is added to the usual epitaxial and growth stresses. Several mechanisms of scale degradation can occur such as spallation, buckling or cracking. For qualitative and quantitative analysis of this oxide scale degradation, the development of specific techniques is required for a better understanding, in our case, of the Ni/NiO creep behaviour and damage to the oxide.

2. Material and Techniques

Samples were taken from a single crystal (111) nickel bar of high purity (residual content of impurities lower than 10 ppm) prepared by Johnson Matthey Comp. A first vacuum annealing was performed for 24 h at 900°C. Plate specimens were machined with a working length of 11 mm, followed by another annealing of 5 h in a vacuum at 900°C.

A specific set-up has been developed which makes it possible to apply a mechanical load in a controlled atmosphere (16-oxygen, 18-oxygen, vacuum) to specimens [4]. In this the *in situ* measurement of the stresses and strains induced during oxidation is possible (Fig. 1). An oxygen isotope (^{18}O) was chosen in order to study the diffusion [5]. A thermal layer of NiO was first developed without any load under oxygen for 4 h at 550°C. Then, creep took place under the same conditions of oxidation, with stresses varying from 15 to 60 MPa. After testing, the samples were cooled in the same atmosphere at a rate of approximately 10^{2}°C h^{-1}.

18-oxygen diffusion profiles were measured by nuclear microanalysis at the Laboratoire Pierre SÜE (Saclay/France). The $^{18}O(p,\alpha)^{15}N$ nuclear reaction was chosen with an incident proton beam at 795 keV. Alpha particles were detected at a detection angle of 170° with an annular detector. The typical beam size, nearly 5×10 µm^2, enables local examination of the oxygen diffusion process on a single oxide strip, so that oxygen diffusion parameters are not influenced by the existence of the cracks.

The acoustic emission system is a commercial P.A.C. Model AEDSP-32/16 & MISTRAS 2001. The gain was 40 dB and the plug-in filter was band pass (20 and 1200 kHz). Acoustic Emission Signals were detected using a piezoelectric transducer with a large frequency range (200 kHz–1MHz).

Microstructural investigations were performed by optical microscopy and scanning electron microscopy with EDS analysis.

3. Microstructural Study

After a short oxidation stage (4 h), a NiO scale with a duplex morphology is developed on the surface of nickel, as previously reported [6]. The oxide is composed of large columnar grains in the outer zone and small equiaxed grains in the inner one (Fig. 2)

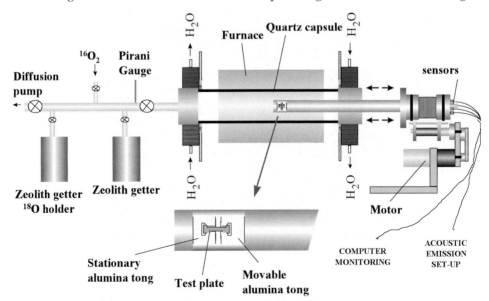

Fig. 1 *Scheme of the deformation–oxidation test set-up with Acoustic Emission equipment.*

[7]. After oxidation for 4 h at 550°C in oxygen the oxide thickness was 0.4 µm for the outer scale and 0.1–0.2 µm for the inner one. The two layers had the same composition and structure, i.e. NiO.

After the nickel samples had been tested in creep in oxygen at 550°C, parallel cracks were observed on the surface of the oxide scale with an orientation perpendicular to the tensile strain axis (Fig. 3) [8,9]. The critical stress for these cracks

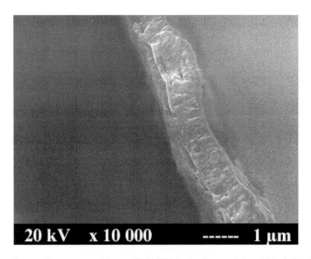

Fig. 2 *Oxide scales on the cross-sections of Ni (111) single crystal oxidised for 4 h at 550°C in oxygen.*

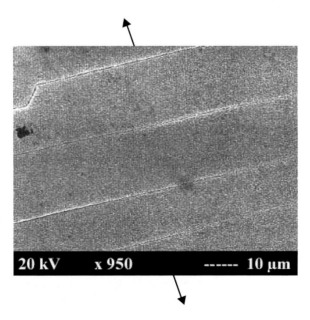

Fig. 3 *Morphology of the oxide scale after creep in oxygen at 550°C. The tensile axis is symbolised by arrows.*

is nearly 10 MPa. The crack spacing decreases when the load is increased. For applied stresses higher than 25 MPa, many cases of oxide spallation and buckling were noticed in the NiO scale. Also, another type of oxidation behaviour was observed for the highest stresses. With these a significant crack healing (or the growth of a new oxide scale) could occur as creep rates close to tensile strain rates for crack healing are reached [10].

4. Acoustic Emission Analysis

The acoustic emission (A.E.) signals which are collected are mainly those emitted by the NiO oxide scale because the number of events is negligible under vacuum. Under oxygen (NiO scale), a significant generation of A.E. (number of events and their intensities) is observed during primary creep. This acoustic activity continues during secondary creep and can be attributed to the development of oxide during creep.

Different mechanisms of oxide scale damage take place during creep and are identified with the help of the acoustic emission technique. The magnitude of the energy measured during A.E. events can be used to identify these mechanisms. During primary creep, a crack generation occurs in the external NiO scale which evolves with the deformation rate. This phenomenon generates A.E. events with a high energy (Fig. 4). The second phase is the spallation and buckling of the scale characterised by a lower energy for A.E. events and a longer time duration. For the higher stresses, cracks are observed in the inner NiO scale, for example with an applied stress of 45 MPa (Fig. 4).

This technique confirms the microstructural results and can be used to follow the type of degradation mechanisms which occur during creep. These first results are promising but require more accurate analysis.

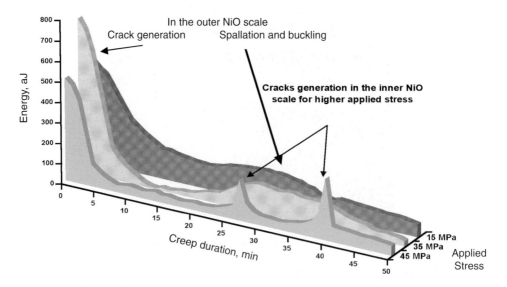

Fig. 4 Evolution of the acoustic emission energy vs creep time.

5. Mechanical Study

Synergistic effects between oxygen diffusion and mechanical loadings may be observed by comparing the mechanical behaviour of nickel in vacuum and in an oxidising atmosphere. This effect modifies the strain rates and the creep laws of nickel samples, depending on the environment.

Under vacuum, the deformation rate increases linearly with respect to the applied stress (slope $n = 1.45$). The evolution of the strain rate vs the applied stress is identified as Harper-Dorn creep [11] (Fig. 5). This creep behaviour corresponds to a low dislocation density, due to the (111) easy glide systems.

Under oxygen, the same tests lead to a different $\dot{\varepsilon}$ vs σ dependence. Now it seems that the slope must be separated into two domains; a first one up to 35 MPa and a second one above this critical applied stress. For low stresses, a diffusion creep mechanism can explain a quasi-linear relationship ($n = 1.2$) between the deformation rate and the applied stress (Fig. 5). A Coble creep mechanism may be considered because of the great number of grain boundaries in the oxide scale. Moreover, boundaries between inner and outer NiO scales and also between oxide and underlying metal must be taken into account.

For high stresses, the mechanical behaviour of the NiO/Ni (111) system follows a power law creep with a power exponent $n \approx 4$, indicating creep resulting from climb and glide of dislocations [11]. This value of 4 is close to the values of 3.5 or 3.8 already reported for nickel [12]. Such a mechanism is most likely to be due to the large number of dislocations in the substrate (induced by high stresses) which control the deformation mechanism.

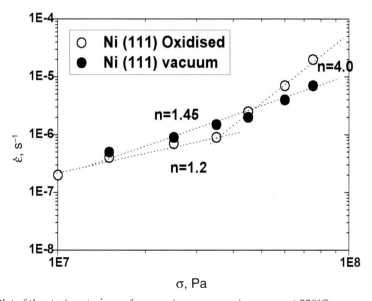

Fig. 5 *Plot of the strain rate $\dot{\varepsilon}$ vs σ for creep in vacuum or in oxygen at 550°C.*

6. Oxygen Diffusion Analysis

Oxygen diffusion analysis gives a complementary investigation of the behaviour of the oxide scales. This study is performed with the use of ^{18}O injected during the secondary creep stage.

A classical model for the diffusion of a species during the growth of its compound on a surface is chosen to study oxygen diffusion [13]. A constant value of the concentration of the diffusing element on the surface is imposed. The mechanism of diffusion of oxygen is dependent on the microstructure of the NiO layer. The diffusion regime is found to correspond to an intergranular type of diffusion via grain boundaries and interfaces (like metal/oxide or external/internal oxide layer interfaces).

The oxygen diffusion parameter increases initially when a stress is applied during oxidation. This increase is high (factor of 100) and would appear to confirm a short-circuit diffusion mechanism. However, the oxygen mobility then decreases with respect to the applied stresses [14] (Fig. 6).

A cyclic behavior can be observed with a new increase for stresses above 40 MPa, probably as a consequence of healing or a new oxide scale growth process, as mentioned in the microstructural study section above.

7. Conclusions

- Nickel oxidation behaviour at 550°C in the presence of an applied stress has been studied with the help of several techniques.

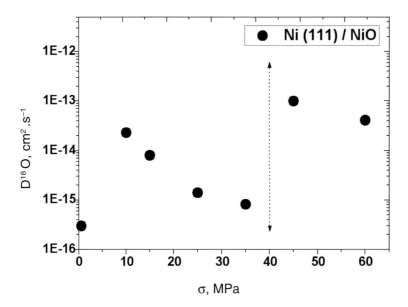

Fig. 6 *Evolution of ^{18}O diffusion parameters vs applied stress for Ni (111) single crystals.*

- The main damage in creep is the generation of periodic cracks, perpendicular to the tensile axis. Increasing the applied stress produces a more defective oxide scale by buckling and spalling. A new oxide layer growth or healing process can occur for specific test conditions. This failure behaviour is confirmed by acoustic emission by continuous monitoring of the damage during creep tests. Most of the degradation processes encountered have been identified by this technique.

- The creep analysis under oxygen and under vacuum at 550°C indicates a different mechanical behaviour for each atmosphere. This can be attributed to the presence of the oxide scale on the surface of the metal. In the oxidising environment, the strain rates are increased and a modification of the creep mechanism is observed.

- Oxygen diffusion analysis has shown a strong influence of the applied stress on the oxidation behaviour. Depending on the load, the oxygen diffusion parameter can increase up to very high values, when short-circuit diffusion via interfaces or grain boundaries appears to be established.

- This article highlights the complementary aspect of acoustic emission, mechanical testing and oxygen diffusion analysis for studying and predicting material degradation submitted to high temperature oxidation.

References

1. J. F. Stringer, High temperature corrosion of aerospace alloys, NATO series, AGARDograph 200, 1975.
2. G. Moulin, C. Mons, C. Severac, C. Haut, G. Rautureau and E. Beauprez, *Oxid. Met.*, 1993, **40**, 85.
3. G. Moulin, P. Arevalo and A. Salleo, *Oxid. Met.*, 1996, **45**, 153.
4. G. Moulin and F. Maurel, Proc. *4th Europ. Conf. on Residual Stresses*, Cluny, 4–6 June, 1996 (S. Denis *et al.*, eds). Paris, 1996, p.537.
5. P. Berger, G. Moulin and M. Viennot, *Nucl. Instr. Meth. Phys. Res. B*, 1997, **130**, 717.
6. A. Atkinson and D. W. Smart, *J. Electrochem. Soc.*, 1988, **11**, 2886.
7. A.W. Harris and A. Atkinson, *Oxid. Met.*, 1990, **32**, 229.
8. P. Hancock and J. R. Nicholls, *Mat. Sci. Tech.*,1988, **4**, 398.
9. M. Nagl, W. T. Evans, D. J. Hall and S. R. J. Saunders, *J. Phys.*, 1993, **3 C9**, 933.
10. M. Schütze, *Protective Oxides Scales and their Breakdown*. John Wiley & Sons, London, 1997.
11. H. J. Frost and M. F. Ashby, in *Deformation Mechanism Map*. Pergamon Press, Oxford, UK, 1982.
12. H. Siethoff, W. Schröter and K. Ahlborn, *Acta. Met.*, 1985, **33**, 443.
13. J. Philibert, *La Diffusion dans les Solides*. Masson Ed., Paris, 1972.
14. L. Gaillet, Ph.D thesis, Université de Technologie de Compiègne, Compiègne, 2000.

List of Abbreviations

The following abbreviations occur in the text and in the Index of contents.

APS	Atmosphere plasma spraying	FE	Finite element
AAS	Atomic absorption spectroscopy		
AE	Acoustic emission	GDS	Glow discharge spectroscopy
AES	Auger electron spectroscopy		
ASSET	Alloy selection system for elevated temperature	HCF	High cycle fatigue
BRITE	Basic research in industrial technologies	ICP-MS	Inductively coupled plasma-mass spectroscopy
BSE	Back scattered electrons	InCF (also ICF)	Intrinsic chemical failure
C.E.C.	Commission of the European Communities	IPS	Inverse problem solution
COSIM	Coating oxidation and substrate interdiffusion model	LEAFA	Life extension of alumina-forming alloys
COSP	Cyclic oxidation simulation programme	LPPS	Low pressure plasma spraying
COST	Cooperation européenne dans la domaine de la recherche scientifique et technique	M	conventional Melting
		MA	Mechanical alloying
CTE	Coefficient of thermal expansion	MICF	Mechanically induced chemical failure
EB-PVD	Electron beam — Physical vapour deposition	NOSH	No self healing
EC	European Commission		
EDS	Energy dispersive X-ray spectroscopy	ODS	Oxide dispersion strengthened
EDX	Energy dispersive X-ray analysis	OM	Optical metallography
EPMA	Electron probe microanalysis	PFBC	Pressurised fluidised bed combustion
ESEM	Environmental scanning electron microscope	PLST	Pulsed laser spallation technique
EURAM	European research in advanced materials	PM	Powder metallurgy

RE	Reactive element
RBS	Rutherford backscattered electron spectroscopy
SEM	Scanning electron microscopy
SIMS	Secondary ion mass spectroscopy
SS	Stainless steel
TBC	Thermal barrier coatings
TEM	Transmission electron microscopy
TGMF	Thermal gradient mechanical fatigue
TGO	Thermally grown oxide
volppm	volume parts per million
XPS	X-ray photoelectron spectroscopy
XRD	X-ray diffraction

Index

Abbreviations are to be found in the Abbreviations list on p.439.
NOTE Generally, alloys in the LEAFA programme, are not separately identified in this Index but can be found under the generic alloy types, for example, FeCrAl(RE)